T0181210

IFIP Advances in Information and Communication Technology

562

Editor-in-Chief

IFIP – The International Federation for Information Processing

IFIP was founded in 1960 under the auspices of UNESCO, following the first World Computer Congress held in Paris the previous year. A federation for societies working in information processing, IFIP's aim is two-fold: to support information processing in the countries of its members and to encourage technology transfer to developing nations. As its mission statement clearly states:

IFIP is the global non-profit federation of societies of ICT professionals that aims at achieving a worldwide professional and socially responsible development and application of information and communication technologies.

IFIP is a non-profit-making organization, run almost solely by 2500 volunteers. It operates through a number of technical committees and working groups, which organize events and publications. IFIP's events range from large international open conferences to working conferences and local seminars.

The flagship event is the IFIP World Computer Congress, at which both invited and contributed papers are presented. Contributed papers are rigorously refereed and the rejection rate is high.

As with the Congress, participation in the open conferences is open to all and papers may be invited or submitted. Again, submitted papers are stringently refereed.

The working conferences are structured differently. They are usually run by a working group and attendance is generally smaller and occasionally by invitation only. Their purpose is to create an atmosphere conducive to innovation and development. Refereeing is also rigorous and papers are subjected to extensive group discussion.

Publications arising from IFIP events vary. The papers presented at the IFIP World Computer Congress and at open conferences are published as conference proceedings, while the results of the working conferences are often published as collections of selected and edited papers.

IFIP distinguishes three types of institutional membership: Country Representative Members, Members at Large, and Associate Members. The type of organization that can apply for membership is a wide variety and includes national or international societies of individual computer scientists/ICT professionals, associations or federations of such societies, government institutions/government related organizations, national or international research institutes or consortia, universities, academies of sciences, companies, national or international associations or federations of companies.

More information about this series at http://www.springer.com/series/6102

Gurpreet Dhillon · Fredrik Karlsson ·
Karin Hedström · André Zúquete (Eds.)

ICT Systems Security and Privacy Protection

34th IFIP TC 11 International Conference, SEC 2019
Lisbon, Portugal, June 25–27, 2019
Proceedings

 Springer

Editors
Gurpreet Dhillon
University of North Carolina
Greensboro, NC, USA

Karin Hedström
Örebro University
Örebro, Sweden

Fredrik Karlsson
Örebro University
Örebro, Sweden

André Zúquete
University of Aveiro
Aveiro, Portugal

ISSN 1868-4238   ISSN 1868-422X   (electronic)
IFIP Advances in Information and Communication Technology
ISBN 978-3-030-22314-4   ISBN 978-3-030-22312-0   (eBook)
https://doi.org/10.1007/978-3-030-22312-0

This Springer imprint is published by the registered company Springer Nature Switzerland AG
The registered company address is: Gewerbestrasse 11, 6330 Cham, Switzerland

Preface

We are honored to bring you this collection of proceedings from the 34th IFIP International Conference on ICT Systems Security and Privacy Protection, which was held in Lisbon, Portugal, during June 25–27, 2019. IFIP SEC conferences are the flagship events of the International Federation for Information Processing (IFIP) Technical Committee 11 on Information Security and Privacy Protection in Information Processing Systems (TC-11).

In this edition, ICT 2019 provided a venue for high-quality papers covering a wide range of research areas in the information security fields. The accepted papers were authored by researchers from different countries. The selection of papers was a highly challenging task. We received 76 submissions in response to our call for papers. From these 76 submissions, we selected 26 full papers to be presented at the conference, based on their significance, uniqueness, and technical quality. Each paper received at least three reviews by members of the Program Committee. Authors received reviewers' comments to be considered for the final camera-ready version of their papers.

We want to record our appreciation of all the contributors who supported in making IFIP SEC 2019 a success. We also acknowledge the authors themselves, without whose expert input there would have been no conference. We would also like to thank members of the Program Committee who devoted significant amounts of their time to evaluate all submissions. The Program Chairs, Fredrik Karlsson, Karin Hedström, and André Zúquete, who coordinated the review process, need a special vote of thanks for their efforts in making this conference a success. Other individuals who deserve special thanks and without whom the current programming and success of the conference would not have been possible include: Sergio Nunes and Jesualdo Fernandes of ISEG, University of Lisbon, Portugal, for all the local arrangements; Ella Kolkowska, Örebro University, Sweden, for organizing the tutorials; Bonnie Anderson, Brigham Young University, USA, and Kane Smith, University of North Carolina Greensboro, USA, for organizing the Work-in-Progress and Emerging Research track (WIPER) of the conference.

We also acknowledge the institutional support for IFIP SEC 2019, which came from the University of North Carolina, Greensboro, USA, Örebro University, Sweden, and ISEG, Universidade de Lisboa, Portugal. Without the support of these institutions, it would not have been possible to organize this conference.

April 2019

Gurpreet Dhillon
Fredrik Karlsson
Karin Hedström
André Zúquete

Organization

Program Committee

Alberto Pinto	Polytechnic Institute of Porto, Portugal
Andrea Kolberger	Upper Austrian University of Applied Sciences, Austria
Angelo Spognardi	Sapienza Università di Roma, Italy
Antonios Tsertsidis	Örebro University, Sweden
Arun Raghuramu	Forescout Technologies Inc.
Audun Jøsang	University of Oslo, Norway
Carlos Rieder	ISEC AG
Christian Damsgaard Jensen	Technical University of Denmark, Denmark
Dagmar Brechlerová	Euromise Prague, Czech Republic
Elham Rostami	Örebro University, Sweden
Ella Kolkowska	Örebro University, Sweden
Fabio Martinelli	IIT-CNR, Italy
Gert Læssøe Mikkelsen	The Alexandra Institute, Denmark
Gilbert L. Peterson	US Air Force Institute of Technology, USA
Gomathi Thangavel	Örebro University, Sweden
Gunnar Klein	Örebro University, Sweden
Gurpreet Dhillon	The University of North Carolina, Greensboro, USA
Heejo Lee	Korea University, South Korea
Helder Gomes	Universidade de Aveiro, Portugal
Ingrid Schaumueller-Bichl	Upper Austrian University of Applied Sciences, Austria
Javier Lopez	Universidad de Málaga, Spain
João Paulo Barraca	University of Aveiro, Portugal
Jonas Hallberg	Swedish Defence Research Agency, Sweden
Kai Rannenberg	Goethe University Frankfurt, Germany
Kai Wistrand	Örebro University, Sweden
Kerry-Lynn Thomson	Nelson Mandela Metropolitan University, South Africa
Lech Janczewski	The University of Auckland, New Zealand
Lingyu Wang	Concordia University, Canada
Lynn Futcher	Nelson Mandela Metropolitan University, South Africa
Mevludin Memedi	Örebro University, Sweden
Miguel Correia	Universidade de Lisboa, Portugal
Miguel Pardal	Universidade de Lisboa, Portugal
Nicola Dragoni	Örebro University, Sweden

Contents

Organizational and Behavioral Security

Crypto and Encryption

Integrity

Intrusion Detection

Hunting Brand Domain Forgery: A Scalable Classification for Homograph Attack

Tran Phuong Thao[1]($\boxtimes$), Yukiko Sawaya[1], Hoang-Quoc Nguyen-Son[1],
Akira Yamada[1], Kazumasa Omote[2], and Ayumu Kubota[1]

[1] KDDI Research, Inc.,Fujimino, Japan
{th-tran,yu-sawaya,ho-nguyen,ai-yamada,kubota}@kddi-research.jp
[2] Tsukuba University,Tsukuba, Japan
omote@risk.tsukuba.ac.jp

Abstract. Visual homograph attack is a way that the attackers deceive victims about what domain they are communicating with by exploiting the fact that many characters look alike. The attack is growing into a serious problem and raising broad attention in reality when recently many brand domains have been attacked such as `apple.com` (Apple Inc.), `adobe.com` (Adobe Systems Incorporated), `lloydsbank.co.uk` (Lloyds Bank), etc. Therefore, how to detect visual homograph becomes a hot topic both in industry and research community. Several existing papers and tools have been proposed to find some homographs of a given domain based on different subsets of certain look-alike characters, or based on an analysis on the registered International Domain Name (IDN) database. However, we still lack a scalable and systematic approach that can detect sufficient homographs registered by attackers with a high accuracy and low false positive rate. In this paper, we construct a classification model to detect homographs and potential homographs registered by attackers using machine learning on feasible and novel features which are the visual similarity on each character and some selected information from Whois. The implementation results show that our approach can bring up to 95.90% of accuracy with merely 3.27% of false positive rate. Furthermore, we also make an empirical analysis on the collected homographs and found some interesting statistics along with concrete misbehaviors and purposes of the attackers.

Keywords: Web security · International domain name · Punycode · Visual homograph attack

1 Introduction

Visual homograph attack was first described by Gabrilovic [1]. To prove the feasibility of this kind of attack, the authors registered a homograph of the brand

Published by Springer Nature Switzerland AG 2019
G. Dhillon et al. (Eds.): SEC 2019, IFIP AICT 562, pp. 3–18, 2019.
https://doi.org/10.1007/978-3-030-22312-0_1

domain `microsoft.com` which incorporated Cyrillic characters. After that, several brand domains were targeted by homograph attacks but this attack was not much attracted. Until April 2017, when the `apple.com` was forged by the homographs [2] such as the one appears under Punycode form `xn-pple-43d.com` which uses the Cyrillic 'a' (U+0430) instead of the ASCII 'a' (U+0061), the attack got mass attention from the media, and thus how to detect visual homographs becomes a significant issue.

Right after the attack on `apple.com` was published, some web browsers disabled the function of automatic IDN conversion. Since the IDNs contain non-ASCII characters (e.g., Arabic, Chinese, Cyrillic alphabet), they are encoded to ASCII strings using Punycode transcription known as *IDNA* encoding and appear under ASCII strings starting with `xn--`; for example, `xn--ggle-0qaa.com` is displayed as `gõõgle.com`. However, there is a big trade-off when a web browser stops supporting the automatic IDN conversion because a huge number of Internet users are using non-English languages with non-Latin alphabets through over 7.5 million registered IDNs in all over the world (by December 2017) [3]. Furthermore, visual homograph not only takes advantage of look-alike Punycode characters in IDNs, but also look-alike Latin characters in even non-IDNs themselves; for example, the homograph `bl0gsp0t.com` was registered targeting to the brand domain `blogspot.com` by replacing 'o' by '0', or the homograph `wlklpedia.org` was registered targeting to the brand domain `wikipedia.com` by replacing 'i' by 'l'. Also, if homograph domains can deceive users before they appear in the address bar of web browsers (e.g., homographs are given from an email or a document under hyper-links) without the users' awareness of the browsers, disabling IDN conversion is not meaningful to prevent users from accessing the homographs. Therefore, the web browsers after that re-enabled the function but are trying to block homographs which can be detected or blacklisted. Then, the problem is still how to detect homographs. Several existing tools and previous papers such as [4–10,17] have been proposed to find homographs of given domains using different inadequate subsets of certain look-alike characters that are defined by themselves, or using the IDN database registered at the time of analysis.

Therefore, our goal is how to propose a scalable, systematic, high-accuracy and low-false-positive-rate approach that can detect sufficient visual homographs not registered by the brand domains' owners to pro-actively protect their brands but by attackers. The research scope in this paper is described as follows. First, there are several types of homographs such as visual-based, semantic-based, top-level-domain (TLD)-based, typosquatting-based but this paper focuses only on the visual-based homograph which is the most popular and serious type (The explanation and the reason why visual-based homograph is the most serious will be described in more details in the background). Second, this paper focuses on finding homographs registered by attackers that can be either phishing (being active and having phishing content) or not phishing yet (being active and not yet have phishing content, but we still consider it as harmful case because of the behavior of registering homograph targeting to brand domains); in other words, our aim is not to detect phishings but homographs.

1.1 Related Work

Many existing tools such as [4–7], or previous work such as [8] by Abawajy et al. have been proposed to find visual homographs on inputting a (brand) domain. Most of these tools simply define a subset of look-alike characters, and then replace each character in the given domain by the look-alike characters in the subset. Some of them such as [7] look up Domain Name System (DNS) to determine whether the homographs are active or not. However, using the tools and previous works cannot find sufficient homographs because the subsets are too small compared with the enormous set of look-alike characters. Furthermore, their approaches cannot distinguish which homographs are registered by owners of the brand domains (to protect their brands), and which homographs are registered by attackers. Also, a formal measure (e.g., Structural Similarity Index (SSIM), Peak Signal-to-Noise Ratio (PSNR) or Mean Squared Error (MSE), etc.) is not used to define the visual similarity instead of subjective feelings on the visual look. There are also some popular tools such as [11,12] but note that the tools are used to generate the other types of homographs like TLD-based or typosquatting, not visual-based as our goal. Tian et al. [13] proposed a method to predict phishing homographs by analyzing HTML content and visual screen-shots. However, using malicious HTML content and similar visual screen-shots does not mean that the phishing has to be homograph in the domain name string. For example, `random.com` (non-homograph) and `faceböök.com` (homograph) are both phishing to the brand *facebook.com* but the former one is a non-homograph and the latter one is a homograph. In other words, the features (i.e., HTML content and screen-shot) are reasonable for detecting phishings but not homographs in term of domain string. Moreover, downloading HTML content and capturing screen-shots require the analyzers to access the domains, and that can lead to malware injections or the cost for setting up a virtual machine. Oliver et al. [10] detect illegitimate links including homographs on a web page but it is not clear which criteria they use to formally define about visual similarity. In their work, they describe "One solution is to introduce a table of characters (i.e., glyphs) that are considered visually similar"; therefore probably their approach is similar to [4–8] as we mentioned above. The same issue is in the work by Tyson et al. [17]. Most recently, Liu et al. [9] have been proposed to detect homographs by using a visual similarity metric for the images of entire domain strings between each of 1.4 million registered IDNs and each of top 1000 popular domains ranked by Alexa. The drawback is that, applying the visual similarity for entire domain strings can lower the accuracy and lead to high false positive rate. When we analyzed their dataset, we found that many domain pairs of the brand domains and sample domains that are too different but have very high SSIM on the entire domain strings; for example, `àa.com` and `ea.com` have 0.952 SSIM and are listed as homographs but actually not. Furthermore, the paper only considers IDNs but as mentioned above, homographs can occur even in Latin alphabet such as ('I', '1', 'l') or ('o', '0'). Also, lots of new domains are registered everyday and thus the method is not scalable.

1.2   Our Work

In this paper, we propose a machine learning based approach which meets the following contributions:

– To the best of our knowledge, ours is the first study proposing a classification of homographs and non-homographs using a visual similarity measure (i.e., SSIM) on each character instead of entire domain string as previous work. This approach can increase the accuracy to 95.36% with merely 2.83% of FPR.
– Although using Whois for detecting phishing is not a new approach, but when it is combined with SSIM on each character, the approach becomes promising to detect homographs and eliminate the domains that look alike to the brand domains but are registered by the brand domains' owners. We do not trivially use entire Whois but select the practical features such as: creation date of homograph is often after that of the brand domain, expiration date of homographs is often before that of the brand domain, register name, and organization of homograph is different from that of the brand domain, along with original creation date and expiration date of homograph and brand domain. The evaluation result shows that our approach can reach to 95.90% of accuracy with merely 3.27% of FPR.
– Last but not least, we make an empirical analysis on the collected homographs and found that a large portion of the domains (44.57%) are for sale or parked domains, 6.38% have the same/related content to the brand domains, 32.45% cannot be accessed or have blank content, 2.13% were created as an education about what is homograph, and 14.47% have completely different content with the brand domains. Interestingly, we figured out several concrete misbehaviors and purposes of the hackers when analyzing these homographs.

Note that, the accuracy when using SSIM only is 95.36% and when using SSIM with Whois is 95.90%. It does not mean that using Whois does not bring much effect (increasing only 0.54% of accuracy because the processes of data labeling in the two cases are different. The important thing here is the high accuracy (over 95%) in both cases.

1.3   Roadmap

The rest of this paper is organized as follows. The backgrounds of homograph attacks, visual similarity measure, and Whois are described in Sect. 2. Our proposed method is presented in Sect. 3. The experiment results are analyzed in Sect. 4. The empirical analysis on the labelled homographs is described in Sect. 5. The discussion of several ideas for future work is given in Sect. 6. Finally, the conclusion is drawn in Sect. 7.

2   Backgrounds

In this section, we present the backgrounds of homograph attacks, visual similarity measures in which SSIM is used in this paper, and Whois information.

2.1 Homograph Attacks

Homograph attack is a way that the attackers deceive victims about what domain they are communicating with by exploiting the fact that many domains look alike. There are several kinds of homographs in the wild, we thus synthesize them into five categories. The first is **visual homograph** which uses different but visually look-alike characters, for example: `facebook.com` and `faceböök.com`. The second is **semantic homograph** which uses synonyms or contextual similar words, for example: `facebook.com` and `mark_zuckerberg_social_network.com`. The third is **TLD homograph** which uses the same main domain names, but different top-level-domain (TLD), for example: `facebook.com` and `facebook.biz`. The fourth is **typosquatting** which relies on mistakes such as typos made by Internet users when typing the domain names, for example: `facebook.com` and `faceboook.com`. The last is the combination of the previous 4 categories. Note that the homographs in which certain characters are inserted or replaced (known as bitsquatting) in the brand domains are also listed in the fourth type (typosquatting homograph); for instance, `travelgoogle.com` targeting to `google.com`. In this paper, we focus on the first that is visual homograph since it is the most popular and serious type.

Visual Homograph Attack. Why visual homograph is the type that is serious the most? First, only the visual homograph can produce a fake domain that is 100% look-alike with the brand domain and even human cannot distinguish. For example, the brand domain `google.com` and the visual homograph `google.com` (encoded by `xn--gogle-m29a.com`) which are completely look-alike with the brand domain, so have very high probability to deceive users. Second, visual homograph not only utilizes the look-alike Punycode characters but also even the look-alike Latin characters such as 'I' (big i), 'l' (el) or '1' (one). For example, the visual homograph `ad0be.com` targeting to the brand domain `adobe.com` by replacing '0' by 'o'.

2.2 Visual Similarity Measure

Visual similarity is a method for measuring the similarity between two images. In this paper, we use the state-of-the-art metric called Structural Similarity Index (SSIM) [14] which is perceptual measure based on visible structures in the images. Meanwhile, the traditional methods such as Peak Signal-To-Noise Ratio (PSNR) and Mean Squared Error (MSE) estimate absolute errors only. The SSIM between two images x and y of the same size $N \times N$ is:

$$SSIM(x,y) = \frac{(2\mu_x\mu_y + c_1)(2\sigma_{xy} + c_2)}{(\mu_x^2 + \mu_y^2 + c_1)(\sigma_x^2 + \sigma_y^2 + c_2)} \quad (1)$$

μ_x and μ_y represent the averages of x and y respectively. σ_x^2 represents the covariance of x and y. σ_x^2 and σ_y^2 represent the variances of x and y respectively.

[Domain Name]   KDDI.JP
[Registrant]   KDDI CORPORATION
[Name Server]   dns101.dion.ne.jp
[Name Server]   dns102.dion.ne.jp
[Name Server]   dnsa01.kddi.ne.jp
[Name Server]   dnsa02.kddi.ne.jp
[Creation on]   2001/04/16
[Expires on]   2017/04/30
[Status]   Active
[Last Updated]   2016/05/01 01:05:12 (JST)
Contact
Information:
[Name]   KDDI CORPORATION
[Email]   kt-tanaka@kddi.com
[Web page]
[Postal code]   163-8003
[Postal Address]   3-2, Nishishinjuku 2-chome, Shinjuku-ku
[Phone]   03-3347-5818
[Fax]

Fig. 1. An example of Whois: the Whois of the domain "kddi.jp"

$c_1 - (k_1 L)^2$ and $c_2 - (k_2 L)^2$ represent the variables to stabilize the division with weak denominator where L is the dynamic range of the pixel-values and is typically set to $L = 2^{\#bits\_per\_pixel} - 1$ and $k_1 = 0.01, k_2 = 0.03$ by default. SSIM values $[-1, 1]$ where 1 indicates perfect similarity.

2.3   Whois

Whois [15] is a protocol that is used for querying databases that store the information of the registered domains such as domain name, registrar, creation date, expiration date, organization, email, etc. Nowadays, there are many competitive services supporting Whois queries by web portal or API. An example of Whois is given in Fig. 1.

3   Our Proposed Method

In this section, we present our method including data collection, data labelling, feature extraction and selection, and learning process.

3.1   Data Collection

The most adequate method to collect homographs is to use the Confusable Unicode table [16] defined by Unicode, Inc. This table (for example version 11.0.0 updated on 2018-05-25) contains 6,296 pairs of confusable characters. However, it is impossible since using the entire set of the confusable pairs, then get all permutations for each position in the domains, and finally query Whois for each permutation are too inefficient (also, the number of non-homographs are extremely dominant compared with the number of homographs). Instead, the way we collected homographs is as follows.

First, in 6,296 Unicode confusable pairs mentioned above, we use only the pairs that have the targeting characters such as A–Z, 1–9 and the hyphen (-) because most of the brand domains (top Alexa ranking) are non-international-domain-name (non-IDN) and thus contains these characters only. We generated 26,021 homographs, then queried their Whois but only got 37 registered domains. Second, we use some tools in the wild such as [4–7] to generate 12,338 homographs which are different from the previous 26,021 generated ones (these tools use different subsets of similar characters), then we queried Whois and got 129 registered domains. Third, thanks to the authors of [9] for sharing us their 1,516 homographs that they matched 1,000 top popular domains ranked by Alexa with their 1.4 million registered IDN out of 300+ million registered domains; we then filter out the overlapping domains with the previous generated domains, re-queried Whois and got 1,006 registered domains (perhaps, at the time that we re-queried Whois, the other 510 domains already expired).

Totally, we got $39 + 129 + 1,006 = 1,174$ unique domains that are (at this time) called "temporary" homographs because we will annotate/label them again later. For non-homographs, from each of unique brand domains in the previous 1,174 "temporary" homographs, we generated at least one domain that is non-IDN and completely look different, and then got 1,969 domains. In summary, there are 3,143 domains in which 1,174 "temporary" homographs and 1,969 "temporary" non-homographs.

3.2 Data Labelling

Although the data we collected was "temporarily" labelled as homographs or non-homographs, we still need to re-label them again by the human because different persons will have different opinions about the visual homographs. For example, person A thinks ess.com is a homograph of ass.com but person B does not think that. For lowering the bias, we employed three analyzers to label the 3,143 domains with the same process and our final decision is based on majority rule. For example, 2/3 analyzers think ess.com is a homograph of ass.com so we finally label it as homograph. For different implementations below, we have different processes for data labelling:

Case 1. This case checks if using SSIM for each character can perform better than using SSIM for entire domain string. Thus, the three analyzers labelled them based on visual look only as Fig. 2. As a result, 3137 domains (99.81%) got the same label from the three analyzers; and the labels of the remaining 6 domains (0.19%) were decided based on majority (the labels decided by 2/3 analyzers). Finally, we got 1060 domains labelled as homograph and 2083 domains labelled as non-homograph.

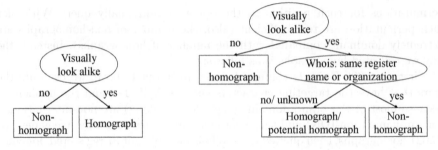

Fig. 2. Data labelling for case 1   **Fig. 3.** Data labelling for case 2

Case 2. This case checks if using SSIM for each character and the Whois can classify the homographs (including potential homographs). The labelling process is described in Fig. 3. We consider potential homographs here because the register names and organizations of some domains (even brand domains or sample domains) are hidden to protect the privacy of domain owners' information. For such pairs of the brand and sample domains that have unknown/hidden register name and organization, we treat them as potential homographs. Also, we cannot separate these potential homographs from actual homograph because we cannot confirm that they are actual or potential homographs. In the case 1, the visual looks are different from each person, the data labelling was thus done by the human. However, in this case 2, the Whois information is obvious, we thus only re-use the labelling result from case 1 and create a program to combine it with the condition of whether the register name and organization are different. In total 3143 domains, we finally got 940 domains labelled as homograph/potential homograph, and 2203 domains labelled as non-homograph.

3.3   Feature Extraction and Selection

For each case mentioned in the data labelling, the process of feature extraction and selection are described as follow:

Case 1. For each pair of the sample domain and its brand domain, we compute SSIM for each character and get the average. More concretely, we first separate the domains into each character. For each pair of characters in order, we parse them into images and compute SSIM for the images. Finally, the average of all SSIMs for every pair of characters which is corresponding to every position in the domain string. For example, SSIM of the two domains 'ab.jp' and 'xy.vn' is the average of the following five pairs: SSIM(image('a'), image('x')), SSIM(image('b'), image('y')), SSIM(image('.'), image('.')), SSIM(image('j'), image('v')), and SSIM(image('p'), image('n')). For the pairs of homographs and brand domains that have a different number of characters (domain string lengths), they are labelled as non-homograph in Sect. 3.2 without the need to compute the SSIM. There are a few examples when the homographs and brand domains have a different string such as 'rn' and 'm' but they are very rare so we do not consider.

Case 2. For this case, besides the SSIM computed on each character in the domain strings as the first case, the Whois of each brand and sample domains is also queried. More concretely, we extract the register name, organization, creation date, and expiration date. We finally use the following 15 features: (1) the average of SSIM on each character; (2) whether register of the brand domain is different from that of sample domain (1 if yes, 0 if no and 2 if both are none or hidden, we do not need to consider the case when one of them (not both) is none or hidden because as long as they are different, they cannot be non-homographs); (3) whether organization of the brand domain is different from that of sample domain (the values are the same as (2)); (4) whether creation date of the brand domains is before that of sample domains (1 if yes and 0 if no); (5) whether expiration date of the brand domain is after that of sample domains (1 if yes and 0 if no); (6) creation year of the brand domain; (7) creation month of the brand domain; (8) expiration year of the brand domain; (9) expiration month of the brand domain; (10) lifetime of the brand domain (the number of days between creation date and expiration date); (11) creation year of sample domain; (12) creation month of sample domain; (13) expiration year of sample domain; (14) expiration month of sample domain; and (15) lifetime of sample domain (the number of days between creation date and expiration date).

3.4   Learning

Since the homograph/non-homograph samples are collected consecutively for each given brand domain, the data must be randomly shuffled at first in order for reducing variance and making sure that models remain general and over-fit less. Then, we apply 7 popular supervised machine learning algorithms for training process including Support Vector Machine, Naive Bayes, Decision Tree, Neural Network, Stochastic Gradient Descent, Nearest Neighbors and Logistic Regression. We use k-fold cross validation (k is set to 10 in our implementation) and compute the accuracy, false positive rate, and true positive rate for our model. Also, the ROC curves are drawn to depict the comparison of previous and our approaches.

4   Experiment

The programs written in Python 2.7.11 on a computer Intel(R) core i7, RAM 16.0 GB, 64-bit Windows 10. The Whois is extracted using the *python-whois* package version 0.6.3. The machine learning algorithms are applied using *scikit-learn* package version 0.18. The SSIM is computed using the *skimage* package version 0.15.dev0.

4.1 Parameters

For each model, different parameters are used. For the Support Vector Machine (SVM), 3 parameters used are SVC, NuSVC which support different kernels and LinearSVC which supports only a linear kernel. For the Naive Bayes, 3 parameters used are GaussianNB which implements the Gaussian Naive Bayes, MultinomialNB which implements the Naive Bayes for multinomially distributed data, and BernoulliNB which implements the Naive Bayes for data distributed according to multivariate Bernoulli distributions. For the Nearest Neighbors, 3 parameters used are KNeighborsClassifier which implements learning based on the n_neighbors nearest neighbors of each query point, n_neighbors = 5 is set by default, RadiusNeighborsClassifier which implements learning based on the number of neighbors within a fixed radius of each training point, radius = 1.0 is set by default), and NearestCentroid which represents each class by the centroid of its members. For the Decision Tree, only 1 parameter used is DecisionTreeClassifier (note that Decision Tree has several algorithms such as ID3, C4.5, CART, CHAID, MARS, and Conditional Inference Tree but only the optimized version of the CART is used in this experiment. For the Neural Network, only 1 parameter used is MLPClassifier which implements a multi-layer perceptron that trains using Backpropagation. For the Stochastic Gradient Descent, only 1 parameter used is SGDClassifier which implements a plain stochastic gradient descent learning routine which supports different loss functions and penalties for classification. Finally, for the Logistic Regression, only 1 parameter used is LogisticRegression.

4.2 Results

The result for each case is described as follows:

Case 1. The results are described in Table 1. For the approach of using SSIM on the entire string, the best accuracy is 86.73% (with FPR = 5.41%) performed by KNeighborsClassifier. For our approach of using SSIM on each character, the best accuracy is 95.35% (with FPR = 2.83%) performed also by KNeighborsClassifier. We achieve 8.62% higher accuracy and 2.58% lower FPR than the previous approach. In this case, only the SSIM is used as the feature and what we expect is that the samples are classified as homograph if its SSIM is larger than a threshold and vice versa. There is only KNeighbors algorithm that classifies an object by a majority vote of its neighbors, with the object being assigned to the class most common among its k nearest neighbors. That is why the KNeighborsClassifier performs the best. The ROC curves of our and previous approaches are depicted in Fig. 4. The light curves mean the curves in each of $k = 10$ folds and the unique bold curve means the average for all the folds.

Case 2. In this case, we tried almost the same parameters for each algorithm as the case 1 except only the RadiusNeighborsClassifier. Since the value of year is much larger than the other features' values, the radius is set larger (i.e., 1400) to avoid the case when some outlier samples do not have any neighbor within the given radius. The results are described in Table 2. Note that, we implemented the case of SSIM on the entire string with Whois for fairly comparing with our achievement that is using SSIM on each character with Whois; there is no previous work using SSIM on the entire string with Whois but SSIM on the entire string only. The result shows that, for the approach of using SSIM on the entire string with Whois, the best accuracy is 92.43% (with FPR = 6.01%) performed by DecisionTreeClassifier. For our approach of using SSIM on each character with Whois, the best accuracy is 95.90% (with FPR = 3.27%) performed also by DecisionTreeClassifier. In this case 2, the way we label the samples is using SSIM at first to filter out the samples that are look alike to the brand domains, and if some domains satisfy this condition, the Whois is then used to filter out the ones which have the same owner with the brand domains. Therefore, it is reasonable why DecisionTree can perform the best because only this algorithm works based on decision rules in a flowchart structure. The ROC curves of the two approaches are depicted in Fig. 5. The explanation for the light curves and the unique bold curves is the same as that in case 1 above.

Why the ROC Curves Strictly Change Direction at the Cut-Point. In both case 1 and 2 and also in both case of using SSIM on each character and case of using SSIM on the entire domain string, the curves do not increase regularly with the same acceleration, but strictly change direction at the cut-point of each curve. This is because applying SSIM on text for a classification has only two groups which are when the SSIM is under a threshold and when the SSIM is over the threshold. It is different from applying SSIM on more complex images (such as landscape or person portrait) which has multiple groups because other elements are considered such as the color grayscale (the amount of light or color intensity).

5   Empirical Analysis on Labelled Homographs in Case 2

In this section, we make an empirical analysis on 940 domains labelled as homograph/potential homograph in the case 2 and found the following results. 60 domains (6.38%) that have the same or related content to the brand domains. 305 domains (32.45%) cannot be accessed or have blank content. 20 domains (2.13%) were created as an education about what is homograph. 419 (44.57%) domains are for sale or parked domains (Parked domains are a kind of domains that are registered without being associated with any services. Instead, these idle domains are used to display relevant advertisements; and every time a consumer clicks on one of the advertisements, the owner can earn money). Finally, 136 domains (14.47%) have completely different content compared with that of the brand domains. Interestingly, while analyzing these 136 domains, we figure out some unusual misbehaviors of the hackers:

- Several homographs were created not for any harmful purpose but just for saying some pointless words. For example, `xn--pple-koa.com` (displayed as äpple.com) targeting to the brand domain `apple.com` has a web content like "the art of killing yourself" or "why is everybody so stupid".

– Several homographs that redirect user's accesses to a safe webpage for the purpose of selling product only. For example, `xn--yotbe-1vab.com` (displayed as yoütübe.com) redirects user to a page on `amazon.com` that is selling a Kindle E-reader.
– Several homographs were created with the purpose of increasing pageview. For example, `xn--youtbe-6ya.com` (displayed as youtübe.com) is just a music video on Youtube with almost 5 million views and 2.3 thousand subscribers. This is also a way to make money (Youtube pays users based on the number of video views).
– Several homographs were created with the purpose of advertising the hackers themselves. For example, `xn--facebok-q0a.com` (displayed as faceboók.com) describes a hacker's profile with a message "If you'd like to hire me ...".
– Several homographs were created to claim (or lower the reputation of) the brand domains for the hacker's demands. For example, `xn--microsftnline-1kdc.com` targeting to `microsoftonline.com` claims Microsoft to support Cyrillic alphabet in its keyboard.

Table 1. Evaluation result for case 1

Algorithms	SSIM on entire string			SSIM on each character		
	Acc(%)	FPR(%)	TPR(%)	Acc(%)	FPR(%)	TPR(%)
svm.SVC	84.98	10.62	76.35	94.18	6.6	95.64
svm.NuSVC	85.14	10.38	76.35	94.02	6.83	95.64
svm.LinearSVC	86.35	8.37	75.96	94.21	5.9	94.45
GaussianNB	84.98	10.62	76.35	94.21	6.55	95.64
MultinomialNB	66.27	0.00	0.00	66.27	0.00	0.00
BernoulliNB	66.27	0.00	0.00	66.27	0.00	0.00
NearestCentroid	82.06	15.62	77.47	92.91	8.75	96.07
KNeighbors	86.73	5.41	71.22	95.35	2.83	91.79
RadiusNeighbors	66.27	0.00	0.00	66.27	0.00	0.00
DecisionTree	82.18	13.72	74.04	94.85	4.05	92.61
MLPClassifier	86.29	8.03	75.14	94.21	5.28	93.21
SGDClassifier	81.39	13.63	70.84	92.02	8.95	94.1
LogisticRegression	66.27	0.00	0.00	66.27	0.00	0.00

Abbreviation: Acc (accuracy), FPR (false positive rate), TPR (true positive rate)

6   Discussion

This section describe several challenges for future work to improve the accuracy.

Average SSIM for Different Characters and Other Measures for SSIM. In this paper, we compute the average SSIM for all the characters. For example, for the pair of domains foo.jp and föö.jp, the SSIM currently used is the average of

Table 2. Evaluation result for case 2

Algorithms	SSIM on string + Whois			SSIM on character + Whois		
	Acc(%)	FPR(%)	TPR(%)	Acc(%)	FPR(%)	TPR(%)
svm.SVC	82.12	2.64	46.28	82.12	2.64	46.28
svm.NuSVC	81.36	1.69	41.48	81.36	1.69	41.48
svm.LinearSVC	70.44	19.74	45.41	68.04	24.28	50.55
GaussianNB	78.87	26.10	90.47	78.84	26.15	90.47
MultinomialNB	80.66	21.61	85.91	80.66	21.61	85.91
BernoulliNB	70.09	0.00	0.00	70.09	0.00	0.00
NearestCentroid	76.27	29.50	89.77	76.27	29.50	89.77
KNeighbors	85.27	10.01	74.08	85.27	10.01	74.08
RadiusNeighbors	85.08	11.25	76.41	85.08	11.25	76.41
tree.DecisionTree	92.43	6.01	88.68	95.90	3.27	93.93
MLPClassifier	79.22	18.35	74.03	78.66	14.85	62.89
SGDClassifier	79.32	9.02	51.90	76.74	22.74	75.09
LogisticRegression	84.63	12.38	77.61	84.63	12.38	77.61

Abbreviation: Acc (accuracy), FPR (false positive rate), TPR (true positive rate)

SSIM of each pair ('f', 'f'), ('o', 'ö'), ('o', 'ö'), ('.', '.'), ('j', 'j'), ('p', 'p'). However, for the domains that have a small number of different characters compared with the number of all characters, using average SSIM cannot reflect clearly the visual difference; instead, the average SSIM on the different characters only perhaps can improved the accuracy. Concretely, a promising SSIM is the average of SSIM of two pairs ('o', 'ö'), ('o', 'ö'). Furthermore, besides using the *average* for SSIM, other measures such as the *median* (the middle value of a range), the *covariance* (the expected value of variations of two random variates from their expected values), or the *correlation* (the expected value of two random variates) can be considered.

Additional Features. Several other features may also help. First, homograph domains have the search engine's result count is less than that of the brand domains. Second, homograph domains often target to the brand domains that are hot topics such as crypto-currency or payment. We analyzed a dataset of phishing sites downloaded from PhishTank, and found that the top three categories with a dominant number of homographs are crypto-currency (38.3%), payment system (20.7%) and online game (10.2%), not bank or online shopping as we thought. Third, the time distance between the creation date/expiration date in the Whois from the present should be also considered as the features because the real brand domains have longer age (as the present) than the homograph domains.

Removing Diacritical Marks. Diacritical marks (also diacritical sign or accent) is a glyph added to a letter. For example, yáhoo.com (encoded by xn--yhoo-5na.com, targeting to the brand domain yahoo.com) have the diacritical acute mark ''' over the letter 'a'. By removing the diacritical marks as a pre-process before data labelling, the accuracy may be improved but probably it can also lead to higher false positive rate; thus a re-implementation is necessary.

Other Implementations. The implementation of the case 2 should be divided into two sub-implementations. The first one uses only SSIM and the second one uses both SSIM and Whois as the current implementation even though they have the same data-labelling process. These two sub-implementations can help figure out how **fairly effective** when Whois is used without comparing it to the implementation of case 1. Furthermore, even though SSIM is proven to be better than the traditional measures such as PSNR and MSE, an additional implementation should be done to confirm whether our approach when using SSIM outperforms it when using PSNR or MSE.

Extending Research Scopes. This paper currently deals with visual homograph but how to thoroughly deal with all types of homographs described in Sect. 2.1 becomes a great demand. The TLD-based is the most trivial since we can straightforwardly replace the TLD with the un-large entire set of available TLDs (1,535 TLDs). The typosquatting may be not difficult since there are several tools that can be used to create the samples such as [7,11,12]. However, the semantic-based is the most challenge since as far as we know, there is no existing methods which can automatically collect enough a number of samples.

Whether Homographs Are Caused By Fonts? Someone may question that if we compare characters in Arial and Times, the characters are probably slightly different. However, homographs are not caused by the font differences but Unicode code differences. Even if 'A' in Arial and Times fonts are visually look different, they have the same Unicode code that is "U+0041". Furthermore, in any web browser, the font (and font size) can be changed in the web body interface but not in the address bar.

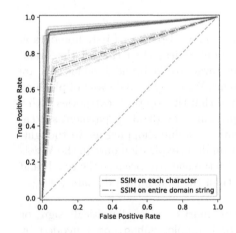
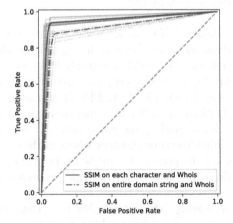

Fig. 4. ROC curves in case 1   **Fig. 5.** ROC curves in case 2

7   Conclusion

This paper proposes the first classification of homographs registered by attackers and non-homographs by using the state-of-the-art visual similarity metric that is SSIM on each character along with some reasonable selected information from Whois to increase the accuracy up to 95.90% with merely 3.27% of FPR. Two implementation cases are analyzed to explain how the approach works when applying only SSIM and applying the combination of SSIM and Whois (with the different labelling processes). An empirical analysis on labelled homographs is also taken place to find the ratio between different contents of the homographs, and to understand certain concrete purposes of the attackers.

Acknowledgement. This research was carried out as part of WarpDrive: Web-based Attack Response with Practical and Deployable Research InitiatiVE, the Commissioned Research of the National Institute of Information and Communications Technology (NICT), JAPAN.

References

1. Gabrilovic, E., Gontmakher, A.: The homograph attack. Commun. ACM **45**(2), 128 (2002)
2. Xudong, Z.: Phishing with Unicode Domains (2017). https://www.xudongz.com/blog/2017/idn-phishing/?_ga=2.53371112.1302505681.1542677803-1987638994. 1542677803
3. IDN World Report: Internationalised Domains show negative growth in 2017. https://idnworldreport.eu/
4. Idn-homograph-attack. https://github.com/timofurrer/idn-homograph-attack
5. EvilURL. https://github.com/UndeadSec/EvilURL
6. Homographs. https://github.com/dutchcoders/homographs
7. Dnstwist. https://github.com/elceef/dnstwist
8. Abawajy, J., Richard, A., Aghbari, Z.A.: Securing websites against homograph attacks. In: Lin, X., Ghorbani, A., Ren, K., Zhu, S., Zhang, A. (eds.) SecureComm 2017. LNICSSITE, vol. 239, pp. 47–59. Springer, Cham (2018). https://doi.org/10.1007/978-3-319-78816-6_4
9. Liu, B., et al.: A reexamination of internationalized domain names: the good, the bad and the ugly. In: 48th Annual IEEE/IFIP International Conference on Dependable Systems and Networks (DSN 2018) (2018)
10. Hunt, O.J., Krstic, I.: Preventing URL Confusion Attacks. Patent US9203849B2 United States. Apple Inc. (2013)
11. Instant Domain Search: Domain Name Generator. https://instantdomainsearch.com/domain/generator/
12. DN Pedia: Search Domain Zones. https://dnpedia.com/tlds/search.php
13. Tian, K., Jan, S., Hu, H., Yao, D., Wang, G.: Needle in a haystack: tracking down elite phishing domains in the wild. In: Internet Measurement Conference (IMC 2018), pp. 429–442 (2018)
14. Wang, Z., Bovik, A., Sheikh, H., Simoncelli, E.: Image quality assessment: from error visibility to structural similarity. IEEE Trans. Image Process. **13**(4), 600–612 (2004)

15. Daigle, L.: WHOIS protocol specification. In: RFC 3912 Internet Society (2004). https://tools.ietf.org/html/rfc3912
16. Unicode Inc: Unicode Security Mechanisms for UTS #39. https://www.unicode.org/Public/security/latest/confusables.txt
17. Tyson, M., Peter, H., Greg, B.: The 2017 homograph browser attack mitigation survey. In: 15th Australian Information Security Management Conference. https://doi.org/10.4225/75/5a84f5a495b4d

GanDef: A GAN Based Adversarial Training Defense for Neural Network Classifier

Guanxiong Liu[1]([✉]), Issa Khalil[2], and Abdallah Khreishah[1]

[1] New Jersey Institute of Technology, Newark, NJ 07102, USA
{gl236,abdallah}@njit.edu
[2] Qatar Computing Research Institute, Doha, Qatar
ikhalil@hbku.edu.qa

Abstract. Machine learning models, especially neural network (NN) classifiers, are widely used in many applications including natural language processing, computer vision and cybersecurity. They provide high accuracy under the assumption of attack-free scenarios. However, this assumption has been defied by the introduction of adversarial examples – carefully perturbed samples of input that are usually misclassified. Many researchers have tried to develop a defense against adversarial examples; however, we are still far from achieving that goal. In this paper, we design a Generative Adversarial Net (GAN) based adversarial training defense, dubbed **GanDef**, which utilizes a competition game to regulate the feature selection during the training. We analytically show that GanDef can train a classifier so it can defend against adversarial examples. Through extensive evaluation on different white-box adversarial examples, the classifier trained by GanDef shows the same level of test accuracy as those trained by state-of-the-art adversarial training defenses. More importantly, **GanDef-Comb**, a variant of GanDef, could utilize the discriminator to achieve a dynamic trade-off between correctly classifying original and adversarial examples. As a result, it achieves the highest overall test accuracy when the ratio of adversarial examples exceeds 41.7%.

Keywords: Neural network classifier · Generative Adversarial Net · Adversarial training defense

1 Introduction

Due to the surprisingly good representation power of complex distributions, NN models are widely used in many applications including natural language processing, computer vision and cybersecurity. For example, in cybersecurity, NN based classifiers are used for spam filtering, phishing detection as well as face recognition [1,18]. However, the training and usage of NN classifiers are based on

© IFIP International Federation for Information Processing 2019
Published by Springer Nature Switzerland AG 2019
G. Dhillon et al. (Eds.): SEC 2019, IFIP AICT 562, pp. 19–32, 2019.
https://doi.org/10.1007/978-3-030-22312-0_2

an underlying assumption that the environment is attack free. Therefore, such classifiers fail when adversarial examples are presented to them.

Adversarial examples were first introduced in [21] in the context of image classification. It shows that a visually insignificant modification with specially designed perturbations can result in a huge change of prediction results with nearly 100% success rate. Generally, adversarial examples can be used to mislead NN models to output any aimed prediction. They could be extremely harmful for many applications that utilize NNs, such as automatic cheque withdrawal in banks, traffic speed detection, and medical diagnosis in hospitals. As a result, this serious threat inspires a new line of research to explore the vulnerability of NN classifiers and develop appropriate defensive methods.

Recently, a plethora of methods to countermeasure adversarial examples has been introduced and evaluated. Among these methods, adversarial training defenses play an important role since they (1) effectively enhance the robustness, and (2) do not limit adversary's knowledge. However, most of them lack the trade-off between classifying original and adversarial examples. For applications that are sensitive to misbehavior or operate in risky environment, it is worth to enhance defenses against adversarial examples by sacrificing performance on original examples. The ability to dynamically control such trade-off makes the defense even more valuable.

In this paper, we propose a GAN based defense against adversarial examples, dubbed **GanDef**. GanDef is designed based on adversarial training combined with feature learning [10,12,24]. As a GAN model, GanDef contains a classifier and a discriminator which form a minimax game. To achieve the dynamic trade-off between classifying original and adversarial examples, we also propose a variant of GanDef, **GanDef-Comb**, that utilizes both classifier and discriminator. During evaluation, we select several state-of-the-art adversarial training defenses as references, including Pure PGD training (**Pure PGD**) [13], Mix PGD training (**Mix PGD**) [7] and **Logit Pairing** [7]. The comparison results show that GanDef performs better than state-of-the-art adversarial training defenses in terms of test accuracy. Our contributions can be summarized as follows:

- We propose the defensive method, GanDef, which is based on the idea of using a discriminator to regularize classifier's feature selection.
- We mathematically prove that the solution of the proposed minimax game in GanDef contains an optimal classifier, which usually makes correct predictions on adversarial examples by using perturbation invariant features.
- We empirically show that the trained classifier in GanDef achieves the same level of test accuracy as that in state-of-the-art approaches. Adding the discriminator, GanDef-Comb can dynamically control the trade-off on classifying original and adversarial examples and achieves the highest overall test accuracy when the ratio of adversarial examples exceeds 41.7%.

The rest of the paper is organized as follows: Sect. 2 presents background material, Sect. 3 details the design and mathematical proof of GanDef, Sect. 4 shows evaluation results, and Sect. 5 concludes the paper.

2   Background and Related Work

In this section, we introduce high-level background material about threat model, adversarial example generators and defensive mechanisms for the better understanding of concepts presented in this work. We also provide relevant references for further information about each topic.

2.1   Threat Model

The adversary aims at misleading the NN model utilized by the application to achieve a malicious goal. For example, adversary adds adversarial perturbation to the image of a cheque. As a result, this image may mislead the NN model utilized by the ATM machine to cash out a huge amount of money. During the preparation of adversarial examples we assume that adversary has full knowledge of the targeted NN model, which is the white-box scenario. Also, we assume that adversary has limited computational power. As a result, the adversary can generate iterative adversarial examples but cannot exhaustively search all possible input perturbation.

2.2   Generating Adversarial Examples

The adversarial examples could be classified into white-box and black-box attacks based on adversary's knowledge of target NN classifier. Based on the generating process, they could be also classified as single-step and iterative adversarial examples.

Fast Gradient Sign Method (FGSM) is introduced by Goodfellow et al. in [6] as a single-step white-box adversarial example generator against NN image classifiers. This method tries to maximize the loss function value, $\mathcal{L}$, of NN classifier, $\mathcal{C}$, to find adversarial examples. The function $\mathcal{F}$ is used to ensure that the generated adversarial example is still a valid image.

$$\underset{\delta}{\text{maximize}} \quad \mathcal{L}(\hat{z} = \mathcal{C}(\hat{x}), t) \quad \text{subject to} \quad \hat{x} = \mathcal{F}(\bar{x}, \delta) \in \mathbb{R}^m_{[0,1]}$$

To keep visual similarity and enhance generation speed, this maximization problem is solved by running gradient ascent for one iteration. It simply generates adversarial examples, $\hat{x}$, from original images, $\bar{x}$, by adding small perturbation, δ, which changes each pixel value along the gradient direction of the loss function. As a single step adversarial example generator, FGSM can generate adversarial examples efficiently. However, the quality of the generated adversarial examples is relatively low due to the linear approximation of the loss function landscape.

Basic Iterative Method (BIM) is introduced by Kurakin et al. in [8] as an iterative white-box adversarial example generator against NN image classifiers. In the algorithm design, BIM utilizes the same mathematical model as FGSM. But, different from the FGSM, BIM is an iterative attack method. Instead of making the adversarial perturbation in one iteration, BIM runs the gradient ascent algorithm multiple iterations to maximize the loss function. In each iteration, BIM applies smaller perturbation and maps the perturbed image through the function $\mathcal{F}$. As a result, BIM approximates the loss function landscape by linear spline interpolation. Therefore, it generates stronger adversarial examples than FGSM within the same neighboring area.

Projected Gradient Descent (PGD) is another iterative white-box adversarial example generator recently introduced by Madry et al. in [13]. Similar to BIM, PGD also solves the same optimization problem iteratively with the projected gradient descent algorithm. However, PGD randomly selects an initial point within a limited area of the original image and repeats this several times to generate an adversarial example. With this multiple time random initialization, PGD is shown experimentally to solve the optimization problem efficiently and generate more serious adversarial examples since the loss landscape has a surprisingly tractable structure [13].

2.3   Adversarial Example Defensive Methods

Many defense methods have been proposed recently. In the following, we summarize and present representative samples from three major defense classes.

Augmentation and Regularization aims at penalizing overconfident prediction or utilizing synthetic data during training. One of the early ideas is the defensive distillation. Defensive distillation uses the prediction score from original NN, usually called teacher, as ground truth to train another smaller NN, usually called student [16,17]. It has been shown that the calculated gradients from the student model become very small or even reach zero and hence become useless to the adversarial example generator [16]. Some of the recent works that belong to this set of methods are referred to as Fortified Network [9] and Manifold Mixup [23]. Fortified Network utilizes denoising autoencoder to regularize the hidden states. Manifold Mixup also focuses on the hidden states but follows a different way. During the training, Manifold Mixup uses interpolations of hidden states and logits during training to enhance the diversity of training data. Compared with adversarial training defenses, this set of defenses has significant limitations. For example, defensive distillation is vulnerable to Carlini attack [4] and Manifold Mixup can only defend against single step attacks.

Protective Shell is a set of defensive methods which aim at using a shell to reject or reform the adversarial examples. An example of these methods is introduced by Meng et al. in [14] which is called MagNet. In this work, the authors design two types of functional components: the detector and the reformer. Adversarial examples are either rejected by the detector or reformed to eliminate the perturbations. Other recent works such as [11] and [19] try to utilize different methods to build the shell. In [11], authors inject adaptive noise to input images which breaks the adversarial perturbations without significant decrease of classification accuracy. In [19], a generator is utilized to generate images that are similar to the inputs. By replacing the inputs with generated images, it achieves resistance to adversarial examples. However, this set of methods usually assume the shell itself is black-box to the adversary and the work in [2] has already found ways to break such an assumption.

Adversarial Training is based on a straightforward idea that treats adversarial examples as blind spots of the original training data [25]. Through retraining with adversarial examples, the classifier learns the perturbation pattern and generalizes its prediction to account for such perturbations. In [6], the adversarial examples generated by FGSM are used for adversarial training and the trained NN classifier can defend single step adversarial examples. Later works in [13] and [22] enhance the adversarial training method to defend examples like BIM and PGD. A more recent work in [7] requires that the pre-softmax logits from original and adversarial examples to be similar. Authors believe this method could utilize more information during adversarial training. A common problem in existing adversarial training defenses is that the trained classifier has no control of the trade-off between correctly classifying original and adversarial examples. Our work achieves this flexibility and shows the benefit.

3   GanDef: GAN Based Adversarial Training

In this section, we present the design of our defensive method (GanDef) as follows. First, the design of GanDef is introduced as a minimax game of the classifier and discriminator. Then we conduct a theoretical analysis of the proposed minimax game in GanDef. Finally, we conduct experimental analysis to evaluate the convergence of GanDef.

3.1   Design

Given the training data pair $\langle x, t \rangle$, where $x \in \cup(\bar{X}, \hat{X})$, we try to find a classification function $\mathcal{C}$ that uses x to produce pre-softmax logits z such that:

$$t_i = f(z_i) = \frac{e^{z_i}}{\sum_{z_j} e^{z_j}} \quad \text{The mapping between } z \text{ and } t \text{ is the softmax function.}$$

Since x can be either original example $\bar{x}$ or adversarial example $\hat{x}$, we want the classifier to model the conditional probability $q_C(z|x)$ with only non-adversarial features. To achieve this, we employ another NN and call it discriminator $\mathcal{D}$. $\mathcal{D}$ uses the pre-softmax logits z from $\mathcal{C}$ as inputs and predicts whether the input to classifier is $\bar{x}$ or $\hat{x}$. This process can be performed by maximizing the conditional probability $q_D(s|z)$, where s is a Boolean variable indicating the source of x is original or adversarial. Finally, by combining the classifier and the discriminator, we formulate the following minimax game:

$$\min_{\mathcal{C}} \max_{\mathcal{D}} J(\mathcal{C}, \mathcal{D})$$

$$\text{where } J(\mathcal{C}, \mathcal{D}) = \mathop{\mathbb{E}}_{x \sim X, t \sim T} \{-log[q_C(z|x)]\} - \mathop{\mathbb{E}}_{z \sim Z, s \sim S} \{-log[q_D(s|z = \mathcal{C}(x))]\}$$

Table 1. Summary of notations

$\mathcal{L}$	Loss function of NN classifier
$\mathcal{F}$	Function which regularize pixel value of generated example
$\bar{x}$, $\bar{X}$; $\hat{x}$, $\hat{X}$; x, X	Original, adversarial and all training examples
$\bar{t}$, $\bar{T}$; $\hat{t}$, $\hat{T}$; t, T	Ground truth of original, adversarial and all training examples
$\bar{z}$, $\bar{Z}$; $\hat{z}$, $\hat{Z}$; z, Z	Pre-softmax logits of original, adversarial and all training examples
$\bar{s}$, $\bar{S}$; $\hat{s}$, $\hat{S}$; s, S	Source indicator of original, adversarial and all training examples
δ	Adversarial perturbation
$\mathcal{C}$, $\mathcal{C}^*$	NN based classifier
$\mathcal{D}$, $\mathcal{D}^*$	NN based discriminator
J, J'	Reward function of the minimax game
Ω, $\Omega_{\mathcal{C}}$, $\Omega_{\mathcal{D}}$	Weight parameter in the NN model
γ	Trade-off hyper-parameters in GanDef

In this work, we envision that the classifier could be seen as a generator that generates pre-softmax logits based on selected features from input images. Then, the classifier and the discriminator engage in a minimax game, which is also known as *Generative Adversarial Net* (GAN) [5]. Therefore, we name our proposed defense as "GAN based Adversarial Training" (GanDef). While other defenses ignore or only compare $\bar{z}$ and $\hat{z}$, utilizing discriminator with z adds a second line of defense when the classifier is defeated by adversarial examples.

The pseudocode of GanDef training is summarized in Algorithm 1 and is visualized in Fig. 1. A summary of the notations used throughout this work is available in Table 1.

Algorithm 1. GanDef Training

Input: training examples X, ground truth T, classifier $\mathcal{C}$, discriminator $\mathcal{D}$
Output: classifier $\mathcal{C}$, discriminator $\mathcal{D}$
1: Initialize weight parameters Ω in both classifier and discriminator
2: **for** the global training iterations **do**
3:     **for** the discriminator training iterations **do**
4:         Randomly sample a batch of training examples, $\langle x, t \rangle$
5:         Generate a batch of boolean indicator, s, corresponding to training inputs
6:         Fix weight parameters $\Omega_{\mathcal{C}}$ in classifier $\mathcal{C}$
7:         Update weight parameters $\Omega_{\mathcal{D}}$ in discriminator $\mathcal{D}$ by stochastic gradient descent
8:     **end for**
9:     Randomly sample a batch of training examples, $\langle x, t \rangle$
10:     Generate a batch of boolean indicator, s, corresponding to training inputs
11:     Fix weight parameters $\Omega_{\mathcal{D}}$ in discriminator $\mathcal{D}$
12:     Update weight parameters $\Omega_{\mathcal{C}}$ in classifier $\mathcal{C}$ by stochastic gradient descent
13: **end for**

3.2   Theoretical Analysis

With the formal definition of our GanDef, we perform a theoretical analysis in this subsection. We show that under the current definition where J is a combination of log likelihood of $Z|X$ and $S|Z$, the solution of the minimax game contains an optimal classifier which can correctly classify adversarial examples. It is worth noting that our analysis is conducted in a non-parametric setting, which means that the classifier and the discriminator have enough capacity to model any distribution.

Proposition 1. *If there exists a solution $(\mathcal{C}^*, \mathcal{D}^*)$ for the aforementioned minmax game J such that $J(\mathcal{C}^*, \mathcal{D}^*) = H(Z|X) - H(S)$, then $\mathcal{C}^*$ is a classifier that can defend against adversarial examples.*

Proof. For any fixed classification model $\mathcal{C}$, the optimal discriminator can be formulated as

$$\mathcal{D}^* = \arg\max_{\mathcal{D}} J(\mathcal{C}, \mathcal{D}) \quad = \arg\min_{\mathcal{D}} \mathop{\mathbb{E}}_{z \sim Z, s \sim S} \{-log[q_D(s|z = \mathcal{C}(x))]\}$$

In this case, the discriminator can perfectly model the conditional distribution and we have $q_D(s|z = \mathcal{C}(x)) = p(s|z = \mathcal{C}(x))$ for all z and all s. Therefore, we can rewrite J with optimal discriminator as J' and denote the second half of J as a conditional entropy $H(S|Z)$

$$J'(\mathcal{C}) = \mathop{\mathbb{E}}_{x \sim X, t \sim T} \{-log[q_C(z|x)]\} - H(S|Z)$$

For the optimal classification model, the goal is to achieve the conditional probability $q_C(z|x) = p(z|x)$ since z can determine t by taking softmax transformation. Therefore, the first part of $J'(\mathcal{C})$ (the expectation) is larger than or equal to $H(Z|X)$. Combined with the basic property of conditional entropy that $H(S|Z) \leq H(S)$, we can get the following lower bound of J with optimal classifier and discriminator

$$J(\mathcal{C}^*, \mathcal{D}^*) \geq H(Z|X) - H(S|Z) \geq H(Z|X) - H(S)$$

This equality holds when the following two conditions are satisfied:

- The classifier perfectly models the conditional distribution of z given x, $q_C(z|x) = p(z|x)$, which means that $\mathcal{C}^*$ is an optimal classifier.
- S and Z are independent, $H(S|Z) = H(S)$, which means that adversarial perturbations do not affect pre-softmax logits.

   In practice, the assumption of unlimited capacity in classifier and discriminator may not hold and it would be hard or even impossible to build an optimal classifier that outputs pre-softmax logits that are independent from adversarial perturbation. Therefore, we introduce a trade-off hyper-parameter γ into the minimax function as follows:

$$\mathop{\mathbb{E}}_{x \sim X, t \sim T} \{-log[q_C(z|x)]\} - \gamma \mathop{\mathbb{E}}_{z \sim Z, s \sim S} \{-log[q_D(s|z = \mathcal{C}(x))]\}$$

When $\gamma = 0$, GanDef is the same as traditional adversarial training. When γ increases, the discriminator becomes more and more sensitive to information of s contained in pre-softmax logits, z.

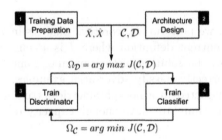

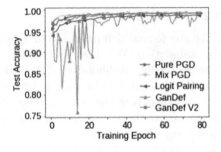

Fig. 1. Training with GanDef   **Fig. 2.** Convergence experiments

3.3   Convergence Analysis

Beyond the theoretical analysis, we also conduct an experimental analysis of the convergence of GanDef. Based on the pseudocode in Algorithm 1, we train a classifier on MNIST dataset. In order to compare the convergence, we also implement Pure PGD, Mix PGD and Logit Pairing and present their test accuracies on original test images during different training epochs.

As we can see from Fig. 2, the convergence of GanDef is not as good as other state-of-the-art adversarial training defenses. Although all these methods converge to over 95% test accuracy, GanDef shows significant fluctuation during the training process.

In order to improve the convergence of GanDef, we carefully trace back the design process and identify the root cause of the fluctuations. During the training of the classifier, we subtract the penalty term $\underset{z \sim Z, s \sim S}{\mathbb{E}} \{-log[q_D(s|z = \mathcal{C}(x))]\}$ which encourages the classifier to hide information of s in every z. Compared with Logit Pairing which requires similar z from original and adversarial examples, our penalty term is too strong. Therefore, we modify the training loss of the classifier to:

$$\underset{x \sim X, t \sim T}{\mathbb{E}} \{-log[q_C(z|x)]\} - \gamma \underset{\hat{z} \sim \hat{Z}, \hat{s} \sim \hat{S}}{\mathbb{E}} \{-log[q_D(\hat{s}|\hat{z} = \mathcal{C}(\hat{x}))]\}$$

Recall that $\hat{x}$, $\hat{z}$ and $\hat{s}$ represent the adversarial example, its pre-softmax logits, and the source indicator, respectively. It is also worth to mention that this modification is only applied to the classifier. Therefore, it does not affect the consistency of the previous proof. During convergence analysis, we denote the modified version of our defensive method as GanDef V2 and its convergence results are also shown in Fig. 2. It is clear that GanDef V2 significantly improves the convergence and stability during the training. Moreover, its test accuracy on the original as well as several different white-box adversarial examples is also higher than the initial design. Due to these improvements, we use it as the standard implementation of GanDef in the rest of this work.

4   Experiments and Results

In this section, we present comparative evaluation results of the adversarial training defenses introduced previously.

4.1   Datasets, NN Structures and Hyper-parameter

During evaluation, we conduct experiments for classifying original and adversarial examples on both MNIST and CIFAR10 datasets. To ensure the quality of evaluation, we utilize the standard python library (CleverHans [15]) and run all experiments on a Linux Workstation with NVIDIA GTX-1080 GPU. We choose the adversarial examples introduced in Sect. 2 and denote them as FGSM, BIM, PGD-1 and PGD-2 examples. For MNIST dataset, PGD-1 represents 40-iteration PGD attack while PGD-2 corresponds to 80-iteration PGD attack. Moreover, the maximum perturbation limitation is 0.3. The per step perturbation limitations for BIM and PGD examples are 0.05 and 0.01. For CIFAR10 dataset, these two sets of adversarial examples are 7-iteration and 20-iteration PGD attack. The maximum perturbation limitation is $\frac{8}{255}$ while per step perturbation limitation for BIM and PGD is $\frac{2}{255}$.

During the training, the vanilla classifier only uses original training data while defensive methods utilize original and PGD-1 examples except for Pure PGD which only requires the PGD-1 examples. For the testing part, we generate adversarial examples based on test data which was not used in training. These adversarial examples together with original test data form the complete test dataset during the evaluation stage. To make a fair comparison, defensive methods and vanilla classifier share the same NN structures which are (1) LeNet [13] for MNIST, and (2) allCNN [20] for CIFAR10. Due to the page limitation, the detailed structure is shown in the Appendix. The hyper-parameter of existing defensive methods are the same as the original papers [7,13]. During the training of Logit Pairing on CIFAR10, we found that using the same trade-off parameter as MNIST lead to divergence. To resolve the issue, we try to change the optimizer, learning rate, initialization and weight decay. However, none of them work until the weight of logit comparison loss is decreased to 0.01.

To validate the NN structure as well as the adversarial examples, we utilize the vanilla classifier to classify original and adversarial examples. Based on the results in Table 2, the test accuracy of the vanilla classifier on original examples matches the records of benchmarks in [3]. Moreover, the test accuracy of the vanilla classifier on any kind of adversarial examples has significant degeneration which shows the adversarial example generators are working properly.

4.2   Comparative Evaluation of Defensive Approaches

As the first step, we compare the GanDef with state-of-the-art adversarial training defenses in terms of test accuracy on original and white-box adversarial examples. The results are presented in Fig. 3 and summarized in Table 2.

On MNIST, all defensive methods achieve around 99% test accuracy on original examples and Pure PGD is slightly better than others. In general, the test accuracy of defensive methods are almost the same and does not go lower than that of the vanilla model. On CIFAR10, we can see that the test accuracy of defensive methods on original data is around 83% and these of Logit Pairing

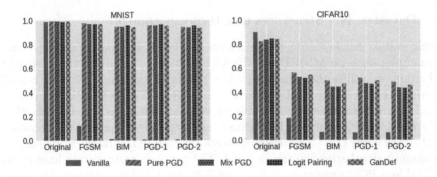

Fig. 3. Visualization of test accuracy on original and adversarial examples

and GanDef are slightly higher than others. Compared with the vanilla classifier, there is about 5% decrease in test accuracy. Similar degeneration is also reported in previous works on Pure PGD, Mix PGD and Logit Pairing [7,13].

During the evaluation on MNIST, there are no significant differences among defensive methods and each could achieve around 95% test accuracy. The Pure PGD method is the best on the evaluation of FGSM and BIM examples, while the Logit Pairing is the best on the evaluation of PGD-1 and PGD-2 examples. Based on the evaluation results from CIFAR10, we can see the differences between defensive methods are slightly larger. On all four kinds of white-box adversarial examples, Pure PGD is the best method and the test accuracy ranges from 48.33% (PGD-1) to 56.18% (FGSM). In the rest of defensive methods, GanDef is the best choice with test accuracy ranges from 45.62% (PGD-1) to 54.14% (FGSM).

Based on the comparison as well as visualization in Fig. 3, it is clear that the proposed GanDef has the same level of performance as state-of-the-art adversarial training defenses in terms of the trained classifier's test accuracy on original and different adversarial examples.

4.3   Evaluation of GanDef-Comb

In the second phase of evaluation, we consider GanDef-Comb which is a variant of GanDef. This variant utilizes both classifier and discriminator trained by GanDef. As we show in Sect. 3, the discriminator could be seen as a second line of defense when the trained classifier fails to make correct predictions on adversarial examples. By setting different threshold values for the discriminator, GanDef can dynamically control the trade-off between classifying original and adversarial examples. In current evaluation, the threshold is set to 0.5.

On MNIST, the test accuracy of GanDef-Comb on original, FGSM and BIM examples is the same as that of GanDef. On PGD-1 and PGD-2 examples, the test accuracy of GanDef-Comb has a small degeneration (less than 0.3%). This is because MNIST dataset is so simple such that the classifier alone can provide near optimal defense. Those misclassified corner cases are hard to be patched by utilizing discriminator. In common cases, the classifier has much larger degeneration on classifying adversarial examples. For example, on the CIFAR10, the

Table 2. Summary of test accuracy on original and adversarial examples

		Vanilla	Pure PGD	Mix PGD	Logit Pairing	GanDef	GanDef-Comb
MNIST	Original	98.70%	99.15%	99.17%	98.50%	99.10%	99.10%
	FGSM	12.15%	97.60%	96.89%	97.00%	96.85%	96.85%
	BIM	1.07%	94.75%	94.58%	95.83%	94.28%	94.28%
	PGD-1	0.87%	95.60%	95.56%	96.34%	95.31%	95.21%
	PGD-2	0.93%	94.14%	93.99%	95.42%	93.62%	93.38%
CIFAR10	Original	89.69%	82.06%	83.70%	84.21%	84.05%	63.97%
	FGSM	18.43%	56.18%	52.21%	51.63%	54.14%	87.61%
	BIM	6.76%	49.21%	44.39%	44.09%	46.64%	76.02%
	PGD-1	6.48%	51.51%	47.11%	46.53%	49.21%	80.39%
	PGD-2	6.44%	48.33%	43.48%	43.28%	45.62%	73.56%

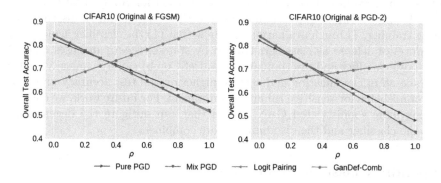

Fig. 4. Visualization of test accuracy under different ratio of adversarial examples

benefit of utilizing discriminator is obvious due to such degeneration. From the results of test accuracy, GanDef-Comb is significantly better than state-of-the-art adversarial training defenses on mitigating FGSM, BIM, PGD-1 and PGD-2 examples. Based on the comparison, GanDef-Comb enhances test accuracy by at least 31.43% on FGSM, 26.81% on BIM, 28.88% on PGD-1 and 25.23% on PGD-2. Although the test accuracy of GanDef-Comb on original examples has about 20% degeneration, the enhancement on defending adversarial examples benefits the overall test accuracy when the ratio of adversarial examples exceeds a certain limit.

To show the benefit of being able to control the trade-off, we design two experiments on CIFAR10 dataset. We form test dataset with original and adversarial examples (FGSM examples in the first experiment and PGD-2 examples in the second one). The ratio of adversarial examples, ρ, changes from 0 to 1. Giving similar weight losses in classifying original and adversarial examples, ρ represents the probability of receiving adversarial examples. Or, giving similar probabilities of receiving original and adversarial examples, ρ represents the weight of correctly classify adversarial examples ($1 - \rho$ for original examples).

These two evaluations are designed for risky or misbehavior-sensitive running environments, respectively.

The results of the overall test accuracy under different experiments are shown in Fig. 4. It can be seen that GanDef-Comb is better than state-of-the-art defenses in terms of overall test accuracy when ρ exceeds 41.7%. In real applications, we could further enhance the overall test accuracy through changing the discriminator's threshold value. When ρ is low, GanDef-Comb gives less attention to discriminator (high threshold value) and achieves similar performance as that of the state-of-the-art defenses. When ρ is high, GanDef-Comb relies on discriminator (low threshold value) to detect more adversarial examples.

5   Conclusion

In this paper, we introduce a new defensive method for Adversarial Examples, GanDef, which formulates a minimax game with a classifier and a discriminator during training. Through evaluation, we show that (1) the classifier achieves the same level of defense as classifiers trained by state-of-the-art defenses, and (2) using both classifier and discriminator (GanDef-Comb) can dynamically control the trade-off in classification and achieve higher overall test accuracy under the risky or misbehavior-sensitive running environment. For future work, we consider utilizing more sophisticated GAN models that can mitigate the degeneration when the classifier and the discriminator are combined.

Appendix Classifier Structures

(See Tables 3 and 4).

Table 3. MNIST LeNet classifier structure

Layer	Kernel size	Strides	Padding	Init
Convolution	$5 \times 5 \times 32$	1×1	Same	Default
MaxPool	2×2	2×2	-	-
ReLU	-	-	-	-
Convolution	$5 \times 5 \times 64$	1×1	Same	Default
MaxPool	2×2	2×2	-	-
ReLU	-	-	-	-
Flatten	-	-	-	-
Dense	1024	-	-	Default
ReLU	-	-	-	-
Dense	10	-	-	Default

Table 4. CIFAR10 allCNN classifier structure

Layer	Kernel size	Strides	Padding	Init
Dropout	0.2 (drop rate)	-	-	-
Convolution	$3 \times 3 \times 96$	1×1	Same	He
ReLU	-	-	-	-
Convolution	$3 \times 3 \times 96$	1×1	Same	He
ReLU	-	-	-	-
Convolution	$3 \times 3 \times 96$	1×1	Same	He
ReLU	-	-	-	-
MaxPool	2×2	2×2	-	-
Dropout	0.5 (drop rate)	-	-	-
Convolution	$3 \times 3 \times 192$	1×1	Same	He
ReLU	-	-	-	-
Convolution	$3 \times 3 \times 192$	1×1	Same	He
ReLU	-	-	-	-
Convolution	$3 \times 3 \times 192$	1×1	Same	He
ReLU	-	-	-	-
MaxPool	2×2	2×2	-	-
Dropout	0.5 (drop rate)	-	-	-
Convolution	$3 \times 3 \times 192$	1×1	Valid	He
ReLU	-	-	-	-
Convolution	$1 \times 1 \times 192$	1×1	Same	He
ReLU	-	-	-	-
Convolution	$1 \times 1 \times 192$	1×1	Same	He
ReLU	-	-	-	-
GlobalAvgPool	-	-	-	-
Dense	10	-	-	Default

References

1. Abu-Nimeh, S., Nappa, D., Wang, X., Nair, S.: A comparison of machine learning techniques for phishing detection. In: Proceedings of the Anti-Phishing Working Groups 2nd Annual eCrime Researchers Summit, pp. 60–69. ACM (2007)
2. Athalye, A., Carlini, N., Wagner, D.: Obfuscated gradients give a false sense of security: circumventing defenses to adversarial examples. arXiv preprint arXiv:1802.00420 (2018)
3. Benenson, R.: Classification datasets results (2018). http://rodrigob.github.io/are_we_there_yet/build/classification_datasets_results.html. Accessed 06 Apr 2018
4. Carlini, N., Wagner, D.: Towards evaluating the robustness of neural networks. In: 2017 IEEE Symposium on Security and Privacy (SP), pp. 39–57. IEEE (2017)
5. Goodfellow, I., Bengio, Y., Courville, A., Bengio, Y.: Deep Learning, vol. 1. MIT Press, Cambridge (2016)

6. Goodfellow, I.J., Shlens, J., Szegedy, C.: Explaining and harnessing adversarial examples. In: International Conference on Learning Representations (2015)
7. Kannan, H., Kurakin, A., Goodfellow, I.: Adversarial logit pairing. arXiv preprint arXiv:1803.06373 (2018)
8. Kurakin, A., Goodfellow, I., Bengio, S.: Adversarial machine learning at scale. In: International Conference on Learning Representations (2017)
9. Lamb, A., et al.: Fortified networks: improving the robustness of deep networks by modeling the manifold of hidden representations. arXiv preprint arXiv:1804.02485 (2018)
10. Lample, G., Zeghidour, N., Usunier, N., Bordes, A., Denoyer, L., et al.: Fader networks: manipulating images by sliding attributes. In: Advances in Neural Information Processing Systems, pp. 5969–5978 (2017)
11. Liang, B., Li, H., Su, M., Li, X., Shi, W., Wang, X.: Detecting adversarial examples in deep networks with adaptive noise reduction. arXiv preprint arXiv:1705.08378 (2017)
12. Louppe, G., Kagan, M., Cranmer, K.: Learning to pivot with adversarial networks. In: Advances in Neural Information Processing Systems, pp. 982–991 (2017)
13. Madry, A., Makelov, A., Schmidt, L., Tsipras, D., Vladu, A.: Towards deep learning models resistant to adversarial attacks. arXiv preprint arXiv:1706.06083 (2017)
14. Meng, D., Chen, H.: MagNet: a two-pronged defense against adversarial examples. In: Proceedings of the 2017 ACM SIGSAC Conference on Computer and Communications Security, pp. 135–147. ACM (2017)
15. Papernot, N., et al.: Technical report on the CleverHans v2.1.0 adversarial examples library. arXiv preprint arXiv:1610.00768 (2018)
16. Papernot, N., McDaniel, P.: Extending defensive distillation. arXiv preprint arXiv:1705.05264 (2017)
17. Papernot, N., McDaniel, P., Wu, X., Jha, S., Swami, A.: Distillation as a defense to adversarial perturbations against deep neural networks. In: 2016 IEEE Symposium on Security and Privacy (SP), pp. 582–597. IEEE (2016)
18. Rowley, H.A., Baluja, S., Kanade, T.: Neural network-based face detection. IEEE Trans. Pattern Anal. Mach. Intell. **20**(1), 23–38 (1998)
19. Samangouei, P., Kabkab, M., Chellappa, R.: Defense-GAN: protecting classifiers against adversarial attacks using generative models. arXiv preprint arXiv:1805.06605 (2018)
20. Springenberg, J.T., Dosovitskiy, A., Brox, T., Riedmiller, M.: Striving for simplicity: the all convolutional net. In: International Conference on Learning Representations (2017)
21. Szegedy, C., et al.: Intriguing properties of neural networks. In: International Conference on Learning Representations (2014)
22. Tramèr, F., Kurakin, A., Papernot, N., Goodfellow, I., Boneh, D., McDaniel, P.: Ensemble adversarial training: Attacks and defenses. arXiv preprint arXiv:1705.07204 (2017)
23. Verma, V., Lamb, A., Beckham, C., Courville, A., Mitliagkis, I., Bengio, Y.: Manifold mixup: encouraging meaningful on-manifold interpolation as a regularizer. arXiv preprint arXiv:1806.05236 (2018)
24. Xie, Q., Dai, Z., Du, Y., Hovy, E., Neubig, G.: Controllable invariance through adversarial feature learning. In: Advances in Neural Information Processing Systems, pp. 585–596 (2017)
25. Xu, W., Qi, Y., Evans, D.: Automatically evading classifiers. In: Proceedings of the 2016 Network and Distributed Systems Symposium (2016)

Control Logic Injection Attacks on Industrial Control Systems

Hyunguk Yoo[1] and Irfan Ahmed[1,2(✉)]

[1] University of New Orleans, New Orleans, LA 70148, USA
`hyoo1@uno.edu`
[2] Virginia Commonwealth University, Richmond, VA 23221, USA
`iahmed3@vcu.edu`

Abstract. Remote control-logic injection attacks on programmable logic controllers (PLCs) impose critical threats to industrial control system (ICS) environments. For instance, Stuxnet infects the control logic of a Siemens S7-300 PLC to sabotage nuclear plants. Several control logic injection attacks have been studied in the past. However, they focus on the development and infection of PLC control logic and do not consider the stealthy methods of transferring the logic to a PLC over the network. This paper is the first effort to explore the packet manipulation of control logic to achieve stealthiness without modifying PLC firmware to support new (obfuscation) functionality. It presents two new control logic injection attacks: (1) Data Execution and (2) Fragmentation and Noise Padding. *Data Execution* attack subverts signatures (based-on packet-header fields) by transferring control logic to the data blocks of a PLC and then, changes the PLC's system control flow to execute the attacker's logic. *Fragmentation and Noise Padding* attack subverts deep packet inspection (DPI) by appending a sequence of padding bytes in control logic packets while keeping the size of the attacker's logic in packet payloads significantly small. We implement the attacks on two industry-scale PLCs of different vendors and demonstrate that these attacks can subvert intrusion detection methods successfully, such as signature-based intrusion detection and Anagram-based DPI. We also release the training and attack datasets to facilitate research in this direction.

Keywords: Control logic · Code injection attack ·
Programmable logic controller · Ladder logic · Critical infrastructure

1 Introduction

Programmable logic controllers (PLCs) in industrial control systems (ICS) are directly connected to physical processes such as wastewater treatment plants, gas pipelines, and electrical power grids. They are equipped with control logic that define how to control and monitor the behavior of the processes [7]. Since ICS environments are increasingly connected to corporate network and Internet,

ⓒ IFIP International Federation for Information Processing 2019
Published by Springer Nature Switzerland AG 2019
G. Dhillon et al. (Eds.): SEC 2019, IFIP AICT 562, pp. 33–48, 2019.
https://doi.org/10.1007/978-3-030-22312-0_3

they are susceptible to cyber attacks [5,6]. In particular, attackers target the control logic of a PLC over the network to manipulate the behavior of a physical process. Stuxnet, for instance, infects the control logic of a Siemens S7-300 PLC to modify the motor speed of centrifuges periodically from 1,410 Hz to 2 Hz to 1,064 Hz and then over again. Since the discovery of Stuxnet in 2010, which sabotaged Iran's nuclear facilities, the number of ICS vulnerabilities reported each year has been dramatically increased [1].

To develop defensive solutions for control logic injection attacks, it is required to explore new attack vectors that target the control logic of a PLC to sabotage a physical process. In the past, different types of control logic injection attacks have been studied [9,12,15,16]. For instance, the analysis of Stuxnet reveals that it compromises the STEP 7 engineering software in a control center to establish the communication with a target PLC in a field site and then transfers a prebuilt malicious control logic to the PLC. These attacks focus on the development and infection of malicious control logic. However, they do not consider the stealthy methods of transferring control logic to a PLC over the network.

This paper is the first effort in this direction and proposes two new control logic injection attacks: (1) Data Execution, and (2) Fragmentation and Noise Padding. These attacks manipulate control logic packets without disrupting the transfer of control logic to a PLC or modifying PLC firmware to support new (obfuscation) functionality.

Data Execution attack subverts the signatures that are based on packet-header fields by transferring the (compiled) code of control logic to the data blocks of a PLC. The data blocks are used to exchange data such as sensor measurement values and states of PLC variables (e.g, `timers` and `counters`). Thus, the signatures must not block their packets. After transferring the logic to a PLC, the attack further modifies the PLC's system control flow to execute the logic located in data blocks. Note that PLCs do not enforce data execution prevention (DEP), thereby allowing the logic to execute.

Fragmentation and Noise Padding attack subverts deep packet inspection (DPI) by generating write request packets that contain a small size of a code fragment with substantial padding of noise. The ICS protocols often have address/offset fields in their headers, which are utilized by the attack to make the PLC discard the noise padding.

We implement the attacks on two PLCs used in industrial setting (i.e., Schneider Electric's Modicon M221 and Allen-Bradley's MicroLogix 1400) and evaluate them against two well-known network intrusion detection methods, i.e., signature-based intrusion detection and DPI (or payload-based anomaly detection [17]). Our evaluation results show that both intrusion detection methods fail to detect the attacks and therefore, warrant more research efforts. We create and release the training and attack datasets to facilitate research in this direction.

Contributions. The contributions of the paper are summarized as follows:

- We propose two new stealthy methods to subvert the detection of control logic code in ICS network traffic.

- We implement and demonstrate the attacks successfully on two industry-scale PLCs of different vendors supporting two proprietary protocols, Allen-Bradley's PCCC and Schneider Electric's M221.
- We evaluate the attacks against both signature-based intrusion detection and DPI to provide evidence of stealth.
- To facilitate research on the defensive solutions for the control logic injection attacks, we create and release the normal and attack traffic datasets[1].

2   Background and Related Work

2.1   PLC and Control Logic

A PLC is an embedded device with multiple inputs/outputs connected to sensors and actuators, used for real-time control and automation of physical processes such as assembly lines and gas pipelines. The PLC primarily provides closed-loop control to maintain the desired state of a physical process. *Control logic* is a program that is executed repeatedly by the PLC in a *scan cycle* consisting of three steps: (1) the sensor inputs are copied to the I/O image table of the PLC, (2) the control logic is executed based on the input values and the state from the previous scan cycle, (3) the result of control logic execution is reflected in the I/O image table from where the output values are transmitted to the connected actuators. PLC vendors provide *engineering software* to program and compile source code of control logic[2]. The engineering software communicates with a PLC to transfer control logic through various interfaces such as RS-232.

Generally, we can divide control logic into four types of blocks that are transferred to/from a PLC: configuration blocks, code blocks, data blocks, and information blocks. The *configuration blocks* contain information about the other blocks such as the address and size of a block, PLC's IP address, etc. The *code blocks* contain the *compiled* control logic code that runs by a PLC. The *data blocks* maintain PLC variables (e.g, `input`, `output`, `timer`, `counter`, etc.) used in the code blocks. The *information blocks* are only used by the engineering software to recover the original project file from decompiled source code when the control logic is retrieved from a PLC.

2.2   Attacks on PLCs

We categorize the existing control logic attacks on PLCs into two groups: (1) traditional control logic attacks, and (2) through firmware modification.

Traditional Control Logic Injection Attacks. They involve modifying the original control logic running on a target PLC by engaging its engineering software, typically employing a man-in-the-middle attack [9,16]. The primary vulnerability involved in this type of attack is the lack of authentication measures

---

[1] https://gitlab.com/hyunguk/control-logic-attack-datasets.

[2] IEC 61131-3 standard defines five PLC programming languages: ladder diagram, instruction list, functional block diagram, structured text, and sequential flow chart.

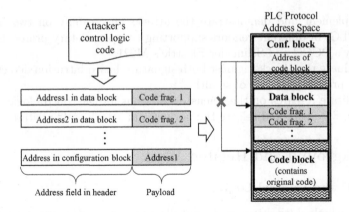

Fig. 1. Data execution attack on protocol header inspection

in the PLC protocols. In many cases, authentication is not supported for control logic download/upload operations or is supported in only one direction, either download or upload. When authentication measures are supported by PLCs, they are rarely used in practice. This section will discuss attacks from literature.

Stuxnet [9] is a representative example of this type of attack, which infects Siemens's STEP 7 (engineering software) and downloads malicious control logic to target PLCs (Siemens S7-300) by utilizing the infected engineering software.

Senthivel *et al.* [16] presents three control logic injection attack scenarios referred to as denial of engineering operations (DEO) attacks where an attacker can interfere with the normal engineering operation of downloading/uploading of PLC control logic. In DEO I, an attacker sitting in a man-in-the-middle position between a target PLC and its engineering software injects malicious control logic to the PLC and replaces it with normal (original) control logic to deceive the engineering software when uploading operation is requested.

DEO II is similar to DEO I except that it uploads malformed control logic instead of the original control logic to crash the engineering software. DEO III does not require a man-in-the-middle position, in which the attacker just injects specially crafted malformed control logic to the target PLC. The malicious control logic is specially crafted in a way that it can be run in the PLC successfully, but the engineering software can not decompile the control logic.

Firmware Modification Attacks. This type of attack [10] infect a PLC at firmware level. Since the firmware provides the control logic with an interface to the hardware such as inputs and outputs, malicious firmware can manipulate a connected physical process without tampering the control logic, which makes it difficult to detect the infection from engineering software or human-machine interfaces (HMIs) [10]. However, infecting PLC firmware would be a challenging task in a real ICS environment. Most PLC vendors protect their PLCs from unauthorized firmware update by cryptographic methods (e.g., digital signature) or allowing firmware update only by local access (e.g., SD cards and USB).

3   Stealthy Control Logic Injection Attacks

We propose two new attacks on PLC control logic in industrial control systems: (1) Data Execution, and (2) Fragmentation and Noise Padding, with the goal of subverting both signature-based intrusion detection and DPI.

3.1   Data Execution Attack

We present a typical signature-based approach to detect the packets of control logic attacks, identify the vulnerability that is exploitable and then, present the data execution attack based on the vulnerability.

Signatures for Detecting Control Logic Attacks. Stuxnet presents a typical control logic injection attack. It compromises the STEP 7 engineering software in a control center to transfer a prebuilt malicious control logic to a target PLC. Generally, control logic injection attacks must transfer a code block of malicious control logic to a PLC. These attacks can be prevented by blocking the packets containing a code block over the network. The PLC protocol header has fields that indicate payload type and are utilized to detect the code-block packets via accurate signatures. For instance, Digital Bond's Quickdraw provides signatures for Snort IDS to monitor ICS traffic including the transfer of control logic [14]. An example signature monitors the network traffic of Schneider Electric's Modicon PLC and raises an alert on the Modbus function code 90 for the uploading/downloading of control logic [3]. Generally, the signatures can be based on any protocol header field that indicates code blocks such as the address field that indicates a restricted address range for code blocks, and the payload type field that identifies code blocks in packet payload.

Vulnerability. We make two observations about PLC communication that cause exploitable vulnerability. First, the data blocks cannot be blocked by the signatures because they are required to exchange the current state of a physical process (being controlled by a PLC) with the control center services such as HMI. Second, the PLCs do not enforce data execution prevention (DEP). Thus, PLCs may be manipulated to execute data blocks.

Subverting Signatures via Data Execution Attack. Figure 1 shows a high-level concept of the Data Execution attack that can subvert the signatures for preventing code blocks over the network. The attack consists of two steps: *First*, the attacker transfers (malicious) control logic code to an arbitrary address of a target PLC, which is not in the address range for code block, and thus, is not blocked by the signatures (e.g., transferring it to a data block).

Second, after sending the entire control logic, the attacker targets a special pointer in configuration block, that indicates the start address of code block and is used by the PLC to execute the control logic. The attacker modifies the pointer to the base address of the malicious control logic code, which in turn, redirects the PLC's system control flow to the malicious logic and forces the PLC to start executing it.

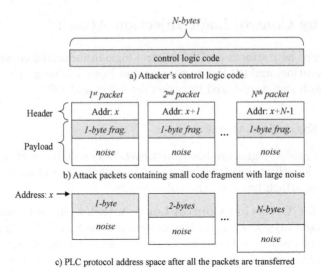

Fig. 2. Fragmentation and noise padding attack on deep packet inspection

Note that while updating control logic do not occur frequently in real-world ICS environment, data values or configuration setting can be exchanged more frequently between PLCs and ICS services (such as HMI and engineering software) and thus, cannot be blocked by the signatures, making this attack stealthy for signature-based detection.

3.2   Fragmentation and Noise Padding Attack

We first discuss payload-based anomaly detection (DPI) that is suitable for ICS traffic monitoring for control logic attacks, identify an exploitable vulnerability and then, present the fragmentation/noise padding attack on the vulnerability.

Deep Packet Inspection for Control Logic Attacks. The semantics of ICS packet payload are often proprietary and unknown. A practical approach for the DPI of ICS traffic is to utilize byte level features without involving semantics. These approaches have been studied extensively in the literature such as PAYL [18] and Anagram [17]. Generally, they obtain n-gram (a sequence of consecutive n byte) frequency as features from packet payloads and then, apply statistics or machine learning algorithms for anomaly detection.

Vulnerability. Hadziosmanovic *et al.* [11] report that the aforementioned DPI techniques cannot detect attack packets that contain significantly small-size attack payload because these packets tend to blend with normal packets. In other words, the smaller the attacker's code in the payload is, the harder it is to detect.

DPI Subversion via Fragmentation and Noise Padding Attack. We utilize Hadziosmanovic *et al.* [11]'s findings in the control logic traffic by fragment-

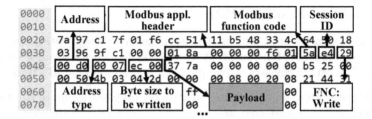

Fig. 3. The write request message format of the M221 protocol

ing the payload of the attacker's control logic and combining it with noise data to make it stealthy against byte-level features for DPI. We notice that some protocols of PLC such as Modicon M221 may allow an attacker to reduce the control logic fragment size to one or two bytes.

Figure 2 describes the *Fragmentation and Noise Padding* attack. Each *write request* packet contains only one-byte fragment of the attacker's code with a large noise of data (i.e., a sequence of padding bytes). To ensure that a target PLC does not use the noise and only execute actual control logic code in packet payloads, the attacker manipulates the address field stating the start location of the write operation in PLC memory for *write request* messages.

The core idea is that the next *write request* overwrites the noise data of the previous *write request* in PLC memory. In other words, the address of each write request packet increases by one (which is the size of an actual code fragment per packet). It keeps overwriting the noise data one byte per packet with the actual code bytes in contiguous memory locations. After all the packets are transferred to the target PLC, the whole code is placed from the address x to $x + N - 1$ where x is the address of the first write request and N is the size of attacker's code.

Note that the attacker can evade both signatures and DPI by using both attacks together. The attacker transfers control logic code to a non-code block in PLC (i.e., data execution), while keeping the size of code to one byte per packet blended with noise data (i.e., fragmentation and noise padding).

4   Implementation

We have implemented the attacks on two industry-scale PLCs: Schneider Electric's Modicon M221 and Allen-Bradley's MicroLogix 1400. To demonstrate the stealthiness of the attacks, we have also implemented two well-known network intrusion detection methods for proof-of-concept, (1) Scapy-based signatures, and (2) Anagram-based DPI [17]. This section further discusses the implementation details for each PLC.

4.1 Attacks on Modicon M221

We have implemented both attacks (i.e., data execution, fragmentation and noise padding) together for Modicon M221 as one stealthy attack against signatures and DPI. The attack consists of two phases: (1) preparation and (2) execution.

Preparation Phase. In this phase, an attacker prepares *compiled* control logic code, which will be transferred to a target PLC, in a lab environment.

Getting Compiled Code Block. The attacker acquires compiled control logic code in the following way. She programs control logic using her engineering software and then, captures network packets when the control logic is being downloaded to a M221 PLC. From the captured packets, compiled logic code can be extracted by assembling the packet payloads containing logic code, indicated by the address fields of write request messages (refer to Fig. 3)[3].

Execution Phase. The attack on Modicon M221 is conducted in five steps.

Step (1) Getting the address space layout of the target PLC. There are four types of block (configuration, code, data, and information) which are transferred to/from a PLC in the normal engineering operation of control logic. We refer to other areas in the address space of the PLC as *unknown areas* which are not occupied by the four types of blocks. Most of the unknown areas are filled with zero but they also contain the chunks of binary data which could be firmware.

To select an injection area, the attacker needs to know the address ranges for each control logic block. In the Modicon M221 PLC, there are two configuration blocks which contain the start addresses and the sizes of other blocks. By reading the first configuration block[4], the attacker can locate the second configuration block and an information block. By further reading the second configuration block, the information of other blocks (i.e., two data blocks and one code block) can be acquired.

Step (2) Select an injection area. To evade signature-based header inspection, the attacker injects her code into an area that does not belong to code block. We have discovered that the address ranges for code block always fall between 0xe000 and 0xfed4, which can be used as a signature to detect control logic over the network. Thus, the attacker can select an injection area except that address range. More specifically, the attacker selects an injection area among the information block, the data blocks, and the unknown areas, since there is no enough *available space* in the configuration blocks to fit logic code.

Step (3) Check the availability of the target injection area. Before injecting the logic code, the attacker needs to check if the target area is *available*, namely, overwriting the area does not affect the operations during the PLC's scan

---

[3] By reverse engineering the M221 protocol (referred it to the PLC's model in this paper), we have figured out that the code block is always written between the address range 0xe000 ∼ 0xfed4.

[4] The first configuration block has a fixed start address (0xfed4) and size (300 bytes).

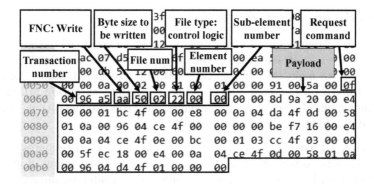

Fig. 4. The write request message format of the PCCC protocol

cycle. If the target area falls into the information block, it is not required to check the availability because the information block is only parsed in the engineering software, after being retrieved from a PLC. On the other hand, the attacker has to examine whether the target area is not currently used in the target PLC for the other locations (i.e., the data blocks and the unknown areas). Overwriting critical data/code of the unknown areas (e.g., firmware) will render the PLC inoperable. Similarly, if the attacker's code is written at the area for the I/O table which is maintained in the data blocks, the code will be overwritten with some arbitrary values at the next scan cycle of the PLC because the I/O table will be updated based on the inputs and outputs.

As a heuristic method, the availability of target area is tested by checking the entire area is filled with zero, based on our observation that the large amount of space in the data blocks and the unknown areas are just filled with zero. Our experimental results have shown that this method is actually effective since all the injected logic code is successfully run on a PLC.

Step (4) Transferring attacker's code using the fragmentation and noise padding. The attacker's logic code (i.e., the compiled code acquired in the preparation phase) is transferred to the target PLC using the fragmentation and noise padding attack. Each write request packet of the attack contains only one-byte of code fragment followed by 235 bytes of zero padding[5]. To perform the data execution attack simultaneously, the addresses of the write request messages start from the target address (i.e., the start address of the injection area) which was selected in the previous step.

Step (5) Change the pointer to code block. Lastly, the attacker modifies the pointer that points to the original logic code to point to the attacker's logic code. The pointer is two-byte size and located at the address 0xff90, which is included in the first configuration block. Consequently, the PLC executes the attacker's code instead of the original code for each PLC scan cycle.

---

[5] The maximum payload size of the M221 protocol is 236 bytes.

4.2   Attacks on MicroLogix 1400

Allen Bradley's MicroLogix PLC family uses the PCCC protocol to communicate with its engineering software [16]. Unlike the attack on Modicon M221, we could not find a way to make the MicroLogix 1400 PLC execute attacker's logic code which is stored in a non-code block and thus the data execution attack could not be implemented. We have implemented only the fragmentation and noise padding attack for the MicroLogix 1400 PLC.

Preparation Phase. In the preparation phase, an attacker prepares captured packets (i.e., a `pcap` file) which are generated in the downloading of attacker's control logic to a PLC from engineering software, in a lab environment.

Packets of Attacker's Control Logic. The MicroLogix 1400 PLC requires a legitimate sequence of communication including all the control logic blocks[6] for updating the logic of the PLC, whereas an attacker could transfer only a code block to the Modicon M221 PLC. To transfer an attacker's control logic to the MicroLogix 1400 PLC, the attacker uses the packets captured when the logic is being downloaded to the PLC from engineering software.

Execution Phase. We have developed an attack tool which takes as input a `pcap` file that is generated in the preparation phase, and replays all the PCCC request messages to the target PLC without modification except the write request messages containing logic code. For the messages indicated to contain logic code (i.e., write request messages with file number 0x02 and file type 0x22 [16]), their payloads (i.e., logic code) are fragmented into two-byte pieces[7] and a padding consisting of zero is appended to each code fragment. Then, new write request messages with code fragments and padding are sent to the target PLC, instead of the original messages.

When manipulating the messages of logic code, the payload size field next to the function code field (refer to Fig. 4) should be fixed according to the modified payload size. Similarly, the sub-element number field which indicates an offset to write within a file (or a block) should be adjusted to the total size of attacker's code fragments transferred previously, to overwrite the padding.

4.3   Network Intrusion Detection Systems (NIDS)

We have implemented two proof-of-concept tools for network intrusion detection, one represents signature-based detection for ICS protocols and other is DPI based on Anagram [17], which is a popular payload-based anomaly detection approach that uses byte-level (n-gram) features. We will show how they effectively detect a traditional control logic injection attack which does not involve any evasion techniques, and evaluate the proposed attacks against them, in Sect. 5.

---

[6] Control logic blocks are referred to as files in MicroLogix PLC family.

[7] Two-byte is the smallest size of payload in the PCCC protocol since the basic data unit of the PLC is 16-bit word.

Scapy-Based Signatures Using Packet Header Fields. We have developed a signature-based control logic detection tool in Python using the Scapy library. In the Modicon M221 PLC, control logic code always exists in the address range between 0xe000 and 0xfed4. Based on this feature, we configure a signature that checks if the address field of write request message is greater than 0xe000 and less than 0xfed4, which means the write request message contains logic code.

Anagram-Based Deep Packet Inspection Method. Anagram [17] is a payload-based anomaly detector that models a higher-order n-grams $(1 < n)$ analysis, which is one of the most effective payload-based anomaly detector for ICS network traffic [11], not requiring semantic knowledge about packet payloads. In a nutshell, it stores n-grams $(1 < n)$ observed in the payloads of normal datasets (i.e., training datasets) in a bloom filter and then, scores each packet by measuring the number of n-grams that were not observed in the normal datasets (i.e., the n-grams which are not stored in the bloom filter), to classify abnormal packets in the detection phase.

In our implementation of Anagram, we models the payload of logic code instead of the payload of normal packets. Remember that our goal is not to detect abnormal packets but to detect packets containing control logic code. Verifying control logic is out of the scope of this work, which has been studied in the literature such as TSV [13]. Note that the detection of control logic must be done to verify it.

In the training phase, we store all the n-grams seen in the payloads of logic code in a bloom filter, based on a training dataset. We examine different n-gram sizes from 2-gram to 20-gram, to find the optimal size of n-gram for each PLC. From our experiment, we have found that 4-gram is the optimal size for Modicon M221 whereas 8-gram for MicroLogix 1400.

In the detection phase, each packet is scored by counting the n-grams in its payload that are present in the bloom filter. If the score is equal or greater than a threshold, it is classified as containing logic code. We use 4 as thresholds for both PLCs with which false positive rates are 0% as discussed in Sect. 5.

5 Evaluation

5.1 Experimental Settings

We evaluate the proposed attacks on two different PLCs used in industrial setting, Schnider Electric's Modicon M221 and Allen-Bradley's MicroLogix 1400, against two different intrusion detection methods, signature-based intrusion detection and DPI. For each PLC, we use its engineering software, SoMachine-Basic v1.6 (Modicon M221) and RSLogix 500 v9.05.01 (MicroLogix 1400), to generate training datasets and in the preparation phase of attack.

The system implementing the proposed attacks runs on Ubuntu 16.04 virtual machine, which communicate with the PLCs over Ethernet. We also place a traditional attacker in the network who utilizes the engineering software (running on Windows 7 virtual machine) to inject control logic to the PLCs. Then, we

Table 1. Description of the datasets for Modicon M221

Datasets	Size (MB)	# of control logic	# of M221 packets	# of write request packets	# of packets w/code
Training	2.1	22	10,148	1,101	27
Code injection w/o evasion	3.7	29	11,092	1,535	38
Data execution & Noise padding	2.2	29	8,168	5,362	3,865

Table 2. Description of the datasets for MicroLogix 1400

Datasets	Size (MB)	# of control logic	# of PCCC packets	# of write request packets	# of packets w/code
Training	19.9	52	71,824	4,084	646
Code injection w/o evasion	39.7	75	168,736	5,465	684
Noise padding	38.6	75	238,657	29,647	24,866

capture the network traffic from each attack to generate attack datasets, and examine if our intrusion detection systems can detect the packets of attacker's logic code.

5.2 Datasets

For the evaluation, we developed 51 different programs of control logic for the Modicon M221 PLC (22 programs for training and 29 programs for attack), and 127 programs for the MicroLogix 1400 PLC (52 programs for training and 75 programs for attack), referred to Tables 1 and 2. The programs are written for different physical processes such as water tanks, gas pipelines, and traffic light involving various complexity.

Training Datasets. The training datasets are generated by capturing network traffic when control logic is being downloaded to a PLC. Based on the training datasets, we have generated the bloom filters of n-grams for each PLC (4-gram for Modicon M221 and 8-gram for MicroLogix 1400).

Traditional Attack Datasets. Generally, traditional attackers remotely targeting the control logic of PLCs engage engineering software typically in a way of man-in-the-middle attack as they did in Stuxent [9] and the DEO attacks [16]. To generate the traditional attack datasets, therefore, we bring an attacker who utilize engineering software to inject control logic to a PLC. Note that the traditional attack does not include any evasion techniques.

Table 3. Detection rates of signatures and DPI against the attacks on Modicon M221

Attacks	# of write request packets	# of packets w/code	TPR	FPR
Code injection w/o evasion	1,535	38	100% (38/38)	0% (0/1497)
Data execution & Noise padding	5,362	3,865	0% (0/3865)	0% (0/1497)

Table 4. Detection rates of DPI against the attacks on MicroLogix 1400

Attacks	# of write request packets	# of packets w/code	TPR	FPR
Code injection w/o evasion	5,465	684	96.78% (662/684)	0% (0/4781)
Noise padding	29,647	24,866	0% (0/24866)	0% (0/4781)

Attack Datasets for Data Execution/Noise Padding. For the datasets of the proposed attacks, we transfer the same set of control logic used in the traditional attack to a PLC but in different ways, as described in Sect. 4. The attack dataset of Modicon M221 includes both the data execution and noise padding attacks[8] whereas the attack dataset of MicroLogix 1400 includes only the noise padding attack.

5.3   Evaluation Results

We present the evaluation results on the traditional attack and the proposed attacks against our implementations of signatures and DPI, described in Tables 3 and 4. The packets of the traditional attack targeting the Modicon M221 PLC is detected on both signatures and DPI with 100% true positive rate (TPR) and 0% false positive rate (FPR). For the MicroLogix 1400 PLC, the traditional attack is detected on DPI with 96.78% of detection rate and 0% of false alarm[9] This result shows that our implementations of signatures and DPI are very effective to detect traditional control logic injection attacks which do not involve any evasion techniques.

Data Execution Attack. We evaluate the effectiveness of the data execution attack on the Modicon M221 PLC against the signature-based IDS. Table 3 shows that the IDS was fully-functional to detect the control logic packets that do not involve the data execution attack. However, the IDS was subverted completely and could not detect a single attack packet when data execution attack was employed.

---

[8] Short for the fragmentation and noise padding attack.

[9] With the classification threshold of 3 for MicroLogix 1400 (instead of 4), the detection rate can be slightly increased (97.95%) in return for increased FPR (0.08%).

Fragmentation and Noise Padding Attack. We evaluate the effectiveness of the fragmentation and noise padding attack on both Modicon M221 and MicroLogix 1400 against the Anagram-based DPI. Tables 3 and 4 show the evaluation results. We noticed a significant increase on the number of packets due to the high fragmentation of control logic i.e., one or two bytes of control logic code per packet. We also found that the noise padding attack subverted the DPI completely. Recall that the DPI was fully functional and detected the control logic packets successfully when the noise padding attack was not used.

6   Countermeasures

This section discusses the countermeasures against the proposed attacks in the following four categories.

User Authentication. Authentication for control logic download operation (or write requests) can defeat the proposed attacks, although it is not supported in many PLCs or just rarely used in practice. When authentication is supported in a PLC, an attacker may find a vulnerability to bypass it [2].

Data Execution Prevention. Data execution prevention (DEP) technique is a well-known protection mechanism used in modern desktop/server environments to prevent code injection attacks to stack or heap. It enforces $W \oplus X$ on all memory regions using a memory management unit (MMU), which is usually not present on microcontrollers. To enforce no-execution of code on non-code regions of a PLC, a memory protection unit (which is present in many microcontrollers) can be utilized as studied in [8].

Control Flow Integrity. Control-flow integrity (CFI) mechanism is a protection technique to detect control-flow hijacking attacks by restricting control-flow transfers based on a pre-determined control-flow graph. Most of CFI solutions designed for general-purpose computers may occur significant performance overhead which can be a critical problem for hard real-time PLCs. Nonetheless, a specially designed CFI solution such as [4], which consider availability and real-time requirements of the ICS systems, could be used.

Robust Network Intrusion Detection Systems. The previously discussed countermeasures require significant change/update on PLC design. It is not expected that those solutions could be widely deployed in the near future, considering the long life-cycle of ICS systems. On the other hand, network-based IDS solutions have the advantages of being deployed without significantly affecting existing systems. Although we have demonstrated that the existing approaches can be deceived by the proposed attacks, there is a room for research to subvert the attacks. For instance, the fragmentation and noise padding attack could be subverted if a NIDS keeps the state of PLC's address space using a mirrored space. Even though there is only small amount of code fragment in an individual attack packet (which is undetectable), it could detect attacker's code by scanning the mirrored space instead of individual packet.

7   Conclusion

We presented two new stealthy control logic injection attacks to demonstrate that an attacker can subvert both signature-based header inspection and DPI to transfer control logic to a PLC successfully. The attacks were implemented on two industry-scale PLCs and demonstrated their effectiveness against two well-known intrusion detection approaches, Scapy-based Signature and Anagram-based DPI. Our evaluation results showed that the attacks were stealthy and warrants for more research on IDS for ICS network traffic monitoring.

References

1. ICS-CERT Annual Vulnerability Coordination Report. Report, National Cybersecurity and Communications Integration Center (2016)
2. ICS-CERT Advisory on Modicon M221 (ICSA-18-240-01) (2018). https://ics-cert.us-cert.gov/advisories/ICSA-18-240-01. Accessed 15 Dec 2018
3. Quickdraw Snort - modicon.rules (2018). https://github.com/digitalbond/Quickdraw-Snort/blob/master/modicon.rules. Accessed 15 Dec 2018
4. Abbasi, A., Holz, T., Zambon, E., Etalle, S.: ECFI: asynchronous control flow integrity for programmable logic controllers. In: Annual Computer Security Applications Conference (ACSAC) (2017)
5. Ahmed, I., Obermeier, S., Naedele, M., Richard III, G.G.: SCADA systems: challenges for forensic investigators. Computer **45**(12), 44–51 (2012). https://doi.org/10.1109/MC.2012.325
6. Ahmed, I., Obermeier, S., Sudhakaran, S., Roussev, V.: Programmable logic controller forensics. IEEE Secur. Priv. **15**(6), 18–24 (2017). https://doi.org/10.1109/MSP.2017.4251102
7. Ahmed, I., Roussev, V., Johnson, W., Senthivel, S., Sudhakaran, S.: A SCADA system testbed for cybersecurity and forensic research and pedagogy. In: Proceedings of the 2nd Annual Industrial Control System Security Workshop, ICSS, pp. 1–9. ACM, New York (2016). https://doi.org/10.1145/3018981.3018984
8. Clements, A.A., et al.: Protecting bare-metal embedded systems with privilege overlays. In: 2017 IEEE Symposium on Security and Privacy (SP) (2017)
9. Falliere, N., Murchu, L.O., Chien, E.: W32.Stuxnet dossier. White paper, Symantec Corp., Security Response **5**(6), 29 (2011)
10. Garcia, L., Brasser, F., Cintuglu, M.H., Sadeghi, A.R., Mohammed, O., Zonouz, S.A.: Hey, my malware knows physics! Attacking PLCs with physical model aware rootkit. In: Network and Distributed System Security Symposium (NDSS) (2017)
11. Hadžiosmanović, D., Simionato, L., Bolzoni, D., Zambon, E., Etalle, S.: N-gram against the machine: on the feasibility of the N-gram network analysis for binary protocols. In: Balzarotti, D., Stolfo, S.J., Cova, M. (eds.) RAID 2012. LNCS, vol. 7462, pp. 354–373. Springer, Heidelberg (2012). https://doi.org/10.1007/978-3-642-33338-5_18
12. Kalle, S., Ameen, N., Yoo, H., Ahmed, I.: CLIK on PLCs! Attacking control logic with decompilation and virtual PLC. In: Binary Analysis Research (BAR) Workshop, Network and Distributed System Security Symposium (NDSS) (2019)
13. McLaughlin, S.E., Zonouz, S.A., Pohly, D.J., McDaniel, P.D.: A trusted safety verifier for process controller code. In: Network and Distributed System Security Symposium (NDSS) (2014)

14. Peterson, D.: Quickdraw: generating security log events for legacy SCADA and control system devices. In: IEEE Conference For Homeland Security (2009)
15. Senthivel, S., Ahmed, I., Roussev, V.: SCADA network forensics of the PCCC protocol. Digit. Invest. **22**, S57–S65 (2017)
16. Senthivel, S., Dhungana, S., Yoo, H., Ahmed, I., Roussev, V.: Denial of engineering operations attacks in industrial control systems. In: ACM Conference on Data and Application Security and Privacy (CODASPY) (2018)
17. Wang, K., Parekh, J.J., Stolfo, S.J.: Anagram: a content anomaly detector resistant to mimicry attack. In: Zamboni, D., Kruegel, C. (eds.) RAID 2006. LNCS, vol. 4219, pp. 226–248. Springer, Heidelberg (2006). https://doi.org/10.1007/11856214_12
18. Wang, K., Stolfo, S.J.: Anomalous payload-based network intrusion detection. In: Jonsson, E., Valdes, A., Almgren, M. (eds.) RAID 2004. LNCS, vol. 3224, pp. 203–222. Springer, Heidelberg (2004). https://doi.org/10.1007/978-3-540-30143-1_11

An Efficient and Scalable Intrusion Detection System on Logs of Distributed Applications

David Lanoë[1,2], Michel Hurfin[1(✉)], Eric Totel[2], and Carlos Maziero[1,3]

[1] Univ Rennes, Inria, CNRS, IRISA, 35000 Rennes, France
michel.hurfin@inria.fr
[2] CentraleSupélec, Inria, CNRS, IRISA, 35000 Rennes, France
{eric.totel,david.lanoe}@centralesupelec.fr
[3] Univ Federal Paraná, Curitiba 81531-980, Brazil
maziero@inf.ufpr.br

Abstract. Although security issues are now addressed during the development process of distributed applications, an attack may still affect the provided services or allow access to confidential data. To detect intrusions, we consider an anomaly detection mechanism which relies on a model of the monitored application's normal behavior. During a model construction phase, the application is run multiple times to observe some of its correct behaviors. Each gathered trace enables the identification of significant events and their causality relationships, without requiring the existence of a global clock. The constructed model is dual: an automaton plus a list of likely invariants. The redundancy between the two sub-models decreases when generalization techniques are applied on the automaton. Solutions already proposed suffer from scalability issues. In particular, the time needed to build the model is important and its size impacts the duration of the detection phase. The proposed solutions address these problems, while keeping a good accuracy during the detection phase, in terms of false positive and false negative rates. To evaluate them, a real distributed application and several attacks against the service are considered.

Keywords: Anomaly detection · Distributed application · Models

1 Introduction

Sensitive applications such as e-commerce or file systems are now running on large scale distributed systems, being prime targets for attackers who can misuse the service, steal information, compromise service availability or data integrity. The usual approach to detect incorrect behaviors in distributed systems consists in monitoring each process with a local intrusion detection system (IDS). Events

This work is partially funded by the French Ministry of Defense (DGA).

generated locally are stamped with a global clock and sent to a central correlation engine. This engine orders all events using their timestamps and analyzes their sequence, to detect predefined patterns corresponding to attack scenarios [8]. This approach can only detect known attacks, for which a "signature" of events can be specified. In addition, a global clock is needed to build the total order over all the events, which may be unfeasible in large distributed applications.

Anomaly-based intrusion detection systems look for deviations from a normal behavior reference model; any deviation is interpreted as an observable consequence of an attack. No previous knowledge about the possible attacks is required. In addition, recent works [15] shown that it is possible to build effective models without relying on a precise observation of the events' order. Instead of a total event ordering, weaker but easier to observe partial order relations, like Lamport's "happened before" [7], may be used. In a previous paper [15], we proposed a method to build a behavior model for a distributed application. This method takes as input an execution trace to build a lattice of consistent cuts [4], which is then used for deducing two complementary sub-models: a list of likely invariants and an automaton (a state machine). A single execution of the distributed application may not capture all its correct behaviors; those not observed could be futurely rejected by the model as false positives, during the detection phase. To face this problem, several executions traces are used, generating distinct sub-models of the same application. They may then be merged [1] and generalized [10,15], to produce models accepting additional behaviors that have not been observed. The main challenge is to build an effective model while addressing scalability (both during the construction of the model and the detection phases). As our ultimate goal is to detect anomalies, effectiveness can be measured in terms of false positive and false negative rates. The time needed to build and update the model is important, specially for longer execution traces. Finally, smaller models enable faster attack detection, which is often a major requirement.

The contribution of this paper is fourfold: (1) As the lattice of consistent cuts grows exponentially in the context of large distributed systems [5], we propose solutions that reduce the computing time and generate a more compact initial model. A judicious selection of a few sequences allows us to obtain more quickly a model that is not less efficient; (2) Existing solutions usually build a huge and complex automaton-based model and then reduce it, during a final generalization phase. We propose an iterative approach where the model under construction is repeatedly merged and generalized; (3) To analyze their impacts on the performance but also on the quality of the detection, the proposed solutions were implemented and evaluated using a real distributed file system and some real attacks against it; (4) finally, we show that the two sub-models (the invariant list and the automaton) are complementary. The paper is organized as follows: Sect. 2 presents the state of the art and discusses some related work. Section 3 describes our contributions to reduce the time complexity of the construction phase without sacrificing effectiveness. Finally, Sect. 4 assesses our proposal.

2   State of the Art

It is usual to model the behavior of a single process by the sequence of systems calls it issues or, more generally, by the sequence of actions performed during its computation. In some distributed systems, the existence of a global clock allows us to keep a model based on a single sequence of actions. Otherwise, the analysis has to focus on causal dependency relations between events. Inferring a model of the correct behaviors of an application from a set of traces is an old research topic. Using such a model to design an anomaly detection system is a more recent research area. The behavior model is usually either an automaton [9,11] or a set of temporal properties [2]. As a model learned from a finite set of traces is possibly incomplete, related work includes also generalization techniques [3,10]. Scalability issues are important and addressed in [5,6,16] in ways different from what is proposed in this paper. Rather than providing separate descriptions of incompatible work (*i.e.* conducted with different objectives), we describe in this section a uniform framework based on the concept of lattice of the consistent cuts [4]. Thus, this framework is closely related to two studies [1,15]. During our experiments, an implementation of this framework is used as a reference point.

In an anomaly-based detection approach, a first phase (called the construction phase) is executed once to build a reference model that specifies correct behaviors. Then, this model is used during a detection phase to monitor an execution potentially targeted by attacks. Figure 1 illustrates this global process, already adopted in [15]. Three main modules (grey boxes in Fig. 1) are in charge of (1) generating an intermediate model for each learned execution, (2) merging the intermediate models and generalizing the obtained model, and (3) detecting an anomaly. They are described in the following sub-sections.

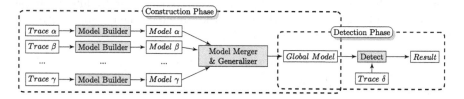

Fig. 1. Overview of the approach

2.1   Building an Intermediate Model from an Execution Trace

During the construction phase, the application is executed a finite number of times (executions α, β, ..., γ) and produces one trace (E^α, E^β, ..., E^γ) per execution. We assume that no attack could occur during the construction phase. The module *Model builder* (see Fig. 1) analyzes each trace separately and generates an intermediate model for each execution. A distributed application consists of n processes (p_1, ..., p_n) running on one or more physical nodes. The number of interacting processes may change from one execution of the application to the other. Therefore, n denotes the maximal number of processes involved during any observed execution. During a computation α, each process produces a sequence of events where an event corresponds to either an internal action or a communication action, namely the sending of a message m on a channel c (denoted $c!m$) or the delivery of a message m received on channel c (denoted $c?m$). The n processes are instrumented to generate log files. During an execution α, each process p_i logs information about its local events in a local log file E_i^α. Thus a trace E^α is a collection of n logs. While events of a same log are totally ordered, events of a same trace are only partially ordered if we consider the well known *happened-before* relation [7], denoted $\prec^\alpha$. We assume that the logs are rich enough to deduce if two events of the trace are causally dependent or not. Either all logged events are stamped with a vector clock or enough information about the communication events is stored, to allow matching the corresponding sending/receiving events in the trace [15]. Based on the causality relation, *consistent cuts* can be defined [4]. A cut is a subset of events. A consistent cut corresponds to a global state that may be reached during the computation.

Definition 1. $C \subseteq E^\alpha$ *is a consistent cut of the distributed computation* α *if and only if* $\forall e \in C, \forall f \in E^\alpha, f \prec^\alpha e \Rightarrow f \in C$.

The set of consistent cuts of a distributed computation forms a lattice. Let us consider the example of an application that involves two processes p_1 and p_2. In Fig. 2, during an execution α, the space-time diagram shows that 7 events occurred in the following order: $<a;\ b;\ c_2!m;\ c_1!m;\ c_1?m;\ c_2?m;\ b>$. But this order is not observable without a global clock. The corresponding lattice is depicted in Fig. 2. Each point (white circle) corresponds to a consistent cut. A path from the empty set of events (the initial cut at the bottom left) to the whole set of events E^α (the final global state at the top right) corresponds to a total order compatible with the observed partial order. There exist 18 different paths and one of them corresponds to the total order in which the events occurred.

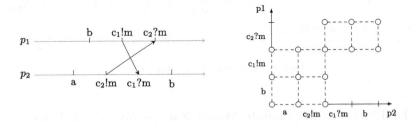

Fig. 2. Space-time diagram of the execution α and its corresponding lattice

An intermediate model, specific to the execution α, is created using the trace E^α. During the construction of the model, each intermediate model (as well as the final one) is dual (*i.e.*, composed of two sub-models). A first sub-model is a list of likely invariants. Assuming that each event has a type (denoted a,

b, ...), usual invariants [2,15] are considered: "a always followed by b" (noted $a \rightarrow b$), "b always preceded by a" ($a \leftarrow b$), and "b never followed by a" ($b \nrightarrow a$). To compute the list of invariants, we adopt a technique proposed in a different context [12]. During the construction of the lattice, for each type of event a, we compute $min(a)$ (respectively $max(a)$) the set of cuts where an event of type a appears for the first time (respectively for the last time) along a path. Each set contains at most n elements and is necessarily a singleton if a is a type of event that can be generated by only one process. For example, during the execution α, $min(c_1!m)$ is equal to $\{\{b, c_1!m\}\}$ and $max(c_1!m)$ is equal to $\{\{b, c_1!m, a, c_2!m\}\}$. To obtain the list, we check the following constraints for each pair (a, b). The coordinates of the cuts in the lattice are used to test the strict inclusion conditions.

$$a \rightarrow b : \forall C_x \in max(a), \exists C_y \in max(b) \text{ such that } C_x \subset C_y$$
$$a \leftarrow b : \forall C_y \in min(b), \exists C_x \in min(a) \text{ such that } C_x \subset C_y$$
$$b \nrightarrow a : \forall C_y \in min(b), \nexists C_x \in max(a) \text{ such that } C_y \subset C_x$$

The second sub-model is an automaton that is trivially deduced from the constructed lattice (see Fig. 3). At this stage, the two sub-models are redundant. Note that scalability issues arise during the construction of each intermediate model: the size of the lattice may grow exponentially [5].

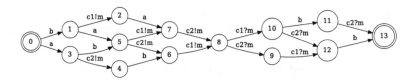

Fig. 3. The intermediate automaton sub-model of the execution α

2.2   Building a Global Model from Multiple Intermediate Models

First the *Model Merger & Generalizer* module (Fig. 1) aims at merging the constructed intermediate models. The final list of invariants must contain only the invariants that have been satisfied during each of the learned executions. Despite the fact that some events are not observed in all executions, merging these lists is rather trivial. Merging the different automata is even easier: it suffices to merge the initial states of the learned automata. At this stage, the obtained automaton accepts only the learned behaviors. As a distributed application may exhibit an infinity of behaviors, a model that captures a finite number of them will necessarily generate false positive alerts during the detection process: correct behaviors that have not been learned will be rejected. The interest of a generalization algorithm is twofold: first, it reduces the size of the model and, second, it extends the model with new behaviors. In fact, generalization algorithms usually introduce loops in the automaton, which allow it to accept infinite behaviors.

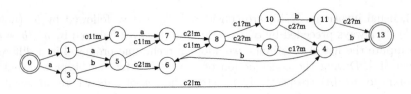

Fig. 4. Generalization of the automaton of execution α with $k = 1$

The *Ktail* algorithm used in [1,15] is a well-known generalization algorithm. It merges states with the same k futures (root subgraphs that are exactly the same up to a distance k).

In Fig. 4, we apply the *Ktail* algorithm on a model learned during the single execution α (see the automaton depicted in Fig. 3). The states 4 and 12 are merged and the resulting automaton now integrates loops and accepts an infinity of behaviors. Among the new behaviors introduced, some are abnormal (For example, in $<a; c2!m; b; c1!m; c2?m; c1?m; b; c1!m; c2?m; c1?m; b>$ three messages are sent and four are received). The automaton sub-model becomes permissive and, during the detection phase, false negative may occur: an anomaly occurs but it is not detected. This explains the interest of keeping two sub-models. Even if the automaton may accept an incorrect behavior, the invariants are still there to possibly reject the behavior and secure the application (In our example, the incorrect behavior described above does not verify $c1!m \nrightarrow c1!m$).

The generalization phase suffers also from a scalability problem. As it occurs after the merge of all the intermediate automata, the number of states is huge.

2.3   Use of the Global Model During a Detection Phase

The *Detection* module (Fig. 1) checks whether the trace generated by a running application is accepted or not by the learned model. Consider that the monitored execution is called δ. Again, the trace E^δ is characterized by several total orders that are compatible with the observed partial order $\prec^\delta$. Like in [15], we use a depth-first search strategy to test all the candidate sequences, until one of them is accepted by the automaton and is in conformity with the set of invariants. An alert is raised if no sequence satisfies both requirements (other strategies may also be considered [13]). Sometimes, another verification can be done during the consumption of the trace E^δ: no receiving event can occur as long as the corresponding sending event has not been executed before. This invariant (denoted InvCom) characterizes the communications and cannot be expressed using invariants between types of events (*i.e.*, $\rightarrow$, $\leftarrow$, and $\nrightarrow$). It may discard candidate sequences (the verifications with the two sub-models are bypassed).

3   Solving Scalability Issues Without Sacrificing Efficiency

3.1   Focusing on a Subset of Specific Sequences in the Lattice

The first proposal modifies the *Model Builder* component (Fig. 1). In [15], for each execution, the entire lattice is built. To face this exponential time and space complexities, we propose to consider only a small subset of x sequences. At both extremes, the n local computations can be carried out sequentially or concurrently. Thus, in the whole lattice, the total number of sequences is between 1 and $\mathcal{S}!/\mathcal{P}$, where $\mathcal{S} = \sum_{i=1}^{n} |E_i^\alpha| = |E^\alpha|$ and $\mathcal{P} = \prod_{i=1}^{n}(|E_i^\alpha|!)$. Moreover, the length of any sequence is equal to $\mathcal{S}$. Among all the possible sequences, a subset of at least one and at most $n!$ sequences defines the concave hull of the lattice (*i.e.*, the envelop corresponding to perimeter of the lattice's geometrical representation). Static priorities between the n processes can be used to determine these sequences. Herein, the notation $p_i \rhd p_j$ indicates that p_i has priority over p_j. Among n processes, $n!$ permutations can be defined. Each permutation is a selection rule that identifies a sequence: given a prefix of the selected sequence (*i.e.* a consistent cut), the rule identifies the next event of the trace that will extend this prefix to obtain the next consistent cut. This deterministic process selects only sequences in the concave hull. It is more appropriate than a random selection of x sequences in the whole lattice because it limits the risk of selecting quite similar sequences. In this work, to reduce again the number of sequences, we consider only a subset of n cyclic permutations where no process appears twice at the same rank in two selected combinations. With $x = n$ selection rules (rather than $x = n!$) we obtain a good approximation of the envelop. In Sect. 4, we show that this drastic reduction in the number of sequences allows to scale without sacrificing efficiency.

The consistent cuts contained in the x selected sequences are elements of the original lattice. As each cut is identified by its coordinates in this lattice, the construction of the automaton (through the enumeration of transitions between cuts) requires no additional cost and can be done separately for each selected sequence. These selected sequences share common consistent cuts (at least the initial and final global states). Thus, in general, thanks to these interleaving points, the number of sequences contained in the built intermediate model is much greater than x. Consider the execution α. If the two selection rules are $p_1 \rhd p_2$ and $p_2 \rhd p_1$ then $<b;\ c_1!m;\ a;\ c_2!m;\ c_2?m;\ c_1?m;\ b>$ and $<a;\ c_2!m;\ b;\ c_1!m;\ c_1?m;\ b;\ c_2?m>$ are the selected sequences. As shown in Fig. 5, these sequences share three consistent cuts. Thus, the automaton accepts two additional valid sequences: $<b;\ c_1!m;\ a;\ c_2!m;\ c_1?m;\ b;\ c_2?m>$ and $<a;\ c_2!m;\ b;\ c_1!m;\ c_2?m;\ c_1?m;\ b>$.

As a subset of sequences is now considered, the size of the automaton (number of states and transitions) can only decrease. The number of likely invariants may increase, but the new list necessarily includes the list corresponding to the whole lattice. Indeed, when some paths are no longer considered, the list of detected invariants can only increase due to the absence of some counterexamples.

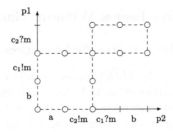

Fig. 5. Two selected sequences (with $p_1 \rhd p_2$ and $p_2 \rhd p_1$) & four sequences in the model

3.2 Model Merging and Generalization

The second proposal suggests to call the *Model Merger & Generalizer* module not only once at the end of the construction phase but repeatedly during this phase. Generalization techniques such as the *Ktail* algorithm require to compare all the pairs of states in the automaton. Even if some optimizations are possible, this is a major time consuming activity, especially when the automaton is the result of learning several long-lasting executions. In the case of an automaton composed of y states, up to $\frac{y(y-1)}{2}$ comparisons may be necessary. The idea is to benefit from the fact that this kind of algorithm takes an automaton as input and outputs an automaton which size is much smaller. We hope to reduce the computation time by calling the generalization algorithm more often but each time with an input automaton of reasonable size (see Fig. 6). A single model under construction is maintained and combined step by step with each new created intermediate model: if z executions are learned, the *Model Merger & Generalizer* module is called $z - 1$ times. The final global automaton may change if the order in which the z executions are learned is modified. Yet, this non-determinism is not a problem.

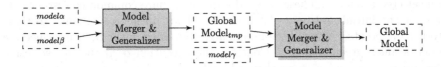

Fig. 6. Repeated calls (R) of the module

4   Assessment of the Approach

We evaluate our approach, using realistic data gathered from real executions of a well-known application which is representative of other distributed applications.

4.1   A Protocol for Conducting Experiments on Traces of XtreemFS

We consider XtreemFS [14], an open source project that provides a general purpose fault-tolerant distributed and replicated file system that can be deployed on cloud infrastructures. The file system service is built by four components: the *Directory Service* (DIR) maintains a database of services and volumes available; the *Metadata and Replica Catalog(s)* (MRC) store and manage files' metadata; the *Object Storage Device(s)* (OSD) store the actual file contents, which may be split and replicated; finally, *clients* request file system operations (volume creation/mount, file creation, etc.).

Multiple execution traces representing the correct behaviors were gathered, varying the behavior of the client and/or the duration of the execution, from 1 min to 5 h. Traces are used to build models and to evaluate the quality of the detection process. The set of traces is divided into v subsets; the construction of the model uses $v-1$ subsets, while the evaluation relies on the whole set of correct traces. During our experiments, the value of v is set to 5. Since there are five possible choices to exclude one subset of traces, we built - for each experiment - five different models. Thus each presented result is an average value obtained from five models that are instances of a given learning strategy. As a strategy is characterized by three parameters, we identify the different models using a naming convention composed of 4 fields separated by the character "-". The first field indicates which model we refer to (Mod stands for the whole model, Aut for the Automaton sub-model, and Inv for the Invariant sub-model). The second field corresponds to the number x of selected sequences (or A, when all the sequences of each lattice are selected). The third field indicates how the merging and generalization process has been performed (S: Single call, R: Repeated calls). The last field indicates the numerical value of k used in *Ktail*.

To evaluate the attack detection capability of each model, we deployed five known attacks against the integrity of a non-secured version of XtreemFS: the *NewFile* attack consists in adding the metadata of a file to the MRC server without adding its content into an OSD server; *DeleteFile* aims at deleting the file metadata on the MRC while keeping the file content on the OSD; the *OsdChange* attack changes an OSD IP address on the DIR database while its IP address did not change; *Chmod* modifies the file access policy on the MRC server without having the permission to do it; finally, the *Chown* attack modifies the file owner in the MRC metadata. A trace was collected for each attack. Regarding the quality of the detection process, we notice that the execution context and the moment the attack occurs are just as important as the nature of the attack. We perform each attack in four different contexts: (c1) while no client is active, (c2) before the client actions, (c3) after the client actions, and (c4) conducted from a source that is not the client address.

4.2   Scalability Issues During the Construction Phase

The first proposed solution (*i.e.* selecting a subset of x sequences) aims to reduce both the time required to build an intermediate automaton and its size.

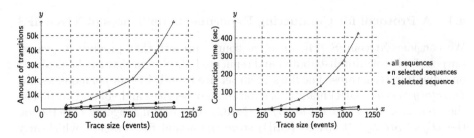

Fig. 7. Construction of an intermediate automaton: space(left)/time(right) complexity

To check its effectiveness, we consider traces with different sizes. Data corresponding to independent constructions of the intermediate models are analyzed. Figure 7 shows the size of the created automaton (number of transitions) and its construction time (seconds) in three cases: (1) all the sequences of the lattice are considered; (2) n of them are selected (here $n = 5 : 1$ DIR + 1 MRC + 2 OSD + 1 client); and (3) only one of them is selected. The rules applied to select sequences in case 2 are based on the following static priorities between processes: $p_1 \triangleright p_2 \triangleright p_3 \triangleright p_4 \triangleright p_5$, $p_2 \triangleright p_3 \triangleright p_4 \triangleright p_5 \triangleright p_1$, $p_3 \triangleright p_4 \triangleright p_5 \triangleright p_1 \triangleright p_2$, $p_4 \triangleright p_5 \triangleright p_1 \triangleright p_2 \triangleright p_3$, and $p_5 \triangleright p_1 \triangleright p_2 \triangleright p_3 \triangleright p_4$. Selecting n sequences avoids the exponential time and space complexity of an approach based on the whole lattice.

Let us now consider the impact on the list of invariants. The number of recognized invariants depends on the number of selected sequences. Table 1 shows the number of invariants discovered during the construction of the corresponding intermediate model, considering the same cases and the same trace sizes as in Fig. 7. When a single sequence is selected ($x = 1$) the model is more restrictive, because more invariants are satisfied by the single selected sequence. Yet the error remains relatively small: the invariant list is always less than 5% bigger whatever the trace size. When only $n = 5$ sequences are selected, the detection is as precise as when considering the whole lattice (the corresponding lines in Table 1 are equal). This non-intuitive and very positive result validates the choice of using a limited number of selected sequences (and appropriate selection rules).

Table 1. Size of an intermediate invariant list

Trace size (events)	212	338	420	581	778	976	1124
1 selected sequence	3226	4721	6389	12230	16332	17295	17392
n selected sequences	3118	4643	6231	12082	16149	17084	17108
All sequences	3118	4643	6231	12082	16149	17084	17108

The Second Proposed Solution (*i.e.* executing repeated calls to the *Model Merger & Generalizer* module rather than a single call) aims at reducing the computation time when either the number of learned executions and/or

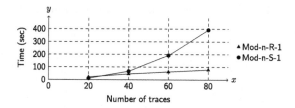

Fig. 8. Merge and generalization: single call versus repeated calls

their duration increase. In Fig. 8, we compare two approaches `Mod-n-R-1` and `Mod-n-S-1`, where R stands for "Repeated calls" and S means "Single call". For distinct numbers of traces, we indicate the time required to compute the whole model. The proposed solution allows us to obtain better results when the number of learned executions increases. The curve corresponding to a single call follows more or less the number of states comparisons $\frac{y(y-1)}{2}$ that are performed during each unique call. In Fig. 8, each intermediate model is built using only $n = 5$ selected sequences. When the whole lattice is considered, the difference (between `Mod-A-R-1` and `Mod-A-S-1`) is even more impressive: taking into account 80 traces requires about 400 s for a single call and only 80 s in the case of repeated calls. Multiple calls are thus far less costly than a single, long lasting call.

4.3   Evaluation of the Accuracy of the Detection Phase

We want to solve scalability issues without impairing detection accuracy.

Table 2 shows detection results for the 5 attacks and 4 contexts described in Sect. 4.1, considering the strategies strategies `Mod-A-S-1` and `Mod-n-R-1`. Sub-models `Aut` and `Inv` are described separately, to show their complementarity and to validate the choice of keeping both. Even if it cannot always be implemented, we indicate with a check mark whether the invariant on communications (see Sect. 2.3) is able to detect the attack. For each strategy, as five models are created, a result $d/5$ means that d models out of five detected the attack ("-" cells mean 0/5). The results show that invariants are useful to detect attacks in contexts c1 and c2, while automata perform better in context c4. The proposed solutions to improve the scalability have no effect on the detection capabilities of the invariants, but they slightly degrade those of the automata. However, our model is dual: an alert is raised when the analyzed trace is rejected by either the `Inv` sub-model or the `Aut` sub-model. The detection based on invariants masks the possible errors of the automata sub-model in 8 out of the 20 cases. In the remaining 12 cases, the results provided by the sub-models `Aut-A-S-1` and `Aut-n-R-1` are the same, except in 2 cases (namely, the *DeleteFile* attack in context c3 and the *Chown* attack in context c2). In these two particular cases, the quality of the detection is quite similar: the scores are 0/5 and 1/5. Thus, in terms of false negative, the proposed solutions have nearly no impact. Differences are small or masked by the fact that the model is dual.

Table 2. Detection of attacks in various contexts using different sub-models

Attack context	NewFile				DeleteFile				OsdChange				Chmod				Chown			
	c1	c2	c3	c4	c1	c2	c3	c4	c1	c2	c3	c4	c1	c2	c3	c4	c1	c2	c3	c4
Aut-A-S-1	-	2/5	-	5/5	-	3/5	1/5	5/5	2/5	5/5	4/5	5/5	-	1/5	-	5/5	-	1/5	-	-
Aut-n-R-1	-	-	-	5/5	-	1/5	-	5/5	4/5	4/5	4/5	5/5	-	-	-	5/5	-	-	-	-
Inv-A-S-1	5/5	5/5	-	-	5/5	5/5	-	-	5/5	5/5	-	-	5/5	5/5	-	-	-	-	-	-
Inv-n-R-1	5/5	5/5	-	-	5/5	5/5	-	-	5/5	5/5	-	-	5/5	5/5	-	-	-	-	-	-
InvCom	✓	✓	✓	✓	✓	✓	✓	✓	✓	✓	✓	✓	✓	✓	✓	✓	-	-	-	-

Table 3. Detection of attacks with different strategies for building the automata

#	Model	NewFile	DeleteFile	OsdChange	Chmod	Chown
1	Aut-A-S-5	100%	100%	100%	100%	100%
2	Aut-A-S-1	35%	45%	80%	30%	5%
3	Aut-A-R-1	30%	35%	50%	30%	5%
4	Aut-n-S-5	100%	100%	100%	100%	100%
5	Aut-n-S-1	25%	25%	85%	25%	0%
6	Aut-n-R-1	25%	30%	85%	25%	0%

Table 3 shows results from 6 distinct strategies to build the automaton sub-model. Each percentage indicates the proportion of attacks detected, knowing that 4 contexts and 5 models were used to test each strategy. As we focus only on the automaton sub-model, the percentages are much lower than those corresponding to the dual model (recall that the invariant sub-model may detect up to half of the first four attacks). When no generalization is done (*i.e.* for a high value of k in *Ktail*), no false negative occur (see line 1 and 4 in Table 3) but any correct behavior that has not been learned raises an alert (false positive). This strategy is not scalable and thus adopting it is unrealistic. In lines 2 and 3 (resp. 5 and 6) an intermediate model is built using the whole lattice (resp. using n sequences).

The model accuracy should be assessed in terms of false negatives but also false positives. Here we focus on the impact of the selection of x sequences on the false positive rate. Figure 9 shows a well-known result: the smaller the value of k in *Ktail*, the better the result. This highlights the importance of the generalization phase. For a given value of k, we observe also differences depending on the number x of selected sequences. When x is too small (equal to 1), less behaviors are accepted. But the best results are obtained when n sequences are selected and not when the whole lattice is used. This good result is again in favor of our proposed solution. It may be explained by the fact that, when $x = n$, the automata that are provided as an input for the merge and generalization phase are neither too simple (not just x paths) nor too complex (the whole lattice). While keeping a wealth of behaviors, these automata have often less nodes and especially less transitions. As a consequence, the *Ktail* algorithm may merge more states and generate a final automaton accepting more behaviors.

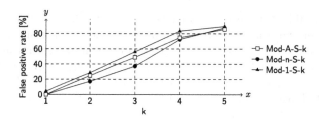

Fig. 9. False positive rate for different values of k and x

4.4   Short Traces and Long Traces

Using short traces, we have shown that a model (Mod-n-R-k) can be produced in a scalable manner without having a significant impact on the detection capabilities. Now we consider longer executions. When the construction phase relies on several short traces (at least 32), long traces are also accepted by the model. So it seems possible to have a learning strategy based mainly on short traces.

Table 4 shown average values allowing to compare the performance of our proposal for short (5 min) and long (5 h) execution traces as inputs, for each phase (construction and merging/generalization). As for the construction phase, the results in the first column roughly correspond to the ones described in Fig. 7 and Table 1. The trace contains more events when the execution lasts 5 h. Thus, the time to construct an intermediate model and the number of transitions are both bigger. Note that the ratios for results in the construction phase are around 35, except for the number of invariants, which is quite similar with short or long traces. This is due to the fact that the activity performed by the tester was similar in both executions: the same types of events were observed.

Table 4. Performance values for each phase

Phase	Construction			Merging & generalization		
Execution duration	5 min	5 hr	Ratio	5 min	5 hr	Ratio
Input trace size (events)	754	26178	34.7	754	26178	34.7
Processing time (seconds)	5	178	35.6	8	386	48.2
Number of transitions	2502	87266	34.9	386	247	0.6
Invariant list size	16643	15239	0.9	2125	2058	0.9
Reduced invariant list size	-	-	-	1190	986	0.8

Regarding the merge & generalization phase, the time needed to merge and generalize the short traces roughly correspond to that indicated in Fig. 8. Of course, it requires much more time to merge and generalize intermediate models obtained after a long execution (due to the bigger number of nodes and transitions). Yet we observe that the sizes of the result automaton and the invariant list are comparable in both cases. This is also due to the fact that, during our tests, the longer executions do not reveal new behaviors. The last line of Table 4 indicates the reduced size of the invariant list after a simple optimization (based on transitivity relations) that does not affect the accuracy of the `Inv-n-R-k` sub-model: for example, when a list contains the invariants $a \rightarrow b$, $a \rightarrow c$, and $b \rightarrow c$, we only have to keep $a \rightarrow b$, and $b \rightarrow c$.

5 Conclusion

We presented a scalable approach to build a behavior model of a distributed application and use it for intrusion detection. This dual model is composed of an automaton and a list of invariants, both learned from traces. Two original propositions have been made in order to scale better in time and space during the construction phase. They were evaluated using a real distributed application and a set of real attacks performed in different contexts. The evaluation showed the interest of the different proposed solutions. These solutions clearly address the scalability issues without having a significant impact on the detection capabilities.

References

1. Beschastnikh, I., Brun, Y., Ernst, M.D., Krishnamurthy, A.: Inferring models of concurrent systems from logs of their behavior with CSight. In: International Conference on Software Engineering, pp. 468–479 (2014)
2. Beschastnikh, I., Brun, Y., Ernst, M.D., Krishnamurthy, A., Anderson, T.E.: Mining temporal invariants from partially ordered logs. ACM SIGOPS Operating Syst. Rev. **45**(3), 39–46 (2012)
3. Biermann, A.W., Feldman, J.A.: On the synthesis of finite-state machines from samples of their behavior. IEEE Trans. Comput. **100**(6), 592–597 (1972)
4. Garg, V.K.: Principles of Distributed Systems. Kluwer Academic Publishers, Dordrecht (1996)
5. Garg, V.K.: Lattice completion algorithms for distributed computations. In: Baldoni, R., Flocchini, P., Binoy, R. (eds.) OPODIS 2012. LNCS, vol. 7702, pp. 166–180. Springer, Heidelberg (2012). https://doi.org/10.1007/978-3-642-35476-2_12
6. Grant, S., Cech, H., Beschastnikh, I.: Inferring and asserting distributed system invariants. In: International Conference on Software Engineering, pp. 1149–1159 (2018)
7. Lamport, L.: Time, clocks, and the ordering of events in a distributed system. Commun. ACM **21**(7), 558–565 (1978)
8. Lanoe, D., Hurfin, M., Totel, E.: A scalable and efficient correlation engine to detect multi-step attacks in distributed systems. In: 37th Symposium on Reliable Distributed Systems (SRDS), pp. 31–40 (2018)
9. Lo, D., Mariani, L., Santoro, M.: Learning extended FSA from software: an empirical assessment. J. Syst. Softw. **85**(9), 2063–2076 (2012)
10. Lorenzoli, D., Mariani, L., Pezzè, M.: Inferring state-based behavior models. In: International Workshop on Dynamic Systems Analysis, pp. 25–32 (2006)

11. Mukund, M., Kumar, K.N., Sohoni, M.: Synthesizing distributed finite-state systems from MSCs. In: Palamidessi, C. (ed.) CONCUR 2000. LNCS, vol. 1877, pp. 521–535. Springer, Heidelberg (2000). https://doi.org/10.1007/3-540-44618-4_37
12. Pfaltz, J.L.: Using concept lattices to uncover causal dependencies in software. In: Missaoui, R., Schmidt, J. (eds.) ICFCA 2006. LNCS (LNAI), vol. 3874, pp. 233–247. Springer, Heidelberg (2006). https://doi.org/10.1007/11671404_16
13. Raguenet, I., Maziero, C.: A fuzzy model for the composition of intrusion detectors. In: Jajodia, S., Samarati, P., Cimato, S. (eds.) SEC 2008. IFIP, vol. 278, pp. 237–251. Springer, Boston, MA (2008). https://doi.org/10.1007/978-0-387-09699-5_16
14. Stender, J., Berlin, M., Reinefeld, A.: XtreemFS: a file system for the cloud. In: Data Intensive Storage Services for Cloud Environments, pp. 267–285. IGI Global (2013)
15. Totel, E., Hkimi, M., Hurfin, M., Leslous, M., Labiche, Y.: Inferring a distributed application behavior model for anomaly based intrusion detection. In: European Dependable Computing Conference (EDCC), pp. 53–64 (2016)
16. Yabandeh, M., Anand, A., Canini, M., Kostic, D.: Finding almost-invariants in distributed systems. In: Symposium on Reliable Distributed Systems (SRDS) (2011)

Access Control

Performance of Password Guessing Enumerators Under Cracking Conditions

Mathieu Valois[✉], Patrick Lacharme, and Jean-Marie Le Bars

Normandie Univ, UNICAEN, ENSICAEN, CNRS, GREYC, 14000 Caen, France
{mathieu.valois,jean-marie.lebars}@unicaen.fr,
patrick.lacharme@ensicaen.fr

Abstract. In this work, we aim to measure the impact of hash functions on the password cracking process. This brings us to measure the performance of password enumerators, how many passwords they find in a given period of time. We propose a performance measurement methodology for enumerators, which integrates the success rate and the speed of the whole password cracking process. This performance measurement required us to develop advanced techniques to solve measurement challenges that were not mentioned before. The experiments we conduct show that software-optimized enumerators like John The Ripper-Markov and the bruteforce perform well when attacking fast hash functions like SHA-1. Whereas enumerators like OMEN and PCFG-based algorithm perform the best when attacking slow hash functions like bcrypt or Argon2. Using this approach, we realize a more in-depth measurement of the enumerators performance, considering quantitatively the trade-off between the enumerator choice and the speed of the hash function. We conclude that software-optimized enumerators and tools must implement academic methods in the future.

Keywords: Password · Hash function · Cracking conditions

1 Introduction

Passwords are for a long time one of the weakest point in digital identity security. NIST guidelines [19] tells us that "Memorized secrets SHALL be salted and hashed using a suitable one-way key derivation function". Such functions are designed to be time and memory costly to hash a password. However, the impact of the hash function on an offline password cracking process has not been studied.

Before our work, even if the time influence has once been mentioned [27], we were not able to get and understand the performance of enumerators and password cracking software depending on the hash function. For example, what is the real impact on enumerators performances when switching from SHA-1 to bcrypt?

Contributions of this work are useful to choose a good hash function knowing the threat in terms of enumerators (defender side) and to choose a good set of

© IFIP International Federation for Information Processing 2019
Published by Springer Nature Switzerland AG 2019
G. Dhillon et al. (Eds.): SEC 2019, IFIP AICT 562, pp. 67–80, 2019.
https://doi.org/10.1007/978-3-030-22312-0_5

enumerators knowing the hash function used to protect passwords (attacker side, password crackers, pentesters).

To measure the impact of the hash function, we aim to measure the performance of the whole offline cracking process, which is composed of the candidate generation by an enumeration algorithm, the candidate hashing and the search for its fingerprint in the targeted dataset. The performance on a period of time is defined as the number of found passwords in that period.

The most popular academic enumerators categories are Probabilistic Context-Free Grammars (PCFG) [15,28,30] and Markov chains models [9,16, 18]. These models are probabilistic, as they assign probabilities to passwords to measure their strength, whereas very fast enumerators are also implemented in free software like John The Ripper (JtR) [21] and Hashcat [24]. These enumerators are optimized for their runtime performance. The faster the hash function is, the more we can test candidates in a period of time. Slowest enumerators generate better quality candidates compared to the fastest ones. It makes sense since slowest enumerators are probabilistic, and propose candidates in decreasing order of probability. There is then a trade-off between the speed of the enumerator and the quality of candidates it generates. We think that the choice of the hash function affects the performance, however it is unclear how the speed of the hash function impacts the performance of the password cracking process. For instance, until which speed of the hash function a naive strategy like bruteforce is viable?

In previous works, probabilistic enumerators were usually compared using the guess number metric [9,27,30]. This metric makes sense as long as enumerators aim to measure the password strength. From the moment when they are used to crack passwords, it becomes inadequate because the candidate generation speed of the enumerator must be taken into account.

Our Contributions:

- We propose a methodology to measure the performance of enumerators in a password cracking context by considering the cost of processing candidates. Since this measurement can not be directly done, we then need to measure separately the success rate and the frequency of the process.
- Measuring the success rate and the frequency of the process are challenging tasks. We present advanced techniques used to conduct these measurements, since there is a sort of Heisenberg effect: measuring a phenomenon has an influence on the value itself. Such techniques have never been presented before and will certainly be useful for future works.
- Experiments on leaked databases with this performance measurement show that fast cryptographic hash functions are unsuitable for password storage, because even the most naive strategies (like bruteforce) perform very well against them. They also show that academic enumerators are very useful against slow hash functions. To make them even more performing, developers should implement them in their password cracking tools.

This paper is organized as following. First, we present the background and related works Sect. 2. Then we introduce our password cracking modelling Sect. 3.

In Sect. 4 we explain how we compute the performance using success rate and frequency measurements. Section 5 contains the results of performance measurements on two publicly leaked passwords using two different hash functions. Then, in Sect. 6, we expose different scenarios in where our work has concrete interests. We conclude in Sect. 7.

2   Background

2.1   Related Works

Password strength evaluation is an active research topic [5–7,26]. Probabilistic passwords models have been studied by Ma et al. in [16] where they introduced a probability-threshold to plot graphs rather than guess-number graphs. However, it is not sufficient since it does not depend on the hash function and enumeration speed, which are required to compute the performance of passwords attacks.

Different cracking strategies have been studied by Ur et al. in [27], where they compare algorithms like PCFG, JtR, Hashcat, Markov from [16] and a strategy from password recovery professionals. They found that professionals are more efficient against complex password policies at a high number of guesses, while automatic approaches perform best at low numbers of guesses. They ended up by providing a Password Guessing Service [25] where they offer to analyze a list of plaintext passwords and return the score for each approach they support. Once again, they do not take care of the enumeration speed nor the hash function to compare enumerators.

2.2   Enumerators

Several enumerators have been proposed by academic researches, but few are currently implemented in the most well-known password cracking software. For example, John The Ripper and Hashcat include a Markov model-based enumerator that has been originally introduced in [18]. However, the remaining proposed enumerators are distributed in standalone versions, meaning they are not actually cracking passwords. Instead, they only generate candidates that might be actual passwords. These candidates should then be gathered by a password cracking software to do the rest of the process: hash them and search for a match.

Bruteforce: This is the most naive and the most known enumerator. Basically, bruteforce will output the words incrementing the characters one by one using a defined alphabet. In this paper, we run our own C implementation of the bruteforce algorithm.

Markov Models: They are probabilistic models that can be applied to words, for which the probability of a character at a given position depends on the previous characters. The software will learn probabilities on a dataset and will then output newly-created words according to these probabilities. In this paper, we consider two enumerators based on Markov models implementations: **John The**

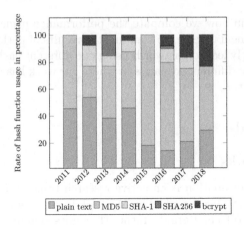

Fig. 1. Distribution of hash functions usage in password leaks since 2011. Source: haveibeenpwned.com

Ripper's Markov mode [1] and **OMEN** [8]. The main differences between them is that OMEN outputs candidates in approximately decreasing probabilities order. Also, OMEN has an adaptive strategy which reorders the sets of next candidates based on the success rate of previous ones.

Probabilistic Context-Free Grammars: Weir et al. [30] are the first to use PCFGs for passcracking. Kelley et al. and Komanduri [13,15] have modified Weir's PCFGs by adding string tokenization and by assigning probabilities to terminals that have not been seen in the training dataset. The implementation used in this paper is the Weir's one and can be found on Github [29]. For easier reading, we will write "PCFG" referring as the Weir et al. algorithm.

In this paper, we make use of these four enumerators: bruteforce, JtR-Markov, OMEN and PCFG. They only generate candidates as they do not hash them or search for them in the targeted database. We don't make use of Hash-cat and John The Ripper password cracking features, only the Markov-based enumerator of John The Ripper.

2.3 Hash Functions

Fast functions like MD5, SHA-1, SHA256 are often used in the leaks that happened the last 10 years. Figure 1 is an overview of the distribution of used hash functions in passwords leaks since 2011: MD5 and SHA-1 remain very used even if we observe an increase of the usage of bcrypt. These figures are not representative as it is more likely that poorly-protected services are also using fast hash functions. However, this provides a lower bound of the bcrypt usage.

Moreover, these hash functions are deterministic, which means that two same words have the same hash. It is unwanted since multiple users having the same password would have the same hash. A common counter-measure is to use a salt [17], a randomly-generated word which is appended to the passwords before hashing, and stored aside the hash value. Attackers are then forced for each candidate to apply the salt value before hashing, slowing down the speed of the attack and making pre-computation-based attacks [11, 20] impracticable. The leaked databases on which we experiment in this paper are not salted. Nevertheless, our model still apply since we don't use pre-computed tables and usage of a salt multiply the cost of the attack by the number of users.

Password-specific hash functions like bcrypt [23], Scrypt [22] or Argon2 [4], winner of the Password Hashing Competition [3], have been proposed to replace fast hash functions. They are designed to break the usage of massive parallel processing units by requiring a big memory and time amounts to compute one hash. Moreover, they handle themselves the salt value, making easier for developers to manage and store passwords.

In this paper we consider two hash functions: the first, SHA-1, due to its wide usage according to publicly leaked passwords datasets (especially in Linkedin). SHA-1 will represent the set of cryptographic hash functions (MD5, SHA-1, SHA2, SHA-3, ...), since these functions are similar for our study. The second is bcrypt, the reference function to store passwords, with a cost factor of 10 which is the default cost factor for PHP's password_hash() function at the time of writing. Since Argon2, Scrypt and bcrypt are for our study very similar, we only consider bcrypt since it is more commonly used than Argon2 and Scrypt.

2.4   Datasets

Here are presented the publicly accessible datasets used in this paper. Note however that there exists many more datasets that can be used for such a study. The ones we used are enough to illustrate our modelling, as it does not aim to be exhaustive. Further works can be conducted on more datasets. A list of recent leaks can be found at [12].

Rockyou: The most used dataset, probably because in plaintext and easily accessible. Rockyou is a company providing services to social networks and video games. The leak happened in 2009. It contains more than 32 million of passwords (14 million unique).

LinkedIn: This leak comes from the LinkedIn social network. The hack happened in 2012. Four years later, the entire database were published, containing more than 160 million of passwords (60 million unique). The passwords were hashed with SHA-1, without salt.

It should be mentioned that the relevance of one password cracking strategy highly depends on the training and target datasets.

3   Cracking Process Modelling

3.1   Context

The cracking process, at the level of a single candidate, can be resumed in 3 steps:

(i) generate a candidate
(ii) hash it
(iii) search for a match in the target database

In practice, searching for a match in step *(iii)* is negligible, even when a fast hash function is used. Indeed, most password cracking tools use a probabilistic structure, similar to a Bloom filter with only one hash function, to store the hashed passwords list in memory. We assume that step *(iii)* takes no time. Then according to the hash function in step *(ii)*, the bottleneck of the process is either step *(i)* or *(ii)*.

3.2   Formalization of the Performance

Even if in practice we compute a discrete version of this password cracking process, it remains a continuous process since the time is continuous, which explains the usage of integrals. Let first consider the performance at the level of a single candidate. For any $i \in \mathbb{N}$, let t^i be the instant where we finish to process the ith candidate of the enumeration, g^i the gain of this candidate, *i.e.* its number of occurrences in the dataset D and c^i the time to process it (steps *(i)*, *(ii)*, *(iii)*), *i.e.* $c^i = t^i - t^{i-1}$. The performance will be $P(t^{i-1}, t^i) = g^i$. Thus

$$P(t^{i-1}, t^i) = c^i \frac{g^i}{c^i} = \int_{t^{i-1}}^{t^i} \frac{g(t)}{c(t)} dt,$$

where $c(t) = c^i$ and $g(t) = g^i$.

Note 1. This model includes the parallelization of the process, since $g(t)$ and $c(t)$ could be measured on multiple cores.

Let t_1 and t_2 be any instants such that $t_1 < t_2$. The performance in the period $]t_1, t_2]$ will be $P(t_1, t_2) = \int_{t_1}^{t_2} \frac{g(t)}{c(t)} dt$, where $c(t) = c^i$ and $g(t) = g^i$, for any $t^{i-1} < t \le t^i$.

The frequency is by definition the number of processed candidates between t_1 and t_2. We get the formula $F(t_1, t_2) = \int_{t_1}^{t_2} \frac{1}{c(t)} dt$. We also define the success rate as the ratio between the number of passwords found in the period $[t_1, t_2]$ and the number of processed candidates in the same period: $S(t_1, t_2) = \frac{P(t_1,t_2)}{F(t_1,t_2)}$.

Comparison Between Enumerators Performance. We show here how to compare two enumerators depending on the speed of the used hash function. For each candidate i, $c^i = c_g^i + c_h^i + c_d^i$ where each term corresponds respectively to one step of the cracking process (generate, hash and search). We show in our experiments that c_d^i is negligible for any enumerator, then $c^i \approx c_g^i + c_h^i$. Let E_1 and E_2 be two enumerators we want to compare. We have two cases:

– **case (a) a slow hash function is used.** Then $c^i = c_g^i + c_h^i \approx c_h^i$ for any i and $F_1(t_1, t_2) \approx F_2(t_1, t_2)$. $P_1(t_1, t_2) > P_2(t_1, t_2)$ when $S_1(t_1, t_2) > S_2(t_1, t_2)$, the enumerator E_1 outperforms E_2 when it has a better success rate.

- **case (b) a fast hash function is used.** Then $c^i = c_g^i + c_h^i \approx c_g^i$ for any i and $P_1(t_1, t_2) > P_2(t_1, t_2)$ when $\frac{S_1(t_1, t_2)}{c_{g,1}} > \frac{S_2(t_1, t_2)}{c_{g,2}}$. The enumerator with the best ratio between the success rate and the enumerator speed has the best performance. We have a trade-off between success rate and speed.

Previous works consider the success rate from the beginning corresponds to the case (a), we have a slow hash function. Indeed in this case, the enumerator speed is not very important and the best performance is obtained with the best success rate. In case (b), when a fast hash function is used, the enumerator E_1 should have a best success rate than the enumerator E_2 but a worse performance if E_2 is a faster enumerator. For instance, if E_2 has twice better success rate than E_1 but a hundred times slower enumerator, the performance of E_2 is fifty times better than E_1, in a same period the enumerator E_2 finds fifty times more passwords than E_2. In that case, measuring only the success rate is clearly insufficient.

3.3 Estimating the Performance

In practice, we can't measure directly the performance, since it would be very expansive in time and uninteresting to compute the gain $g(t)$ for every t. One solution is to estimate the frequency and the success rate in small periods and derive an estimation of the performance. We will perform these measures for each interval of one second: $P(j, j+1) = F(j, j+1) S(j, j+1)$.

Firstly, as we already stated, the measurement of $c_g(t)$ has an impact on its value. We want then to have the less measurements possible while keeping a good enough accuracy. $c_g(t)$ does not vary much between j and $j+1$, then we suppose $c_g(t)$ to be constant in that period. For that, we note c_g its value, and take the mean of $c_g(t)$ as its value. Secondly, we can consider c_h to be constant given a hash function, because passwords size is almost always smaller than the input size of the compression function.

If we note $c = c_g + c_h$ in the interval $[j, j+1]$, then we have $F(j, j+1) = \frac{1}{c}$ the frequency of the enumerator during that period.

Let now consider a period of k seconds $[t_1, t_2 = t_1 + k]$, we have

$$P(t_1, t_2) = \sum_{l=0}^{k-1} P(t_1 + l, t_2 + l + 1)$$

as it is computed in the previous researches with the guess number comparisons, called the "Cumulative Distribution Function (CDF)".

Table 1. N value for measuring time for each enumerator

Enumerator	Bruteforce	JtR-Markov	OMEN	PCFG
N	10^9	4×10^6	10^5	10^4

4   How to Measure Performances of the Cracking Process

While the time cost to generate a candidate has too briefly been highlighted in a previous work [27], the computation of the performance has never been considered. We show that S and F must be computed separately. An estimation of these two values is enough. Furthermore, it makes possible the study of them separately, which is a work that can be done in a foreseeable future.

Nevertheless, as mentioned in [27], due to the fact that some enumerators are very fast, computing $S(j, j + 1)$ and $F(j, j + 1)$ in practice is very challenging. The measurements that have been performed are now presented. First of all, none of the enumerators implementations provide the measurement of the time between candidates. OMEN embeds the success rate measurement. For the remaining ones, we need to make our own measurements using different techniques depending on the implementation. Since time is not considered in the success rate measurement, we may use different techniques depending on the enumerator.

Number of Candidates in a Period. The 3 steps ((i), (ii) and (iii)) must be taken into account when measuring the number of candidates. However, steps (ii) and (iii) take constant times given a hash function, we then only need to analyze the generation step. Measuring times between each candidates until reaching one second is a very costly process. We instead estimate this number of candidates by measuring the time to generate a fixed number N of candidates. Then we compute the time spent to hash these candidates by multiplying the number of candidates with the time to hash one candidate. For example, in average, PCFG generates $\approx 8 \times 10^4$ candidates per second, OMEN 10^6, JtR-Markov 2×10^7 and bruteforce 1.6×10^9 (note however that these times are not constant during enumeration, hence are presented here to have a glance on enumerators speed). In our experiments, we chose N such that times required to generate N candidates are about $0.1\,\mathrm{s}$ (at beginning) to have both a good estimation of $C(t)$ and an acceptable number of floating point numbers to store. Our values of N for each enumerator can be found in Table 1.

Success Rate Measurement. To measure the success rate, we want to have for each candidate, its rank and its gain (how many times it appears in the dataset). There are two ways of measuring the success rate: by running the enumerator and counting occurrences in the dataset, or backwardly by computing for each password of the dataset, its rank in the enumeration if the enumerator has an index function. The former can be applied to every enumerators. However, it requires to run them, which is a long process, especially if we want to benchmark them for a long time. The latter however is doable only for few enumerators: bruteforce and JtR-Markov since the rank of a word is predictable and easily computable (index function). Since the time is not considered, using different techniques to measure the success rate of enumerators is not an issue. Once we get all the gains, we can compute the success rate for any period of time using the number of candidates for that period. For that, we search the rank of the first generated candidate of the given period, and compute the sum of

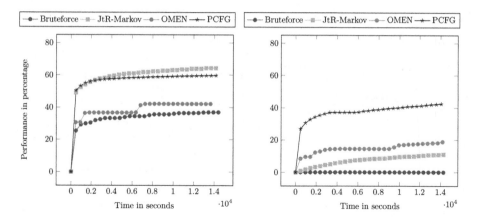

Fig. 2. LinkedIn dataset using SHA-1

Fig. 3. LinkedIn dataset using bcrypt with a cost factor of 10

gains corresponding to candidates of this period. The rank can be computed by summing the number of candidates for all previous periods of one second.

Performance Computation. Once we get the success rate and the number of candidates for every period, the computation of the performance is straight forward $P = F \times S$.

5   Experimental Results

In this section we present the experiments of the performance computation for four enumerators, the bruteforce, John The Ripper-Markov mode, OMEN and the PCFG-based enumerator, over the datasets LinkedIn and Rockyou, and using two hash functions, SHA-1 and bcrypt (cost 10). To be able to compare plots with previous researches and across datasets, rather than plotting the performance, we plot the percentage of cracked passwords from the beginning. We can then compare how they perform between each other over time and across datasets. We still took [10] for hash functions benchmarks. For SHA-1, $1/c_h \approx 12.5 \times 10^9$, and for bcrypt with a cost of 10, which is the default cost factor for PHP's password_hash() function at the time of writing, $1/c_h \approx 700$. Note that the benchmarks in Hashcat uses a cost 5 bcrypt, which means 2^5 rounds of the internal key-derivation function. Thus, c_h for a cost-10 bcrypt is 2^5 times smaller, giving $23 \times 10^3/2^5 \approx 700$.

On Fig. 2, showing how enumerator perform over LinkedIn with the SHA-1 function, JtR-Markov and PCFG cracked about 60% of passwords after four hours, while OMEN cracked around 43% and the bruteforce around 36%. Bruteforce, even though it is the most naive method, still has good results. PCFG is surprisingly good since it exploits the high number of passwords sharing the same grammatical structure in this dataset: 37% of passwords share the top

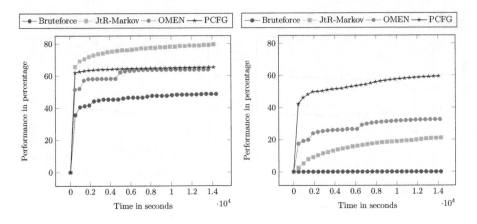

Fig. 4. Rockyou dataset using SHA-1

Fig. 5. Rockyou dataset using bcrypt with a cost factor of 10

five most frequent structures. However using bcrypt (Fig. 3), enumerators performances are completely disrupted. Because of the poor quality candidates of the bruteforce, the whole process spends its time to hash candidates which give mediocre success rates. After four hours, bruteforce cracked less than 1% of the dataset, while PCFG still performed great by cracking 43% of passwords. OMEN is better placed than before, cracking more than 18% of the dataset. Finally, JtR-Markov is worse than OMEN, finding only 11% of passwords after four hours. On Rockyou using SHA-1 (Fig. 4), JtR-Markov cracked 15% passwords more than on LinkedIn after four hours, while OMEN cracked 23% more than on LinkedIn, and the bruteforce 12% more. However, PCFG perform similar than on LinkedIn using SHA-1. On Rockyou using bcrypt (Fig. 5), results are similar than with LinkedIn using bcrypt: even though enumerators cracked not the same number of passwords, their performances are in the same order. The "steps" of the OMEN curves on all figures is due to the fact that it generates sets of candidates of same length. When OMEN switches the length of the candidates, it suddenly cracks many more passwords.

These four graphics highlight the impact of the hash function concerning the performance of the cracking process. They confirm our hypothesis that hacker enumerators are well-performing using fast hash functions while academic enumerators perform better using slow hash functions. One special mention to the PCFG-based enumerator which performs quite well on LinkedIn using SHA-1. Finally, experiments on each dataset can be easily adapted to other hash functions or to other parameters, like the cost of bcrypt. We could also choose the algorithm depending on the used hash function from the beginning. With the rise of memory-hard functions usage like bcrypt or Argon2, it will be even more interesting to have software-optimized academic enumerators implemented in password cracking tools.

It is important to note that the performance of enumerators highly depends on the dataset, and that OMEN becomes better than PCFG when attacking more complex datasets where passwords are longer and having more complex structures, like when keeping only strong passwords (longer than 8 characters, include all four characters classes). We also observed that enumerators perform differently on particular password composition policies, like the basic, complex, longbasic and longcomplex as defined in [27]. Results suggest that the beginning of the cracking session is decisive enough to determine which enumerator will perform the best on the targeted dataset.

6   Impacts of Our Contributions

We present here the usages that can be made of our contribution and results, for different actors of the computer security community.

Password Guessing Researchers. Imagine a scenario where you are a password guessing researcher who wants to build a new enumerator. Since you aim to attack passwords, you have to take care of the time cost of the enumeration algorithms you build. Concretely, you have to measure both their success rate and their frequencies in password cracking conditions, as we did in Sect. 4.

Then, you have a concrete proof that your enumerators are worth implementing them since you have measured their performances as if they were integrated in password cracking software. You could bring that proof in your future publications to encourage password cracking tools developers to implement your solution.

Furthermore, since you independently measured success rate and frequencies during a cracking session, you can independently analyze the behaviors of the success rate or the frequency during the enumeration. Thanks to that, you have a clearer understanding on your enumerator performance.

Then you are able to propose different enumerators settings regarding the attacked dataset. If your measurements show that a set of parameters provides good results against a given dataset, you can be pretty confident that this set of parameters will also provides good results against a similar dataset (for example where the password composition policy is similar). Moreover, you can implement a strategy in your enumerator that adapt the enumerator settings depending on the found passwords. For example, if you found a lot of passwords of length 6, your enumerator can for a while only generate passwords of length 6.

Password Cracking Tools Developers. Imagine once again a scenario where you are a developer of a password cracking tool and want to implement and optimize better enumerators in your tool. Using our comparison methodology, you are able to compare existing enumerators as if they were integrated in your software. For example, on both attacked datasets in Sect. 5, PCFG is not really impacted when using bcrypt instead of SHA-1. Implementing it directly in your cracking tool would make it even more efficient in such contexts.

Therefore, based on the hash function used in the targeted dataset, your tool can select different enumerators at the beginning of the cracking session. You can select those which are known to be more efficient against fast hash functions when you detect such functions, and similarly for slow hash functions. That way, the first steps of a cracking session can be run without user interaction. You can then crack a non-negligible part of weakest passwords automatically. Moreover, you can implement an adaptive strategy to select enumerators during the same single cracking session: based on previous performances of the running enumerator, your tool can change the enumerator for a one that is more likely to be efficient. For example, if a lot of found passwords share the same base structure, your tool benefits to switch to a PCFG-based algorithm since its performance is high on such passwords (provided that this algorithm performs well on the used hash function).

Security Community & CISO. Imagine a scenario where your are a CISO, system administrator of a company or an university, and that you want to improve the security of your infrastructure and users. The results of our research emphasize on the importance of a good hash function to protect passwords. Cryptography-oriented hash functions are unsuitable for password hashing since they allow to most of attacks to be very efficient even if their success rate is low. Therefore, it is essential for you to protect passwords using a dedicated hash function that has been designed for it, like bcrypt or Argon2d. Nevertheless speed performance of the hash function also depends on the hardware [14]. Dedicated hardware (ASIC or FPGA) focuses particularly on hash functions used in cryptocurrencies mining (typically SHA-256 but also Scrypt) [2].

Nonetheless, the results also show that a slow hash function is not enough to offer a very good protection for passwords. For example, PCFG still perform well against bcrypt-protected datasets. The only remaining protection to such attacks is by ensuring a good password strength before registering it in the database. That is why it is important for you to provide, when users register, a satisfying password composition policy that aims to increase the spread of passwords in their universe. Our work can also be used as a leverage to recommend or force the usage of slow hash functions and the usage of password strength meters in organizations services.

7   Conclusion

Our study proves the importance of the speed of the enumerator and the hash function in the performance measurement of a password cracking process. Thanks to that, it becomes possible to evaluate how efficient a slow memory-hard function is against the different enumeration strategies of the literature.

Even if we observe an increase of the bcrypt and Argon2d usage, recent passwords database leaks still confirm the high usage of cryptographic hash functions like SHA-1 and its siblings. We recommend the usage of dedicated slow and memory-hard hash functions to protect passwords.

We bring the technical challenges out when computing the performance of enumerators in password cracking context. Firstly, we highlight the impossibility to compute both the number of found passwords and the time to generate the required candidates without altering their values. Even if we only measure the latter it remains a complicated task. Then we provided a methodology to estimate these values accurately enough.

When the enumerator has an index function, as in [18], we have a big advantage since it becomes possible to compute the success rate of the enumerator without running it. However it has not been considered when recent enumerators have been designed. In the other case, we need to run the algorithm and measure its success rate along the enumeration.

Our experiments over two publicly leaked passwords lists show unexpected results. First, we observe that bruteforce is still useful against fast hash functions. Secondly, the PCFG-based algorithm is nearly as good as JtR-Markov against LinkedIn using SHA-1, meaning that it would be better than JtR-Markov if it was optimized. In OMEN paper [9], authors showed that it was better than PCFG on the Rockyou dataset. However we show that considering the cost of generating candidates, PCFG becomes better in all presented experiments.

We showed that JtR-Markov is really relevant when using a fast hash function like SHA-1, while probabilistic enumerators like OMEN and PCFG-based algorithms perform better than others using slow hash function like bcrypt.

If PCFG and OMEN were optimized for password cracking, results would change only against fast hash functions, where OMEN, PCFG and JtR-Markov would be more distinguishable. We encourage Hashcat and John The Ripper developers to implement such algorithms in further versions of their tools. Hashcat developers have already took a step in that way since the version *v5.0.0* by introducing the feature "slow candidates" which aims to facilitate the integration of slow enumerators proposed by academics.

More generally, the security community, especially in the field of password protection, lack of researches on password enumerator performances. Nowadays, the usage of fast hash functions remains too high. However, slow hash functions are more likely to be used in the near future. Therefore, our study is relevant and should be extended with future works that take into account the password cracking context in order to be closer to the attacker environment.

Acknowledgment. We would like to thank the Région Normandie for supporting this research.

References

1. John the ripper implementation. https://github.com/magnumripper/JohnTheRipper
2. Mining hardware comparison. https://en.bitcoin.it/wiki/Mining_hardware_comparison
3. Aumasson, J.P.: Password hashing competition. https://password-hashing.net/
4. Biryukov, A., Dinu, D., Khovratovich, D.: Argon2: new generation of memory-hard functions for password hashing and other applications. In: 2016 IEEE European Symposium on Security and Privacy (EuroS&P), pp. 292–302. IEEE (2016)
5. Carnavalet, X.D.C.D., Mannan, M.: A large-scale evaluation of high-impact password strength meters. ACM Trans. Inf. Syst. Secur. **18**, 1–32 (2015)

6. Delahaye, J.P., Zenil, H.: Numerical evaluation of algorithmic complexity for short strings: a glance into the innermost structure of randomness. Appl. Math. Comput. **219**, 63–77 (2012)
7. Dell'Amico, M., Filippone, M.: Monte carlo strength evaluation: fast and reliable password checking. In: Proceedings of the 22nd ACM SIGSAC Conference on Computer and Communications Security, pp. 158–169. ACM (2015)
8. Dürmuth, M., Angelstorf, F., Castelluccia, C., Perito, D., Chaabane, A.: OMEN implementation. https://github.com/RUB-SysSec/OMEN
9. Dürmuth, M., Angelstorf, F., Castelluccia, C., Perito, D., Chaabane, A.: OMEN: faster password guessing using an ordered Markov enumerator. In: Piessens, F., Caballero, J., Bielova, N. (eds.) ESSoS 2015. LNCS, vol. 8978, pp. 119–132. Springer, Cham (2015). https://doi.org/10.1007/978-3-319-15618-7_10
10. Gosney, J.: 8x Nvidia GTX 1080 Hashcat benchmarks. https://gist.github.com/epixoip/ace60d09981be09544fdd35005051505
11. Hellman, M.: A cryptanalytic time-memory trade-off. IEEE Trans. Inf. Theory **26**(4), 401–406 (1980)
12. Hunt, T.: Have I been pwned (2017). https://haveibeenpwned.com
13. Kelley, P.G., et al.: Guess again (and again and again): measuring password strength by simulating password-cracking algorithms. In: 2012 IEEE Symposium on Security and Privacy, pp. 523–537. IEEE (2012)
14. Khalil, G.: Password security - thirty-five years later (2014)
15. Komanduri, S.: Modeling the adversary to evaluate password strength with limited samples. Ph.D. thesis, Microsoft Research (2016)
16. Ma, J., Yang, W., Luo, M., Li, N.: A study of probabilistic password models. In: 2014 IEEE Symposium on Security and Privacy, pp. 689–704. IEEE (2014)
17. Morris, R., Thompson, K.: Password security: a case history. Commun. ACM **22**, 594–597 (1979)
18. Narayanan, A., Shmatikov, V.: Fast dictionary attacks on passwords using time-space tradeoff. In: ACM Conference on Computer and Communications Security (CCS), pp. 364–372. ACM (2005)
19. NIST: NIST special publication 800–63b. https://pages.nist.gov/800-63-3/
20. Oechslin, P.: Making a faster cryptanalytic time-memory trade-off. In: Boneh, D. (ed.) CRYPTO 2003. LNCS, vol. 2729, pp. 617–630. Springer, Heidelberg (2003). https://doi.org/10.1007/978-3-540-45146-4_36
21. OpenWall: John the ripper (2017). http://www.openwall.com/john
22. Percival, C.: Stronger key derivation via sequential memory-hard functions (2009)
23. Provos, N., Mazieres, D.: A future-adaptable password scheme. In: USENIX Annual Technical Conference, FREENIX Track, pp. 81–91 (1999)
24. Steube, J.: Hashcat implementation. https://github.com/hashcat/hashcat
25. Ur, B.: Password guessability service (2015). https://pgs.ece.cmu.edu/
26. Ur, B., et al.: Design and evaluation of a data-driven password meter. In: Proceedings of CHI (2017)
27. Ur, B., et al.: Measuring real-world accuracies and biases in modeling password guessability. In: 24th USENIX Security Symposium (USENIX Security 2015), pp. 463–481 (2015)
28. Weir, C.M.: Using probabilistic techniques to aid in password cracking attacks. In: ACM Conference on Computer and Communications Security (CCS) (2010)
29. Weir, M.: PCFG implementation. https://github.com/lakiw/pcfg_cracker
30. Weir, M., Aggarwal, S., De Medeiros, B., Glodek, B.: Password cracking using probabilistic context-free grammars. In: 2009 30th IEEE Symposium on Security and Privacy, pp. 391–405. IEEE (2009)

An Offline Dictionary Attack Against zkPAKE Protocol

José Becerra[1(✉)], Peter Y. A. Ryan[1(✉)], Petra Šala[1,2(✉)],
and Marjan Škrobot[1(✉)]

[1] University of Luxembourg, 6, Avenue de la Fonte,
4364 Esch-sur-Alzette, Luxembourg
{jose.becerra,peter.ryan,petra.sala,marjan.skrobot}@uni.lu
[2] Computer Science Department, École Normale Supérieure,
45 rue d'Ulm, 75230 Paris, France
petra.sala@ens.fr

Abstract. Password Authenticated Key Exchange (PAKE) allows a user to establish a secure cryptographic key with a server, using only knowledge of a pre-shared password. One of the basic security requirements of PAKE is to prevent offline dictionary attacks.

In this paper, we revisit zkPAKE, an *augmented* PAKE that has been recently proposed by Mochetti, Resende, and Aranha (SBSeg 2015). Our work shows that the zkPAKE protocol is prone to offline password guessing attack, even in the presence of an adversary that has only eavesdropping capabilities. Results of performance evaluation show that our attack is practical and efficient. Therefore, zkPAKE is insecure and should not be used as a password-authenticated key exchange mechanism.

Keywords: Password Authenticated Key Exchange ·
Augmented PAKE · zkPAKE · Offline dictionary attack ·
Zero-knowledge Proofs

1 Introduction

Password Authenticated Key Exchange (PAKE) is a primitive that allows two or more users that start only from a low-entropy shared secret – which is a typical user authentication setting today – to agree on the cryptographically strong session key. Since the introduction of PAKE in 1992, a plethora of protocols trying to achieve secure PAKE has been proposed. However, due to patent issues, only recently have PAKEs begun to be considered for a wide-scale use: SRP [28] has been used in password manager called 1Password [2], J-PAKE of Hao and Ryan [14] was used in Firefox Sync [12], while *Elliptic Curve* (EC) version of the same protocol (EC-J-PAKE [11]) has been used to enable authentication and authorization for network access for *Internet-of-Things* (IoT) devices under the Thread network protocol [13].

© IFIP International Federation for Information Processing 2019
Published by Springer Nature Switzerland AG 2019
G. Dhillon et al. (Eds.): SEC 2019, IFIP AICT 562, pp. 81–90, 2019.
https://doi.org/10.1007/978-3-030-22312-0_6

From a deployment perspective, the most significant advantage of using PAKE compared to a typical key exchange protocol is that it avoids dependence on functional *Public Key Infrastructure* (PKI). On the downside, the use of low-entropy secret as the primary means of authentication comes with the price: PAKEs are inherently vulnerable to *online* dictionary attacks. To mount this attack, all an adversary needs to do is repeatedly send candidate passwords to the verifying server to test for their validity. In practice, this type of attack can be relatively easily avoided in a two-party setting by limiting the number of guesses (i.e., wrong login attempts) that can be made in a given time frame.

At the same time, a well-designed PAKE must be resistant against *offline* dictionary attacks. In such attack scenario, the adversary typically operates in two phases: in the first (usually online) phase, the adversary – either by eavesdropping or impersonating a user – tries to collect a function of the password that is being targeted to serve him as the password verifier. Later, in the second (offline) phase, the adversary has to correlate the verifier that has been collected in the first step with offline password guesses to determine the correct password.

Concerning of design, PAKEs can follow a symmetric or asymmetric approach concerning the value that is used as an authenticator. For instance, the first PAKE to be proposed, EKE [6], follows symmetric design strategy: Both client and server are required to know their joint password in clear to successfully run the EKE protocol. Such protocols are usually called *balanced* PAKEs. Over time it has been realized that the risk of losing a large number of passwords in case of a server compromise increases if passwords are kept in the clear. Damage inflicted from such loss could be very high, especially today when most people typically use many online services while authenticating with only a few related passwords.

One way to mitigate such treat is to use asymmetrically designed PAKE, also known as *augmented* PAKE[1]. This type of PAKE guarantees that the password is not stored on the server side as plaintext, but, in fact, as an image of the password. Nevertheless, for long it has been argued, from a theoretical perspective, that augmented PAKEs do not add much benefit over balanced PAKEs, since the brute-force attack on a stolen password file (a list containing password hashes) would quickly yield a number of underlying passwords. With the introduction of *sequential memory-hard hash functions* such as Scrypt [25] and Argon2 [8] and use of *salt*, which can be used to slow down password cracking significantly, this may not be the case anymore.

1.1   Our Contribution

Recently, Mochetti, Resende and Aranha [23] proposed (without exhibiting a security proof) a simple augmented PAKE called zkPAKE, which they claim is suitable for banking applications, requiring the server to store only the image of a password under a one-way function. Their main idea was to use zero-knowledge proof of knowledge (password) to design an efficient PAKE. However, here we

---

[1] For the latest results on augmented PAKE check Jarecki et al. [20].

present an offline dictionary attack against the zkPAKE protocol. In addition, we show that the same attack works on a slight variant of zkPAKE that has been proposed later in [24]. We also provide a prototype and share the benchmarks of the attack to demonstrate its feasibility. Our dictionary attack can be carried out in two ways: passively - by eavesdropping on the zkPAKE protocol execution, or actively - by impersonating the server and having the client attempt to log in.

1.2 Previous Works

Password Authenticated Key Exchange was introduced by Bellovin and Meritt [6] in 1992. Their EKE protocol was first to show that it is possible to bootstrap a low-entropy string into a strong cryptographic key. A few years later, Jablon proposed an alternative - the SPEKE protocol [18]. Over the next 25 years, plenty more PAKE proposals have surfaced [4, 14, 21, 22]. In parallel, augmented versions of different PAKEs were introduced (e.g. A-EKE [7], B-SPEKE [19]). As explained above, augmented PAKEs have an additional security property compared to balanced PAKEs: if implemented well, it is considered to be more resistant to server compromise in a sense that clients' passwords are not immediately revealed once the password file is leaked since the attacker still has to perform password cracking. Finally, a number of them have been standardized in IEEE [16], IETF [15] and ISO [17].

Security of early PAKE proposals was argued only informally by showing that a protocol can withstand all known attacks. Starting from 2000, the two formal models of security for PAKE appeared in [5] and [9]. More specifically, following a game-based approach Bellare, Pointcheval and Rogaway have argued in [5] that a provably secure PAKE protocol must provide the indistinguishability of the session key and satisfy the authentication property. The Real-or-Random (RoR) variant of their model from [3], along with the Universally Composable PAKE model from [10] are considered to be state-of-the-art models that rigorously capture PAKE security requirements.

Since we exclusively deal with an offline dictionary attack on zkPAKE, in this paper, we keep the discussion here short and refer readers to Pointcheval's survey [26] for a more detailed overview of PAKE research field.

1.3 Organization

The rest of the paper is organized as follows. Section 2 describes the zkPAKE protocol and its variant. In Sect. 3, we present an offline dictionary attack against both variants of the zkPAKE protocol. Finally, we conclude the paper in Sect. 4.

2 The zkPAKE Protocol

In this section, we review the zkPAKE protocol. We will start with the variant of zkPAKE from [24] whose description is presented in Fig. 1, and then point out the differences with the original design from [23]. The reason for this order of presentation is because the variant of zkPAKE that is proposed later is slightly more elaborate than the original zkPAKE, so we want to show that zkPAKE does not stand against our attack even with proposed modifications.

2.1 Protocol Description

zkPAKE, as described in [24], is a two-party augmented PAKE protocol meant to provide authenticated key exchange between a server S and a client C.

Initialization Phase. The protocol starts with an enrollment phase, which is executed for every client only once. In this phase, a client and a server (e.g., bank) share a secret value of low entropy that can be remembered by the client. More specifically, in case of zkPAKE, the client must remember the password π, while the server only stores an image of the password R. Before the server computes the corresponding image R, public parameters must be chosen and agreed on: (1) a finite cyclic group $\mathbb{G}$ of prime order q and a random generator g of the group $\mathbb{G}$; (2) Hash functions H_1 and H_2 whose outputs are k-bit strings, where k is the security parameter representing the length of session keys.

Protocol Execution. Once the enrollment phase is executed and the public parameters are established, the zkPAKE protocol (see Fig. 1) will run in three communication rounds as follows:

1. First, the server S chooses a random value n from $\mathbb{Z}_q$, computes N that is supposed to act both as a nonce and Diffie-Hellman value, and sends it to the client C.
2. Now, upon receiving the nonce N, the client C inputs his password, computes the hash of the password - r, chooses a random element v from $\mathbb{Z}_q$, and computes $t := N^v$. Then, C computes $c := H_1(g, g^r, t, N)$ and obtains $u := v - H_1(c)r$ that should lie in $\mathbb{Z}_q$. Next, C computes the session key $sk_c := H_2(c)$ and sends u and $H_1(c)$ to the S.
3. Upon receiving $H_1(c)$ and u, S recovers t' by computing $g^{un}R^{nH_1(c)}$. Then, S calculates $c' := H_1(g, R, t', N)$. Next, S checks if $H_1(c')$ echoes $H_1(c)$. If it does, S computes the session key $sk_s := H_2(c')$ and sends $H_1(sk_s)$ to C. Otherwise; it aborts the protocol.
4. Similarly, upon receiving $H_1(sk_s)$, C checks if $H_1(sk_s)$ and $H_1(sk_c)$ match. If values are equal, C saves computed session key sk_c and terminates.

As we said before, the authors of zkPAKE have presented two variants of it. The original proposal from [23] differs from the follow-up version in two places: Nonce N is left underspecified, and value t on the client side is computed without involving received nonce. This difference also affects the computation of t' from the server side. In more details, the original zkPAKE protocol runs as follows:

Initialization

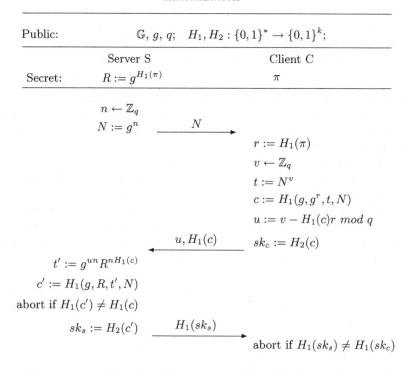

Public:	$\mathbb{G}, g, q;\quad H_1, H_2 : \{0,1\}^* \to \{0,1\}^k;$	
	Server S	Client C
Secret:	$R := g^{H_1(\pi)}$	π

$n \leftarrow \mathbb{Z}_q$
$N := g^n \qquad \xrightarrow{\quad N \quad}$

$r := H_1(\pi)$
$v \leftarrow \mathbb{Z}_q$
$t := N^v$
$c := H_1(g, g^r, t, N)$
$u := v - H_1(c)r \bmod q$
$\xleftarrow{\quad u, H_1(c) \quad} \quad sk_c := H_2(c)$

$t' := g^{un}R^{nH_1(c)}$
$c' := H_1(g, R, t', N)$
abort if $H_1(c') \neq H_1(c)$
$sk_s := H_2(c') \qquad \xrightarrow{\quad H_1(sk_s) \quad}$
abort if $H_1(sk_s) \neq H_1(sk_c)$

Fig. 1. The zkPAKE protocol.

1. The server sends his nonce N to the client C.
2. The client calculates the hash of his password r, chooses a random parameter $v \leftarrow \mathbb{Z}_q$, and computes $t := g^v$. Then, C computes $c := H_1(g, g^r, t, N)$ and obtains $u := v - H_1(c)r$ in $\mathbb{Z}_q$. Next, C computes the session key $sk_c := H_2(c)$ and sends u and $H_1(c)$ to the S.
3. Upon receiving $H_1(c)$ and u, S recovers t' by computing $g^u R^{H_1(c)}$. Then, S calculates $c' := H_1(g, R, t', N)$. Next, S checks if $H_1(c')$ echoes $H_1(c)$. If it does, S computes the session key $sk_s := H_2(c')$ and sends $H_1(sk_s)$ to C. Otherwise, he aborts the protocol.
4. Finally, upon receiving $H_1(sk_s)$, C checks if $H_1(sk_s)$ echoes $H_1(sk_c)$. If values are equal, C saves computed session key sk_c and terminates.

3 Offline Dictionary Attack on zkPAKE

In the next section, we will show how both variants of the zkPAKE protocol are vulnerable to an offline dictionary attack. Our attack exploits the fact that r, which is a hash of clients password, is of low entropy.

3.1 Attack Description

Let the enrollment phase be established and let an attacker $\mathcal{A}$ be allowed only to eavesdrop on the communication between two honest parties. The attack on the version of zkPAKE protocol presented in Fig. 1 proceeds as follows:

Step 1. The execution of the protocol starts and S sends his first message, N. The attacker $\mathcal{A}$ sees the message and stores it in his memory.

Step 2. C does all the computations demanded by the protocol and sends u and $H_1(c)$ in the second transmission to S. $\mathcal{A}$ observes the second message and obtains u and $H_1(c)$.

Step 3. The adversary that now holds N, u and $H_1(c)$ from the first two message rounds may go offline to perform a dictionary attack. His goal is to compute a candidate c' and then use stored $H_1(c)$ as a verifier. The adversary will compute c' by hashing $H_1(g, g^r, t', N)$. Two intermediate inputs to the hash function are obtained by first choosing a candidate password π, and then computing the corresponding r and t'. Note that the adversary can easily compute $t' = N^v$, since $v := u + H_1(c)r$. Finally, the adversary checks if his guess $H_1(c')$ echoes $H_1(c)$.

Step 4. The adversary repeats Step 3 until he guesses the correct password.

As for the original zkPAKE protocol, the same attack works in a very similar way: Steps 1, 2 and 4 are the same while in Step 3 we need to make a minor change:

Step 3a. The adversary that now holds N, u and $H_1(c)$ from the first two message rounds may go offline to perform a dictionary attack. Same as above, the adversary aims to obtain candidate c_i' by computing a hash $H_1(g, g^{r_i}, t_i', N)$. Here the only difference is that $t_i' = g^{v_i}$, while the formula for computing v_i stays the same.

Note that one can mount a similar dictionary attack by impersonating a server. In this case, the only difference with the eavesdropping attack described above is the attacker picks the value of the nonce N. Such knowledge, however, does not additionally help the adversary in our attack. Once the adversary receives clients reply, he can continue with Steps 3 and 4 from the eavesdropping attack.

3.2 Attack Implementation

We implemented a prototype[2] in Python 3 to simulate the attack described above. Our simulation consists of two steps: in the first step, a password is randomly chosen from one of three fixed dictionaries that vary in size and the zkPAKE protocol is executed between two honest parties. Then, in the second step (see Algorithm 1), the adversary is given access to honestly generated values as described in Sect. 3.1. With this information in hand, the adversary can easily perform an offline dictionary attack against chosen password.

---

[2] Available under GPL v3 at https://github.com/PetraSala/zkPAKE-attack.

Input: Values N, u, $H(c)$ and a dictionary of passwords P
Output: pw', K
for *each pw' in dictionary P* **do**
$\quad$ | $\quad c' = hash.sha256(pw')$
$\quad$ | $\quad r = c' \bmod q$
$\quad$ | $\quad v = (u + H(c) * r) \bmod q$
$\quad$ | $\quad t = N^v \bmod p$
$\quad$ | $\quad R = g^r \bmod p$
$\quad$ | $\quad H(c') = hash.sha256(g, g^r, t, N)$
$\quad$ | $\quad$ **if** $H(c) == H(c')$ **then**
$\quad$ | $\quad$ | $\quad K = HKDF(c')$
$\quad$ | $\quad$ | $\quad$ Return $(pw', H(c'), K)$
$\quad$ | $\quad$ **end**
end

Algorithm 1. Offline search algorithm

We performed a set of experiments, using a 224-bit subgroup of a 2048-bit finite field Diffie-Hellman group[3], to determine the time it takes to complete an offline dictionary attack depending on the size of a selected dictionary. Each set of experiments involved mounting the attack by enumerating dictionaries that contain 1000, 10000, and 100000 random password elements. Each experiment was performed 50 times.

Results. The times it took the adversarial algorithm described above to find a matching password for each given dictionary are summarized in Table 1.

Table 1. Results for different dictionary sizes

Dictionary size	Average time until the correct password is found (*ms*)	Std dev
1000	3694	1898
10000	27322	17461
100000	244540	178465

Our results demonstrate that there is a linear relationship between the size of the dictionary and the average time to find a matching password, and shows that an attack is feasible for any adversary with even a small computational power[4]. As expected, the total time for cracking a 100000 password-size dictionary is less than 5 min, and thus we conclude that the attack would be feasible for

---

[3] Selected group parameters, which are originally coming from the standard NIST cryptographic toolbox, are specified in Appendix A.
[4] In all cases the experiments were run under Windows 10 on a 2.8 GHz PC with 8 GB of memory.

dictionaries with significantly more elements. We also note that there are more powerful tools to create more efficient dictionaries, such as HashCat [27] or John the Ripper [1], which would make the offline search more effective.

4 Conclusion

In this paper, we showed that both versions of the zkPAKE protocol [23, 24] are vulnerable to offline dictionary attacks. To make matters worse, the adversary in case of zkPAKE only needs eavesdropping capabilities to mount the attack.

By taking a wider view on zkPAKE, the problem with its design lies in a fact that variable r, which is of low-entropy, is used as a mask for the secret value v. In contrast, in a typical zero-knowledge proof of knowledge, which was used as an inspiration for zkPAKE design, such value is of high entropy. By showing this vulnerability, we hope that in future protocol designers will be more careful in claiming the security of proposed protocols, especially when a proof of security does not back those claims. Since zkPAKE protocol core design is flawed beyond repair and there already exist many mature PAKE alternatives, we do not pursue further study to improve on the zkPAKE protocol.

A Appendix

The group parameters are taken from the NIST cryptographic toolbox using 2048 modulus, and are shown in Table 2.

Table 2. Group parameters

Parameter	Value (Base 16)
Prime modulus	AD107E1E 9123A9D0 D660FAA7 9559C51F A20D64E5 683B9FD1 B54B1597 B61D0A75 E6FA141D F95A56DB AF9A3C40 7BA1DF15 EB3D688A 309C180E 1DE6B85A 1274A0A6 6D3F8152 AD6AC212 9037C9ED EFDA4DF8 D91E8FEF 55B7394B 7AD5B7D0 B6C12207 C9F98D11 ED34DBF6 C6BA0B2C 8BBC27BE 6A00E0A0 B9C49708 B3BF8A31 70918836 81286130 BC8985DB 1602E714 415D9330 278273C7 DE31EFDC 7310F712 1FD5A074 15987D9A DC0A486D CDF93ACC 44328387 315D75E1 98C641A4 80CD86A1 B9E587E8 BE60E69C C928B2B9 C52172E4 13042E9B 23F10B0E 16E79763 C9B53DCF 4BA80A29 E3FB73C1 6B8E75B9 7EF363E2 FFA31F71 CF9DE538 4E71B81C 0AC4DFFE 0C10E64F
Generator	AC4032EF 4F2D9AE3 9DF30B5C 8FFDAC50 6CDEBE7B 89998CAF 74866A08 CFE4FFE3 A6824A4E 10B9A6F0 DD921F01 A70C4AFA 00C29F52 C57DB17C 620A8652 BE5E9001 A8D66AD7 C1766910 1999024A F4D02727 5AC1348B B8A762D0 521BC98A E2471504 22EA1ED4 09939D54 DA7460CD B5F6C6B2 50717CBE F180EB34 118E98D1 19529A45 D6F83456 6E3025E3 16A330EF BB77A86F 0C1AB15B 051AE3D4 28C8F8AC B70A8137 150B8EEB 10E183ED D19963DD D9E263E4 770589EF 6AA21E7F 5F2FF381 B539CCE3 409D13CD 566AFBB4 8D6C0191 81E1BCFE 94B30269 EDFE72FE 9B6AA4BD 7B5A0F1C 71CFFF4C 19C418E1 F6EC0179 81BC087F 2A7065B3 84B890D3 191F2BFA
Subgroup order	801C0D34 C58D93FE 99717710 1F80535A 4738CEBC BF389A99 B36371EB

References

1. John the ripper password cracker. https://www.openwall.com/john/. Accessed 25 Feb 2019
2. 1Password Security Design. https://1password.com/files/1Password%20for %20Teams%20White%20Paper.pdf. Accessed 27 Feb 2018
3. Abdalla, M., Fouque, P.-A., Pointcheval, D.: Password-based authenticated key exchange in the three-party setting. In: Vaudenay, S. (ed.) PKC 2005. LNCS, vol. 3386, pp. 65–84. Springer, Heidelberg (2005). https://doi.org/10.1007/978-3-540-30580-4_6
4. Abdalla, M., Pointcheval, D.: Simple password-based encrypted key exchange protocols. In: Menezes, A. (ed.) CT-RSA 2005. LNCS, vol. 3376, pp. 191–208. Springer, Heidelberg (2005). https://doi.org/10.1007/978-3-540-30574-3_14
5. Bellare, M., Pointcheval, D., Rogaway, P.: Authenticated key exchange secure against dictionary attacks. In: Preneel, B. (ed.) EUROCRYPT 2000. LNCS, vol. 1807, pp. 139–155. Springer, Heidelberg (2000). https://doi.org/10.1007/3-540-45539-6_11
6. Bellovin, S.M., Merritt, M.: Encrypted key exchange: password-based protocols secure against dictionary attacks. In: IEEE Symposium on Research in Security and Privacy, pp. 72–84 (1992)
7. Bellovin, S.M., Merritt, M.: Augmented encrypted key exchange: a password-based protocol secure against dictionary attacks and password file compromise. In: Denning, D.E., Pyle, R., Ganesan, R., Sandhu, R.S., Ashby, V. (eds.) Proceedings of the 1st ACM Conference on Computer and Communications Security, CCS 1993, pp. 244–250. ACM (1993)
8. Biryukov, A., Dinu, D., Khovratovich, D.: Argon2: new generation of memory-hard functions for password hashing and other applications. In: IEEE European Symposium on Security and Privacy, EuroS&P 2016, pp. 292–302. IEEE (2016)
9. Boyko, V., MacKenzie, P., Patel, S.: Provably secure password-authenticated key exchange using Diffie-Hellman. In: Preneel, B. (ed.) EUROCRYPT 2000. LNCS, vol. 1807, pp. 156–171. Springer, Heidelberg (2000). https://doi.org/10.1007/3-540-45539-6_12
10. Canetti, R., Halevi, S., Katz, J., Lindell, Y., MacKenzie, P.: Universally composable password-based key exchange. In: Cramer, R. (ed.) EUROCRYPT 2005. LNCS, vol. 3494, pp. 404–421. Springer, Heidelberg (2005). https://doi.org/10.1007/11426639_24
11. Cragie, R., Hao, F.: Elliptic curve J-PAKE cipher suites for transport layer security (TLS) (2016). https://tools.ietf.org/html/draft-cragie-tls-ecjpake-01
12. Foundation, T.M.: Firefox Sync. https://www.mozilla.org/en-US/firefox/sync/. Accessed 28 Feb 2018
13. Group, T.: Thread protocol. http://threadgroup.org/. Accessed 06 Apr 2017
14. Hao, F., Ryan, P.: J-PAKE: authenticated key exchange without PKI. Trans. Comput. Sci. **11**, 192–206 (2010)
15. Harkins, D.: Dragonfly key exchange. RFC 7664, RFC Editor, November 2015
16. Standard specifications for password-based public key cryptographic techniques: standard. IEEE Standards Association, Piscataway, NJ, USA (2002)
17. ISO/IEC 11770–4:2006/cor 1:2009, Information technology - Security techniques - Key management - Part 4: Mechanisms based on weak secrets. Standard, International Organization for Standardization, Genève, Switzerland (2009)

18. Jablon, D.P.: Strong password-only authenticated key exchange. ACM SIGCOMM Comput. Commun. Rev. **26**(5), 5–26 (1996)
19. Jablon, D.P.: Extended password key exchange protocols immune to dictionary attacks. In: 6th Workshop on Enabling Technologies (WET-ICE 1997), Infrastructure for Collaborative Enterprises, pp. 248–255. IEEE Computer Society (1997)
20. Jarecki, S., Krawczyk, H., Xu, J.: OPAQUE: an asymmetric PAKE protocol secure against pre-computation attacks. In: Nielsen, J.B., Rijmen, V. (eds.) EUROCRYPT 2018. LNCS, vol. 10822, pp. 456–486. Springer, Cham (2018). https://doi.org/10.1007/978-3-319-78372-7_15
21. Katz, J., Ostrovsky, R., Yung, M.: Efficient password-authenticated key exchange using human-memorable passwords. In: Pfitzmann, B. (ed.) EUROCRYPT 2001. LNCS, vol. 2045, pp. 475–494. Springer, Heidelberg (2001). https://doi.org/10.1007/3-540-44987-6_29
22. MacKenzie, P.: The PAK suite: protocols for password authenticated key exchange. DIMACS Technical report 2002-46, (2002)
23. Mochetti, K., Resende, A., Aranha, D.: zkPAKE: a simple augmented PAKE protocol. In: Brazilian Symposium on Information and Computational Systems Security (SBSeg) (2015)
24. Mochetti, K., Resende, A., Aranha, D.: zkPAKE: a simple augmented PAKE protocol (2015). http://www2.ic.uff.br/~kmochetti/files/abs01.pdf
25. Percival, C.: Stronger Key Derivation via Sequential Memory-hard Functions, pp. 1–16. Self-published (2009)
26. Pointcheval, D.: Password-based authenticated key exchange. In: Fischlin, M., Buchmann, J., Manulis, M. (eds.) PKC 2012. LNCS, vol. 7293, pp. 390–397. Springer, Heidelberg (2012). https://doi.org/10.1007/978-3-642-30057-8_23
27. Team Hashcat: hashcat - advanced password recovery. https://hashcat.net/hashcat/. Accessed 25 Feb 2019
28. Wu, T.D.: The secure remote password protocol. In: Proceedings of the Network and Distributed System Security Symposium, NDSS 1998. The Internet Society (1998)

Fine-Grained Access Control in Industrial Internet of Things
Evaluating Outsourced Attribute-Based Encryption

Dominik Ziegler[1]([✉]) [iD], Josef Sabongui[2], and Gerald Palfinger[3]

[1] Know Center GmbH, Inffeldgasse 13, 8010 Graz, Austria
`dominik.ziegler@tugraz.at`
[2] Institute for Applied Information Processing and Communications,
Graz University of Technology, Inffeldgasse 16a, 8010 Graz, Austria
`josef.sabongui@student.tugraz.at`
[3] A-SIT Secure Information Technology Center Austria,
Seidlgasse 22/Top 9, 1030 Vienna, Austria
`gerald.palfinger@a-sit.at`
`http://www.know-center.tugraz.at, https://www.iaik.tugraz.at`
`https://www.a-sit.at`

Abstract. Putting Attribute-Based Encryption (ABE) to the test, we perform a thorough performance analysis of ABE with outsourced decryption. In order to do so, we implemented a purely Java and Kotlin based Ciphertext-Policy Attribute-Based Encryption (CP-ABE) system. We specifically focus on the requirements and conditions of the Industrial Internet of Things (IIoT), including attribute revocation and limited computing power. We evaluate our system on both resource-constrained devices and high-performance cloud instances. Furthermore, we compare the overhead of our implementation with classical asymmetric encryption algorithms like RSA and ECC.

To demonstrate compatibility with existing solutions, we evaluate our implementation in the Siemens MindSphere IIoT operating system. Our results show that ABE with outsourced decryption can indeed be used in practice in high-security environments, such as the IIoT.

Keywords: Fine-grained access control · IIoT · Performance analysis · ABE

1   Introduction

The integration of Internet of Things (IoT) [4] into classical enterprise systems promises fundamental improvements to existing workflows, efficiency gains or optimised decision making. This trend towards interconnected devices in the manufacturing industry is called the Industrial Internet of Things (IIoT) or Industry 4.0 [12]. Its goal is to integrate and connect manufacturing environments via global networks. In fully automated IIoT systems, arbitrary sensors,

© IFIP International Federation for Information Processing 2019
Published by Springer Nature Switzerland AG 2019
G. Dhillon et al. (Eds.): SEC 2019, IFIP AICT 562, pp. 91–104, 2019.
https://doi.org/10.1007/978-3-030-22312-0_7

production facilities or heavy machinery are all communicating and controlling each other independently. As a result, the IIoT will allow enterprises to quickly adapt to customer requirements and individualisation in production processes while maintaining resource and energy efficiency.

However, Sadeghi et al. [20] show that devices in enterprise environments can generate vast amounts of security critical or personal data. Furthermore, processing of such data might be restricted by laws, like the General Data Protection Regulation (GDPR). As a result, these challenges call for an efficient and flexible access-control mechanism, to protect sensitive data from unauthorised access.

A cryptographic solution for fine-grained access control is Attribute-Based Encryption (ABE), first introduced by Sahai and Waters [21]. It represents a generalisation of Identity-Based Encryption (IBE), a concept proposed by Shamir [22]. In contrast to conventional public-key cryptography schemes, ABE defines the recipient of a message as a set of attributes. It does so, by combining ciphertexts with distinctive attributes and access control policies. There exist two main flavours of ABE. Key-Policy Attribute-Based Encryption (KP-ABE) [10] and Ciphertext-Policy Attribute-Based Encryption (CP-ABE) [6]. KP-ABE schemes encrypt messages under a set of attributes. Access control policies, embedded into a party's secret key, determine which ciphertexts a key can decrypt. In CP-ABE constructions the role of ciphertexts and keys are reversed. Access control structures are directly embedded into ciphertexts and attributes are associated with a party's secret key. Parties can only decrypt ciphertexts if their attributes match the embedded policy.

A central issue with classical ABE systems, however, is that they typically do not have strong security guarantees or are based on expensive bilinear pairings and are hence not very efficient. Indeed, Wang et al. [23] show that classical ABE systems should only be used "when the computing device has relatively high computing power and the applications demand low to medium security". In contrast, we typically find a variety of devices with usually limited resources, e.g. sensors with low computing power or memory, in the IIoT. Hence, performance is a critical aspect of ABE in the IIoT. As a result, several approaches aim at improving ABE, in terms of efficiency. For example, approaches based on Elliptic Curve Cryptography (ECC) or Linear Secret Sharing Schemes (LSSSs) [1,19], address performance issues of ABE. Another promising approach for resource-constrained environments is outsourcing heavy computations to cloud servers [11,17,18]. Clients only need to perform lightweight operations. Recent advances in Outsourced Attribute-Based Encryption (OABE), promise applicability in resource-constrained environments, such as the IIoT. Hence, in this paper, we study OABE from two perspectives.

- **How does ABE with outsourced decryption compare to established cryptographic primitives like RSA and ECC, in practice?** We implemented a purely Java and Kotlin based CP-ABE system with outsourced decryption. We specifically focus on industrial requirements such as attribute revocation, key escrow and key exposure. We perform a thorough evaluation of the execution time of operations and overhead on both, resource-constrained devices and powerful cloud servers. Although a fair comparison of these primitives is not possible, our goal is to highlight the overall performance of ABE with outsourced decryption with established primitives.

- **How can ABE efficiently be deployed in IIoT-based systems with high security requirements?** To demonstrate compatibility with existing infrastructure, we successfully deployed our system in the Siemens MindSphere cloud. MindSphere is the cloud-based, open IoT operating system from Siemens for the Industrial Internet of Things. It focuses on data acquisition and access control in IIoT environments. Our evaluation shows that, while adding some computational overhead, ABE can provide fine-grained access control in IIoT, in practice.

Motivated by our findings, we first provide an introduction to ABE and relevant aspects. Next, we evaluate execution time and overhead of our implementation on a Raspberry Pi 3 Model B+ and high-performance cloud instances. We evaluate the performance of different security levels. Furthermore, we show how OABE can successfully be deployed in IIoT environments such as the Siemens MindSphere platform. Subsequently, we highlight relevant work. Finally, we discuss our findings and give an outlook.

2   Background: ABE with Outsourced Decryption

We evaluate how ABE can be used in professional environments in practice. The work of Lin et al. [19] addresses several typical challenges of IIoT systems and was therefore chosen to serve as the foundation of the implemented system, concerning the ABE framework.

In our system, a Client's (CL) private key is split three-ways. To fully decrypt ciphertexts, all key parts have to be used. Since the majority of decryption calculations are outsourced to a cloud server, no expensive bilinear pairings have to be done on the client's side. Furthermore, the server's key-parts have to be retrieved on each decryption request. This ensures that key revocation is enforced instantaneously, by simply invalidating and updating user key-parts. The split private keys also prevent key escrow, since no actor has knowledge of all three key-parts throughout the entire decryption process.

In this section, we first give an introduction to involved actors of the implemented ABE systems with outsourced decryption. Next, we discuss security definitions and required algorithms.

2.1   Actors

The architecture of this ABE system identifies multiple actors, as outlined in this section. Each actor plays a significant role in the environment of this ABE system.

- *Data Owner:* A Data Owner (DO) produces information, which is encrypted using a random secret and an access policy. This initial ciphertext is subsequently sent to the Re-Encryption Server (RS).

- *Re-Encryption Server:* The main responsibility of the Re-Encryption Server (RS) is to re-encrypt initial ciphertexts. The re-encryption process incorporates the attribute groups of a system into the ciphertext. The RS is furthermore involved in the generation of the system parameters.
- *Key Authority:* The Key Authority (KA) is in charge of key management in the system. It creates and updates client key-parts in conjunction with the RS, but remains the sole authority to grant or revoke attributes from key-parts mentioned above. Any modifications to the system parameters, such as additional attributes throughout the systems lifetime, are exclusively performed by the KA.
- *Decryption Server:* The Decryption Server (DS) is capable of partially decrypting ciphertexts, obtained by a client. For this operation, the DS requires the key-parts from KA and RS. This partial decryption performs all expensive bilinear pairing operations, while leaving a simple El-Gamal encrypted ciphertext for the CL to decrypt.
- *Client:* A Client (CL) is an endpoint, which wants to access encrypted data. Since a valid CL key-part is necessary for this operation, the client needs to participate in the update process of key-parts. If the key-parts associated with a CL fulfil the access policy of a ciphertext, it is possible to obtain the plaintext with the help from the DS.

2.2 Definitions

Lin et al. [19] based their protocols on bilinear pairings of Type I. However, Chatterjee et al. [8] claim that Type I pairings are expected to be slower compared to Type II or Type III pairings. Hence, we chose a Type III pairing, specifically the *Ate Pairing over Barreto-Naehrig curves.* We will be using the notion of additive groups instead of multiplicative groups for the two paired groups $\mathbb{G}_1$ and $\mathbb{G}_2$, for illustration purposes. The target group remains a multiplicative group.

Bilinear Maps. Given two additive cyclic groups $\mathbb{G}_1$, $\mathbb{G}_2$ with generators g_1, g_2 respectively and a multiplicative group $\mathbb{G}_T$. Let the pairing map be:

$$e : \mathbb{G}_1 \times \mathbb{G}_2 \to \mathbb{G}_T \tag{1}$$

The prime order of $\mathbb{G}_1$ is $p \in \mathbb{P}$. The hash functions of this system are defined as:

$$\mathbb{H}_0 : \{0, 1\}^* \to \mathbb{G}_1 \tag{2a}$$

$$\mathbb{H}_1 : \mathbb{G}_1 \to \mathbb{Z}_p^* \tag{2b}$$

$$\mathbb{H}_T : \mathbb{G}_T \to \mathbb{Z}_p^* \tag{2c}$$

The system defines attributes, that may be used to build ciphertext policies. This is denoted by the set S. All authorised clients are contained in the set $\mathcal{U}$. Each attribute is associated with an attribute group G, consisting of system clients that hold this attribute. The set of all attribute groups is specified as $\mathcal{G}$. The implemented ABE scheme consists of the following algorithms.

System Setup. The setup of an ABE system consists of three algorithms. They generate the system parameters *params* which are required to be publicly available to all system actors.

– *BaseSetup* $(\lambda, S) \to params_{base}$: On input the security parameter λ and a set S of attributes, the algorithm outputs the base system parameters.
– *KeyAuthoritySetup* $(params_{base}) \to params_{KA}$: On input the base system parameters $params_{base}$, the algorithm selects a random element $q \in \mathbb{Z}_p^*$ as its secret parameter and outputs the KA parameters.
– *ReEncryptionSetup* $(params_{base}) \to params_{RS}$: On input $params_{base}$, the algorithms selects a random element $\alpha \in \mathbb{Z}_p^*$ as its secret parameter and outputs the public parameters $params_{RS}$.

After the RS registers its public parameter with the KA, the final system parameters are created and distributed to all actors in the system:

$$params = \{params_{base}, \ params_{KA}, \ params_{RS}\} \tag{3}$$

Key Creation and Update. In this system, the private key of a client is split 3-ways, where KA, RS and the CL each possess a part of the key. The process for generating these key parts is based on two algorithms:

– *KeyCreation* $(params, S) \to IK_{KA}$: The key creation algorithm takes as input the base system parameters *params* and the CL's set of attributes S. It outputs the initial key IK_{KA}.
– *KeyUpdate* $(params, IK_{KA}) \to (PK_{KA}, PK_{RS}, PK_{CL})$: The key update algorithm takes as input the base system parameters *params* and the initial key IK_{KA}. It outputs the corresponding keys for KA, RS and CL.

Encryption and Decryption. Encryption & decryption consists of four algorithms, discussed in the following:

– *Encryption* $(params, \mathbb{A}, \mathcal{M}) \to CT_{init}$: On input the base system parameters *params*, an access structure $\mathbb{A}$ and a plaintext $\mathcal{M}$ the algorithm outputs an initial ciphertext CT_{init}.
– *Re-Encryption* $(params, CT_{init}) \to CT$: On input the base system parameters *params* and the initial ciphertext CT_{init}, the re-encryption algorithm calculates and outputs the final ciphertext CT.
– *Partial-Decryption* $(CT, PK_{KA}, PK_{RS}) \to \mathcal{M}_{part}$: On input the final ciphertext CT and the keys PK_{KA} and PK_{RS}, the algorithm outputs the partially decrypted message $\mathcal{M}_{part}$.
– *Decryption* $(\mathcal{M}_{part}, PK_{CL}) \to \mathcal{M}$: On input the partially decrypted message $\mathcal{M}_{part}$ and the client's key PK_{CL} the algorithm outputs the plaintext.

3   Evaluation

In ABE the main factors, influencing the performance of cryptographic operations are the number of used attributes and the security parameter. Hence, we evaluate how different security levels and the number of attributes affect the system's performance. As our goal is to evaluate applicability in the IIoT, we first test all cryptographic operations on a cloud-instance. Next, we measure the execution time of encryption and decryption operations on a Raspberry Pi 3 Model B+, representing a low-performance IIoT device.

3.1   Test-Setup

To provide consistent test-results we performed a series of benchmark tests, repeating each test 100 times. Each iteration was run on a single core. We encrypted a random ciphertext of 16 bytes using AES with a 256-bit key. The key is then encrypted via ABE. We used the median to eliminate outliers that could occur due to initialisation operations our scheduling. We relied on the IAIK Provider for the Java^{TM} Cryptography Extension (IAIK-JCE) [15] and the IAIK ECCelerate^{TM} [14] for cryptographic functionality, such as bilinear pairings.

Settings. Since heavy operations are outsourced to a cloud server, we used two Intel(R) Xeon(R) CPU E5-2699 v4 @ 2.20 GHz running a 64-bit Linux Kernel to act as cloud and decryption server. To measure the performance of resource-constrained devices, we used a Raspberry Pi 3 Model B+.

Attributes. To demonstrate the impact of attributes, we perform our tests for a varying range of attributes. However, while Ambrosin et al. [2] claim that 30 attributes are a "range that represents a reasonable choice in real scenarios", Green et al. [11] show that policies can become highly complex in typical use cases. Hence, we tested our implementation for up to $n = 100$ attributes. To reflect this in our experiments, we generated policies in the form $(A_1, (A_2, (\ldots, (A_{n-1}, A_n, AND), AND), AND), AND)$.

Security. To measure the impact of the security level, we evaluated our system for six security levels and associated curves, as recommended in [7]. As shown in [16], the security of pairing-based cryptography depends on the prime order n of the basepoint $P \in E(\mathbb{F}_q)$ and the embedding degree k. We evaluate our implementation with $n = 2^{160}$ to $n = 2^{512}$ and according k.

3.2   Performance

As argued by Wang et al. [23], ABE has a considerable performance overhead. This may lead to long execution times, especially on resource-constrained devices. This section provides an overview of the execution time of the different operations of our implementation, to show the applicability of our approach in an IIoT scenario.

Table 1. Key size recommendations published by ECRYPT-CSA [9] and NIST [5].

	Symmetric	Factoring modulus	Elliptic curve
Legacy	80	1024	160–223
Near term	128	3072	256–383
Long term	256	15360	512+

Cloud-Server. Figure 1 depicts the execution time of different ABE operations using varying security levels and policies with up to 100 attributes. The examined prime order n ranges from 160-bit to 512-bit. This range comprises the recommended sizes for legacy (continued use or already deployed), near-term (at least ten years) and long-term (thirty to fifty years) use cases, as shown in Table 1. The first row shows the timings corresponding to systems with legacy security requirements with prime order of 160-bit and 192-bit. The near-term scenarios are depicted in the three subsequent diagrams, with prime order 224-bit, 256-bit, and 384-bit. The long-term scenario is shown in the last diagram, with prime order 512-bit.

As expected, the execution time of all operations, except for the decryption operation, grows linearly with the number of attributes. As discussed, decryption is constant due to it being a simple El-Gamal operation. Thus, the execution time of the decryption step is independent of the number of attributes and only depends on the security level. Even with prime order 512-bit, the decryption operation in our comparison takes less than 30 ms on our cloud server. While encryption and key-generation have a close to linear growth rate, both are still quite lightweight. For legacy and near-term applications, these operations take just a few tenths of a second, even when using up to 100 attributes. Only when aiming for long-term security, the required time to encrypt or generate keys while using policies with 100 attributes can take slightly longer than half a second. The most expensive operations in our scheme are the partial-decryption and the re-encryption operations. The execution time of partial-decryption exceeds five seconds in the long-term scenario when using a large number of attributes. However, in the more traditional near-term scenario using a limited number of attributes, these two operations still perform well by providing the results in less than a second on our test server.

IoT-Device. In our implementation, the encryption and decryption operations will typically be executed on an IIoT device. To depict such a resource-constrained device, we have chosen a Raspberry Pi 3 Model B+ for evaluation. Table 2 illustrates the execution times of the two operations. We have focused on prime order 256-bit, as such a security level provides near-term security. The encryption operation grows linearly in the amount of time. When using policies with up to 20 attributes, the operation is completed in less than a second and close to 4 seconds for up to 100 attributes. As the decryption operation is a constant time operation, it takes less than two-tenths of a second, irrespective of the number of attributes. Therefore, we propose to rely on OABE in applications which only require near-term security or for systems which do not rely on real-time computing.

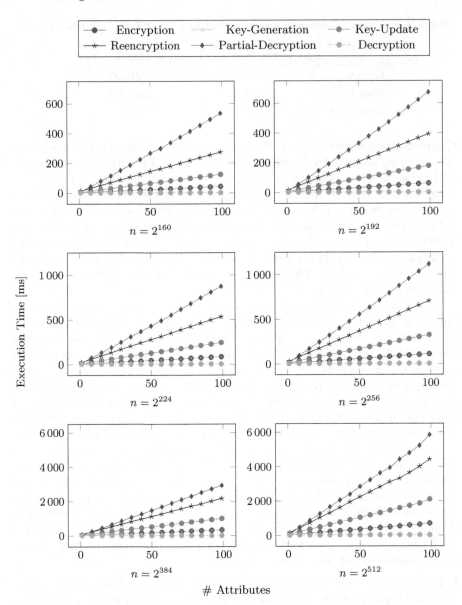

Fig. 1. Execution Time of ABE operations for different prime orders. Tests were executed on an Intel(R) Xeon(R) CPU E5-2699 v4 @ 2.20 GHz.

Table 2. Execution time of ABE En- and decryption on Raspberry Pi 3 Model B+

Security Level	Attributes	Median		Security Level	Attributes	Median
256	1	231 ms		256	1	155 ms
256	20	968 ms		256	20	158 ms
256	40	1 738 ms		256	40	154 ms
256	60	2 580 ms		256	60	158 ms
256	80	3 306 ms		256	80	157 ms
256	100	4 085 ms		256	100	157 ms

(a) Execution Time of ABE Encryption

(b) Execution Time of ABE Decryption

3.3 Data Overhead

ABE introduces additional data overhead, as access structures need to be embedded in the final ciphertext. Hence, we evaluate the additional payload of the resulting ABE ciphertext. Figure 2 illustrates this circumstance. As expected, the ciphertext size grows with the number of used attributes and the security level. As encrypted data is a constant size AES key, data overhead is independent of the underlying plaintext.

Our tests reveal that the ciphertext, generated by a client is approximately 1 kB for 160-bit prime order and 100 attributes. In applications with near-term security, the ciphertext size can increase up to 40 kB when 100 attributes are used. For 512-bit security, ciphertext size can even grow to 60 kB for 100 attributes. In contrast, we compared these results with encryption in RSA. With RSA ciphertexts getting as large as 2 kB for a 15360-bit modulus, at most.

We conclude that while 60 kB, does not seem as large overhead, given today's infrastructure, it can have a significant influence on the network. Especially in IIoT environments, with typically a large amount of small messages in short times, this additional overhead may be larger than the actual message itself. Hence, ABE not only introduces computational overhead but can also put stress on the network itself.

3.4 Comparison: RSA and ECC

In order to get a better understanding of OABE compares to established encryption algorithms, we have conducted a comparison with RSA and ECC. The comparison can be found in Fig. 3. The security level corresponds to the equivalent symmetric security level. While ABE offers a bigger feature set than both RSA and ECC, and, therefore, cannot directly be compared, it still gives an intuition of the overhead ABE introduces. For this experiment, all ABE operations have

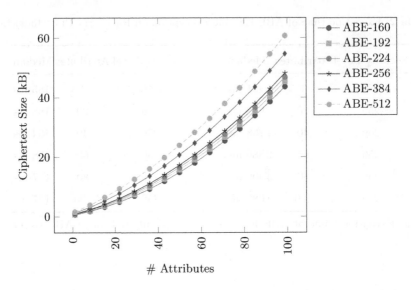

Fig. 2. Ciphertext size of ABE with prime order 160-bit to 512-bit. Tests were executed on an Intel(R) Xeon(R) CPU E5-2699 v4 @ 2.20 GHz.

been executed using a policy with a single attribute. For better illustration, a logarithmic scale is used on the y-axis. The experiment compares all steps needed to generate keys, encrypt, and decrypt ciphertexts. We found that generating a private key for RSA can take a highly variable amount of time, from 2 to 35 min. We attribute this to the fact that two large random prime numbers have to be generated. Therefore, RSA key generation has been omitted from the Figure. As shown in the figure, ECC operations are nearly instant, taking less than a millisecond, even when operating at the highest security level. This is because ECC does not have a direct form of encryption. Instead, it can be used as a key agreement protocol. Hence, the timings for ECC encryption and decryption operation merely represent the timing for the corresponding AES operation. While encryption in RSA is also rather fast, decryption performance in RSA for near-term applications is already comparable to the operations in our ABE scheme. When using even higher RSA key sizes decryption starts to take considerably longer than partial-decryption and re-encryption in our ABE scheme.

3.5 Use-Case Description: Siemens MindSphere

Our experiments reveal that ABE can indeed be used in resource-constrained environments with near and long-term security requirements. Hence, we deployed our scheme to the Siemens MindSphere IOT. We evaluated, how sensitive (sensor) data, can be efficiently shared with multiple users while reducing overhead for IIoT devices. We found that encryption operations can be performed on resource-constrained devices, as long as there are no real-time computing requirements. Additionally, we discovered that attribute revocation, respectively

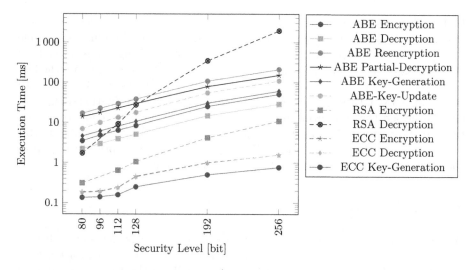

Fig. 3. Comparison of ABE with RSA, and ECC. For ABE, policies with a single attribute have been used. Tests were executed on an Intel(R) Xeon(R) CPU E5-2699 v4 @ 2.20 GHz.

key-update, which is essential in professional environments, does at most take two seconds, even for a high prime order. Hence, ABE can even provide a speed-up in revocation, in comparison to RSA.

4 Related Work

The calculation of pairings constitutes the most expensive part of ABE schemes. This is why one approach is to offload CPU intensive parts to dedicated servers, as can be seen in the work of Green et al. [11]. In their approach, any ABE ciphertext satisfied by that user's attributes is translated into a (constant-size) El Gamal-style ciphertext, regardless of the number of attributes. Hence, the resource-constrained device only has to decrypt the El Gamal-style ciphertext. Lin et al. [19] presented another approach, which offloads decryption operations. Their system is based onHur's [13] architecture but adds another actor, which partially decrypts ciphertexts for users. A scheme proposed by Zhang et al. [24] allows outsourcing key generation, decryption, and encryption. They use two different non-colluding servers for key generation. They evaluate their system and compare it against plain CP-ABE. However, they do not assess their approach on more resource-constrained devices, such as smartphones or IoT devices.

    Wang et al. [23] have conducted a performance evaluation of an ABE implementation on a smartphone and a laptop using an Intel CPU. In their study, all operations were executed solely on the client device, without the help of a dedicated server. They compare different performance metrics while using up to 30 attributes. They examine three different security levels, ranging from an 80 to 128-bit security level. Encryption and decryption operations on the smartphone lead to execution times of multiple seconds, even when using the lowest examined security level. While this scenario on the utilised laptop resulted in a reasonable encryption and decryption time of less than a second, raising the security level also increased the execution time substantially. Ambrosin et al. evaluate the feasibility of ABE on different IoT devices [2] as well as smartphones [3], using both Intel and ARM CPUs. Similar to Wang et al. [23], their evaluation setup also only harnessed the computing power of the device. They also report execution times in the seconds for security levels of 80-bit to 128-bit. In contrast, this paper evaluates the feasibility of ABE with outsourced decryption using up to 100 attributes and security levels from 80-bit to 512-bit.

5 Conclusions

We implemented a purely Java and Kotlin based CP-ABE system with outsourced decryption. We presented performance data regarding execution time and data overhead. To prove the feasibility of ABE in the IIoT we conducted our experience on both resource-constrained devices and high-performance cloud instances. Additionally, we successfully deployed our implementation in the Siemens MindSphere cloud to achieve fine-grained access control.

As expected, our results show that the performance of our ABE implementation scales linearly with the number of used attributes and security level. However, as most of the work is done by cloud servers, clients are left with only lightweight operations. As a result, we achieve constant decryption time and encryption time scaling linearly with the number of attributes and security parameters. Hence, we achieve fast decryption, even on devices with limited computing power. Our experiments further reveal that data overhead of ABE can be up to 60 kB. Thus ABE can introduce an additional burden on the network, in systems with a large number of small messages in a short time.

Summarising, we showed that ABE, while introducing some additional overhead, can indeed be successfully deployed in the IIoT to provide fine-grained access control. In the future we will further explore how our ABE implementation can be integrated into existing IIoT infrastructure and Identity and Access Management (IAM) solutions. We believe that fine-grained access control is crucial for the success of IIoT.

Acknowledgments. This research was conducted in cooperation with Graz University of Technology, the Institute for Applied Information Processing and Communications (IAIK), Know-Center GmbH and Siemens AG Austria, as part of the Siemens MindSphere research project.

References

1. Agrawal, S., Chase, M.: FAME: fast attribute-based message encryption. In: Proceedings of the 2017 ACM SIGSAC Conference on Computer and Communications Security, CCS 2017, pp. 665–682. ACM, New York (2017). https://doi.org/10.1145/3133956.3134014, ISBN 978-1-4503-4946-8
2. Ambrosin, M., et al.: On the feasibility of attribute-based encryption on internet of things devices. IEEE Micro **36**(6), 25–35 (2016). https://doi.org/10.1109/MM.2016.101. ISSN 0272-1732
3. Ambrosin, M., Conti, M., Dargahi, T.: On the feasibility of attribute-based encryption on smartphone devices. In: Proceedings of the 2015 Workshop on IoT Challenges in Mobile and Industrial Systems, IoT-Sys 2015, pp. 49–54. ACM, New York (2015). https://doi.org/10.1145/2753476.2753482, ISBN 978-1-4503-3502-7
4. Atzori, L., Iera, A., Morabito, G.: The internet of things: a survey. Comput. Netw. **54**(15), 2787–2805 (2010). https://doi.org/10.1016/j.comnet.2010.05.010. ISSN 1389-1286
5. Barker, E.: Recommendation for key management part 1: general. Technical report, National Institute of Standards and Technology, Gaithersburg, MD, January 2016. https://doi.org/10.6028/NIST.SP.800-57pt1r4
6. Bethencourt, J., Sahai, A., Waters, B.: Ciphertext-policy attribute-based encryption. In: 2007 IEEE Symposium on Security and Privacy, SP 2007, pp. 321–334, May 2007. https://doi.org/10.1109/SP.2007.11
7. Brown, D.R.L.: SEC 2: recommended elliptic curve domain parameters. Technical report, Standards for Efficient Cryptography, Certicom Research (2010)
8. Chatterjee, S., Hankerson, D., Menezes, A.: On the efficiency and security of pairing-based protocols in the type 1 and type 4 settings. In: Hasan, M.A., Helleseth, T. (eds.) WAIFI 2010. LNCS, vol. 6087, pp. 114–134. Springer, Heidelberg (2010). https://doi.org/10.1007/978-3-642-13797-6_9. ISBN 978-3-642-13797-6
9. ECRYPT - CSA: D5.4 algorithms, key size and protocols report (2018). Technical report, H2020-ICT-2014 - project 645421 (2018)
10. Goyal, V., Pandey, O., Sahai, A., Waters, B.: Attribute-based encryption for fine-grained access control of encrypted data. In: Proceedings of the 13th ACM Conference on Computer and Communications Security, CCS 2006, pp. 89–98. ACM, New York (2006). https://doi.org/10.1145/1180405.1180418, ISBN 1-59593-518-5
11. Green, M., Hohenberger, S., Waters, B.: Outsourcing the decryption of ABE ciphertexts. In: Proceedings of the 20th USENIX Conference on Security, SEC 2011, p. 34. USENIX Association, Berkeley (2011)
12. Kargermann, H., Wahlster, W., Helbig, J.: Umsetzungsempfehlungen für das Zukunftsprojekt Industrie 4.0. Technical report, April 2013
13. Hur, J.: Improving security and efficiency in attribute-based data sharing. IEEE Trans. Knowl. Data Eng. **25**(10), 2271–2282 (2013). https://doi.org/10.1109/TKDE.2011.78. ISSN 1041-4347
14. Institute for Applied Information Processing and Communications (IAIK): IAIK ECCelerate Library (2018). URL https://jce.iaik.tugraz.at
15. Institute for Applied Information Processing and Communications (IAIK): IAIK-JCE (2018). URL https://jce.iaik.tugraz.at
16. Koblitz, N., Menezes, A.: Pairing-based cryptography at high security levels. In: Smart, N.P. (ed.) Cryptography and Coding 2005. LNCS, vol. 3796, pp. 13–36. Springer, Heidelberg (2005). https://doi.org/10.1007/11586821_2

17. Li, J., Chen, X., Li, J., Jia, C., Ma, J., Lou, W.: Fine-grained access control system based on outsourced attribute-based encryption. In: Crampton, J., Jajodia, S., Mayes, K. (eds.) ESORICS 2013. LNCS, vol. 8134, pp. 592–609. Springer, Heidelberg (2013). https://doi.org/10.1007/978-3-642-40203-6_33

18. Li, J., Huang, X., Li, J., Chen, X., Xiang, Y.: Securely outsourcing attribute-based encryption with checkability. IEEE Trans. Parallel Distrib. Syst. **25**(8), 2201–2210 (2014). https://doi.org/10.1109/TPDS.2013.271. ISSN 1045–9219

19. Lin, G., Hong, H., Sun, Z.: A collaborative key management protocol in ciphertext policy attribute-based encryption for cloud data sharing. IEEE Access **5**, 9464–9475 (2017). https://doi.org/10.1109/ACCESS.2017.2707126

20. Sadeghi, A.R., Wachsmann, C., Waidner, M.: Security and privacy challenges in industrial internet of things. In: Proceedings of the 52nd Annual Design Automation Conference, DAC 2015, pp. 54:1–54:6. ACM, New York (2015). https://doi.org/10.1145/2744769.2747942, ISBN 978-1-4503-3520-1

21. Sahai, A., Waters, B.: Fuzzy identity-based encryption. In: Cramer, R. (ed.) EUROCRYPT 2005. LNCS, vol. 3494, pp. 457–473. Springer, Heidelberg (2005). https://doi.org/10.1007/11426639_27

22. Shamir, A.: Identity-based cryptosystems and signature schemes. In: Blakley, G.R., Chaum, D. (eds.) CRYPTO 1984. LNCS, vol. 196, pp. 47–53. Springer, Heidelberg (1985). https://doi.org/10.1007/3-540-39568-7_5

23. Wang, X., Zhang, J., Schooler, E.M., Ion, M.: Performance evaluation of attribute-based encryption: toward data privacy in the IoT. In: 2014 IEEE International Conference on Communications, ICC 2014, pp. 725–730 (2014). https://doi.org/10.1109/ICC.2014.6883405

24. Zhang, R., Ma, H., Lu, Y.: Fine-grained access control system based on fully outsourced attribute-based encryption. J. Syst. Softw. **125**, 344–353 (2017). https://doi.org/10.1016/j.jss.2016.12.018. ISSN 0164–1212

Towards an Automated Extraction of ABAC Constraints from Natural Language Policies

Manar Alohaly[1,2(✉)], Hassan Takabi[1], and Eduardo Blanco[1]

[1] University of North Texas, Denton, TX, USA
ManarAlohaly@my.unt.edu, {Takabi,Eduardo.blanco}@unt.edu
[2] Princess Nourah bint Abdulrahman University, Riyadh, Kingdom of Saudi Arabia

Abstract. Due to the recent trend towards attribute-based access control (ABAC), several studies have proposed constraints specification languages for ABAC. These formal languages enable security architects to express constraints in a precise mathematical notation. However, since manually formulating constraints involves analyzing multiple natural language policy documents in order to infer constraints-relevant information, constraints specification becomes a repetitive, time-consuming and error-prone task. To bridge the gap between the natural language expression of constraints and formal representations, we propose an automated framework to infer elements forming ABAC constraints from natural language policies. Our proposed approach is built upon recent advancements in natural language processing, specifically, sequence labeling. The experiments, using Bidirectional Long-Short Term Memory (BiLSTM), achieved an F1 score of 0.91 in detecting at least 75% of each constraint expression. The results suggest that the proposed approach holds promise for enabling this automation.

Keywords: Access control policy · Attribute-based access control · Natural language processing · Constraints specifications

1 Introduction

Attribute-based access control (ABAC) is an effective model in governing the controlled access to information and resources using the notion of rules. Authorization rules are defined by means of attributes which describe properties or characteristics of entities involved in authorization decisions. Hence, a user may exercise a permission only if the assigned values of the attributes of the user and object satisfy the configuration of the authorization rule [8]. Further, to comply with high-level organizational policies, these attributes often need to be constrained. Hence, a proper selection and specification of the constraints are crucially important for policy adherence.

G. Dhillon et al. (Eds.): SEC 2019, IFIP AICT 562, pp. 105–119, 2019.
https://doi.org/10.1007/978-3-030-22312-0_8

Recently, several studies have focused on the formal specification and enforcement of constraints in the context of ABAC systems [2,3,10,21]. However, defining formal constraints requires a significant amount of training and mathematical sophistication. It also involves a considerable degree of expert manual effort. That is, the security architects of the system have to analyze multiple natural language policy documents to extract constraints-relevant information necessary to provide the formal notion of constraints. Therefore, the effectiveness of the set of formally defined constraints is heavily dependent on the individual's particular level of skill. This fact introduces a need for an automated aid to assist security architects in analyzing natural language policy documents and defining the formal constraints.

To that effect, we present this work, that builds upon the state-of-the-art advances in natural language processing (NLP), in order to aid security architects in extracting information necessary to form ABAC constraints from natural language policies. This automation does not only aid security architects in the constraints specification task, but also enables the traceability between formal constraints expressions and the policies they represent. This in turn, can be used to defend against intentional or unintentional unauthorized attributes assignments when manually handled by the security architect. To the best of our knowledge, there is no work in the literature that addresses the automated extraction of constraints directly from natural language policies.

The main research contributions of this paper are as follows: (1) we design an annotation scheme to capture ABAC constraint expressions within natural language access control policies (NLACPs), (2) we design a framework to extract the constraints from NLACPs, and (3) we present experimental results showing that constraints extraction task can be automated reliably. The remainder of this paper is organized as follows: Sect. 2 provides the background information. In Sect. 3, we discuss our approach to automate the extraction of ABAC constraints from natural language policies. Our dataset and experimental results are discussed in Sect. 4. We review the related work in Sect. 5; and, in Sect. 6, we conclude our study with the recommendations for future work.

2   Background

In the following subsections, we provide background information regarding: (1) ABAC-based constraints specification language that we adopt to encode constraints, (2) natural language parsing strategies and (3) sequence labeling model.

2.1   Overview of Constraints Specifications in ABAC

Since authorization decisions in ABAC are based on the notion of attributes, the constraints are also defined according to the relationships among these attributes. Particularly, constraints in ABAC can be seen as the component that defines what values can or cannot be assigned to attributes of entities involved in the access. For instance, suppose a university system has defined a policy

as "*An adjunct faculty member should not be considered a full-time employee.*"
Two attributes, possibly named position and employmentType, were used in
this system to express the faculty position and employment type information.
In this case, a mutual exclusion constraint over the values of these attribute,
namely adjunct and full-time, has to be defined in order to satisfy the given
policy. Numerous studies have researched how to specify and enforce such constraints in ABAC systems [2,3,7,11]. To the best of our knowledge, the proposal
of Bijon et al. is the most comprehensive work to date on constraint specifications in ABAC [2,3]. Therefore, we adopted their definition of constraints in our
work.

Bijon et al. presented a policy specification language called attribute-based
constraint specification language (ABCL). The key idea of ABCL is to define
constraints using the conflicting relations among attribute values which, in turn,
can be used to express notions such as mutual exclusion, preconditions, and
obligations amongst attribute values. The authors have identified two dimensions
or factors to define four levels of conflict: the number of attributes and the
number of entities involved in the conflict relation. Table 1 defines the four levels
of conflict, referred to as level 0 to 3, and provide an illustrative example for each
level. The primary components of ABCL's notion of constraints are the relation-
sets and the ABCL expressions [2,3]. The relation-sets contain the conflicting
values and the expressions are used to precisely specify constraints based on
these conflicts as demonstrated in the following examples.[1]

S1: A faculty member cannot hold tenure and non-tenure track positions simul-
taneously.
RS: $R1 = \{$'tenure', 'non-tenure'$\}$
C1: $|R1 \cap \text{track}(\mathbf{OE(U)})| \leq 1$
S2: An adjunct faculty member should not be considered a full-time employee.
RS: $R2 = \{$'adjunct', 'full-time'$\}$
C2: $|R2 \cap \text{position}(\mathbf{OE(U)}) \cup \text{employmentType}(\mathbf{OE(U)})| \leq 1$
S3: Only one faculty member can hold the chair position.
RS: $R3 = \{$'chair'$\}$
C3: $|\mathbf{assignedEntities}_{U,position}(\text{'chair'})| \leq 1$
S4: Only one full-time faculty member can hold the chair position.
RS: $R4 = \{$'full-time', 'chair'$\}$
$U_{\text{chair}} = \mathbf{assignedEntities}_{U,position}(\text{'chair'})$
C4: $|\mathbf{assignedEntities}_{U_{chair},employmentType}(\text{'full-time'})| \leq 1$

The goal of this work is to automatically identify the elements of the relation-
set, denoted as R, along with the limitation, if any, on the number of entities
subject to the constraint, as discussed in Sect. 3.

---

[1] In these examples *track*, *position* and *employmentType* are attribute names. According to the constraints specification language, ABCL [2,3], attributeName ($\mathbf{OE(U)}$)
is an expression that computes all possible values of the respective attribute while
assignedEntities returns the entities that satisfy certain attribute values.

Table 1. The four levels of conflict that form ABAC constraints as proposed by [2,3]

Conflict	Definition	Example
Level 0	The conflicts occur among values of a **single** attribute and the constraints apply on each **entity individually**	A faculty member cannot hold **tenure** and **non-tenure** track positions simultaneously
Level 1	The conflicts occur among values of **multiple** attributes and the constraints apply over each **individual entity**	An **adjunct** faculty member should not be considered a **full-time** employee
Level 2	The conflicts occur among values of a **single** attribute and the constraints apply on **multiple entities**	Only one faculty member can hold the **chair** position
Level 3	The conflicts occur among values of a **multiple** attribute and constraints apply on **multiple entities**	Only one **full-time** faculty member can hold the **chair** position

2.2   Natural Language Parsing Strategies

This subsection presents two natural language parsing strategies that are essential for NLACPs analysis.

Dependency parsing: is a way of representing the syntax of a sentence in a tree-like structure [14]. A dependency parse tree is a directed graph in which each node represents a word, whilst edges represent the grammatical relationship between two words.

Semantic role labeling (SRL): also known as semantic parsing, decomposes a sentence into its semantic constituents (arguments). The decomposition occurs with respect to each target element (predicate) in the sentence. Then, the semantic role of each argument is recognized with regard to the same target. In other words, the SRL captures who did what to whom, how, when and where for each predicate regardless of the syntactic structure of the sentence [12].

2.3   Sequential Labeling

In natural language processing, a sequence labeling task consists of assigning labels to a sequence of words [24]. In our case, the sequence of words is the NLACP sentence, and the labels indicate words belonging to constraints. For example, the word sequence $(w_1, w_2, w_i, \ldots, w_n)$ of a policy sentence is labeled with $(l_1, l_2, l_i, \ldots, l_n)$, where l_i is the label assigned to the word w_i. Each label is selected from a tag set, typically {B, I, O}, to denote the Beginning, the Intermediate and the non-constraint (Outside) tokens.[2]

---

[2] A token is usually defined as "an instance of a sequence of characters in some particular document that are grouped together as a useful semantic unit for processing."

Recurrent Neural Networks (RNNs) have shown great promise in this task; we particularly used the Bidirectional LSTM (BiLSTM) model as it has been reported to achieve state-of-the-art performance [9]. More explanation on BiLSTM is given in Sect. 3.

Table 2. Examples of annotated policy sentences using our annotation scheme. Bold type denotes the labeled segment of each constraint

Annotation example	Explanation
A faculty member **cannot hold tenure and non-tenure track positions** simultaneously	The constraint is composed of two conflicting values of a single attribute, name it *track*. Hence, it is a level 0 conflict. The annotated span contains the two conflicting values which are "tenure" and "non-tenure"
An adjunct faculty member **should not be considered a full-time employee**	The constraint is composed of two conflicting values across two attributes, name them *employmentType* and *position*. Hence, it is level 1 conflict. The annotated span contains the portion of conflict that is under constraint. The conflicting values are "full-time" and "adjunct"
Only one faculty member can hold the chair position	The constraint is composed of a single attribute, name it *position*, with the value of "chair". And it applies on multiple entities, i.e., "faculty members." Hence, it is a level 2 conflict. The annotated span is the portion of constraints that expresses the restriction over the number of entities
Only one full-time faculty member can hold the chair position	The conflict occurs between multiple attributes, name them *employmentType* and *position*, and applies on multiple entities. Hence, it is a level 3 of conflict. The annotated span is the portion of constraints that expresses the restriction over the number of entities. The conflicting values are "full-time" and "chair"

3   The Proposed Methodology

On the basis of the work of Bijon et al. [2,3], ABAC constraints are defined using the notion of conflict. Accordingly, to extract constraints from NLACPs one should first extract the conflicting factors that form the constraints (see examples in Subsect. 2.1). Hence, the question becomes how to pinpoint the conflicts in NLACPs. Since different combinations of the conflict factors lead to different realizations of constraints, we define our annotation scheme based on the levels of conflict discussed in [2,3]. Then, we design our constraints extraction framework according to the annotation scheme used in the training data, i.e. policy sentences. Particularly, we divide the extraction task into two main phases: identification and normalization. In the former phase, the effort is focused on identifying the right boundary of conflicting values within a policy sentence. In the normalization phase, the focus is on extracting the actual conflicting values that pose a constraint, i.e., the elements of the relation-set, see Subsect. 2.1. In the rest of this section, we provide detailed description of constraints annotation scheme, constraints identification phase and constraints normalization phase.

3.1 Annotation Scheme

The question here is how should we annotate the NLACPs to capture different levels of conflict that form ABAC constraints. Since it is extremely challenging to come up with a one-fits-all annotation rule that captures different variations of constraint expressions, we define our annotation scheme based on the level of conflict. Particularly, we identify the level of the conflict, ranging from 0 to 3, for each sentence as it would be done manually by a security architect. Then, we select the portion of policy text that informs our decision regarding the conflict level. We further explain our annotation scheme by examples in Table 2. In each example, we encode the presence of constraints by marking the span of text that best expresses the one factor that is likely to indicate the level of conflict, i.e. values that are subject to constraints for levels 0 and 1 or the restriction on the number of entities considering levels 2 and 3.

To codify the presence of constraints in NLACP, each sentence is represented using the BIO format (Beginning, Inside, Outside). A token is labeled with B-label if the token is the beginning of a constraint, I-label if it is inside a constraint but not the first token within the constraint, and O-label otherwise.

3.2 Constraints Identification

This phase focuses on the automated detection of constraint expressions within the policy sentence. Hence, the effort is concentrated on predicting their boundaries in the text. In this sense, the constraints identification step can be regarded as a sequence labeling task leading to the choice of BiLSTM, amongst the RNN family models, as it yields the best results. Figure 1 visualizes the abstract functioning of a BiLSTM model when employed in the constraints identification task. The core of our sequence labeling model consists of:

- **Input layer:** This layer forms the real-valued vector representation of individual words in the NLACP sentence using word embeddings. The embeddings can either be learned as a part of the model or loaded from a pretrained

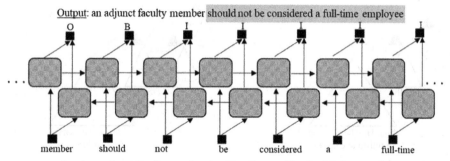

Fig. 1. BiLSTM when applied in the constraints identification task; tokens belonging to the constraint are tagged with the labels B and I. The shaded span of text in the input-output example indicates the output of the BiLSTM model which defines the starting point of the normalization phase as shown in Figs. 2 and 3

word embeddings. Following the recommendation of Chen et al. [4], which calls for the latter approach, we build our model using GloVe pretrained word embeddings. The output of this layer is a matrix $\mathbf{x} = [x_1, x_2, x_t, \ldots, x_n]$. The size of $\mathbf{x}$ is $m \times n$ where m is the dimensionality of the word embedding vectors, and n is the length of the input sequence $\mathbf{x}$.

- **Bidirectional LSTM (BiLSTM) layer:** The BiLSTM layer is an extension of the traditional LSTM layer that further improves the performance on sequential learning problems. The core idea of BiLSTM is that it trains two, instead of one, LSTMs; one learns the sequential information of an input as-is, while the other learns information encoded in the reversed form of the sequence. That is, when a classical LSTM is fed with an ACP sentence, the LSTM computes a representation of the left context of the sentence at each word, denoted as $\overrightarrow{h}_t$. Naturally, generating a representation $\overleftarrow{h}_t$ to capture right context as well using a backward LSTM is likely to add useful information. The combination of both representations results in a vector $h_t = [\overrightarrow{h}_t; \overleftarrow{h}_t]$. This representations effectively encodes the surrounding context of a word, which is useful for numerous sequential learning tasks.
- **Output layer:** In this step, the hidden state vector $\mathbf{h} = [h_1, h_2, h_t, \ldots, h_n]$ that is obtained from the BiLSTM layer is fed into a fully connected layer to predict or assign a label for each word indicating whether it is inside, a beginning or outside of a constraint expression as shown in Fig. 1.

3.3 Constraints Normalization

The input to this phase is a policy sentence in which the sequence of words associated with constraints is identified as discussed in Subsect. 3.2.

The goal is to extract the exact conflicting values as well as the restrictions, if any, over the number of entities that are subject to the constraint. In this work,

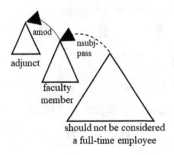

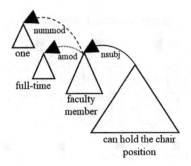

Fig. 2. Collapsed dependency tree representation of "an adjunct faculty member should not be considered a full-time employee." The relation-set R containing the conflicting values is $R = \{\text{adjunct, full-time}\}$

Fig. 3. Collapsed dependency tree representation of "One faculty member can hold the chair position." $R = \{\text{full-time, chair}\}$ second element here is captured using SRL tagger

we address this task using a heuristic-based approach. The heuristics have been designed using a combination of the findings of our earlier work in extracting ABAC attributes from NLACPs [1] and the observations gained from our manual analysis of constrained NLACPs. The proposed heuristics are applied as follows.

First, we analyze the dependency tree (DT) representation of the sentence to determine the level of conflict. If none of the words belonging to the segment identified in the previous phase is marked as a numerical modifier, i.e. nummod, or indicates an exact quantity, e.g. the words single, twice etc., then the conflict is regarded as either level 0 or level 1. Otherwise, the conflict is considered as either level 2 or 3. In the former case, i.e., level 0 or 1, the labeled segment consists of one or more conflicting values as specified in the annotation scheme (see the first two examples in Table 2). To locate the actual values within the segment, we split items combined with commas, "and", "or" or other delimiters. For each split, we search for noun phrases[3] and named entities (NEs)[4]; and, we filter out stop words.

Next, we trace the labeled segment, e.g., "should not be considered a full-time employee," to the corresponding policy element, e.g., "faculty member", that it modifies. The tracing here is done using the DT representation of the sentence as shown in Fig. 2 by the dashed arrow. After identifying the policy element, the goal is to extract its attributes, if any. These attributes are to be added to the set of conflicting values, denoted as R. To address this task, we use the list of patterns, identified in our earlier study of ABAC attributes extraction [1], as

---

[3] A noun phrase is a sequence of words consisting of a head noun and zero or more modifying adjectives and/or nouns.

[4] A named entity is a real-world object, such as persons, locations, organizations, products, etc.

patterns encoding subject-attribute and object-attribute relations. The result of this step is illustrated by the dotted arrow in Fig. 2.

For conflicts of levels 2 or 3, the labeled segment contains the restriction over the number of entities involved in the policy as shown in the last two examples in Table 2. Hence, in this case, the analysis begins with the word marked as the numerical modifier in the DT representation. This word is traced back to the element it modifies as shown in Fig. 3 with the dashed arrow between "one" and "faculty member". With this, we automatically infer restrictions over the number of entities, e.g., (1, faculty member). To extract the attributes, if any, of the policy element under restriction we again use the same set of subjects and objects attributes patterns as shown with the dotted arrow in Fig. 3. To capture other conflicting attribute values, we apply the SRL tagger on the policy sentence. The resulting arguments—i.e., the who, what, how, when and where—that co-occur with the element under restriction are then directly mapped to these values. The intuition is that a conjunction of these arguments needs to be satisfied to trigger the constraint. Consider for instance the policy sentence, "Only one full-time faculty member can hold the chair position". The SRL tagger captures "Only one full-time faculty member" and "the chair position" as the who and what arguments of the predicate "hold". Since "the chair position" co-occurs with the detected portion of the constraint expression, i.e., "Only one full-time faculty member", as arguments to the same predicate, it is considered as its conflicting value; and, so should be added to the relation-set R.

4   Experimental Results and Performance Evaluation

In this section, we present the evaluation conducted to assess the effectiveness of our proposed approach in addressing the constraints identification and normalization tasks. We begin with a description of the our dataset followed by an overview of the general evaluation criteria applied in this study and the experiments and the experimental results.

4.1   Dataset

To evaluate our proposed approach, we constructed a dataset from real-world policy documents from educational institutions, specifically colleges and universities. These documents provide among other things a detailed description of authorizations rules for a wide variety of departments, including Human Resources, Information Technology, Risk Management Services, Faculty and Student Affairs, Administration, Intellectual Property, Technology Transfer, and Equity Development. The documents are written in English. Since not all the sentences in these document express constraints, we manually read through them and collected those that do. The collection process resulted in 801 occurrences of constraints in 747 natural language access control policy (NLACP) sentences. Each sentence contains at least one constraint. Constraints were annotated as discussed in Subsect. 3.1. All resulting constraints expressions are multi-token, of which 687 expressions span 5 or more tokens and some span as many as 15 tokens.

4.2   Evaluation Criteria

To evaluate the performance of the constraints identification phase, we calculate the precision, recall, and F1 score over the tokens correctly classified as part of the constraints and over the predicted constraints. The token-level evaluation enables a more thorough understanding of the per-class performance behavior. It answers questions along the line of: How does the model perform at predicting the beginning and intermediate tokens of a sequence? Or is the model biased towards the majority of the non-constraint tokens? The latter, on the other hand, reflects the predictive power of the model at the constraint-level. For this purpose, we use two different definitions of matching: strict and sloppy matching [6]. With strict matching, a predicted constraint is considered correct if and only if its boundary (i.e., the beginning and end) exactly matches the correct constraint. With sloppy matching, a predicted constraint is considered correct if it overlaps with the correct constraint.

To evaluate the performance of the normalization phase, we also use the notion of precision, recall and F1 score. The evaluation measures are computed based on the count of true positives, false positives and false negatives. These counts are determined by comparing the manually identified conflicting values, posing constraint, in each policy sentence against the automatically extracted ones.

4.3   Experiment and Experimental Results

Next, we present the experiments that we conducted to assess the effectiveness of our constraints extraction framework. Through the experiments, we address the following research questions:

- **RQ1:** How effectively, in terms of precision, recall and F1-score, does the constraint identification phase perform in capturing constraints boundary within NLACP sentences ?
- **RQ2:** How effectively, in terms of precision, recall and F1-score, does the constraint normalization phase perform in extracting the conflicting values that form constraints from NLACP sentences?

To answer RQ1, we split our dataset into three separate sets of training, validation, and testing with the proportion of (70:10:20). Training and validation sets were used to adjust the BiLSTM parameters and to decide on the word embedding method, vector length and embedding's source that better fit our data. The testing set, on the other hand, was used to evaluate the performance of the final model. To configure the model, we set one hyper-parameter value at a time. Our default settings: dropout $= 0$, decay rate $= 0$, number of BiLSTM cells (i.e., layers) $= 1$, and GloVe (Common crawl) with 300 dimensions.

To determine which word embeddings better capture the semantics of words in our data, we conducted several experiments using pretrained word vectors provided by Mikolov et al. [16] and Pennington et al. [20]. Mikolov's vector representations were trained on Google News using Word2Vec approach whereas Pennington's embeddings were trained on Common Crawl and Wikipedia using GloVe method. Our model performs the best on the validation set when using the word embeddings trained on Wikipedia with 300 dimensions; thus, we use this setting in our final model. Similarly, we ran several experiments to adjust the number of BiLSTM cells and the size of each one, dropout, decay rate and activation function and ultimately set these values as 3, 100 units, 0.1, 0.00001 and tanh, respectively. The detailed results on the validation set were omitted due to space limit.

Once the BiLSTM parameters were configured, we tested the resultant model on our test set. To further understand the difficulty of the problem and to better assess the predictive power of our model, we established a comparative baseline using a conditional random field (CRF) sequence labeler [15]. Note that the CRF sets a strong baseline as it has been successfully applied to sequence labeling problem in many fields including natural language processing [13]. The L1 and L2 regularization parameters of the CRF were set to 0.0001 and 0.0, respectively, following the same tuning procedure applied on the BiLSTM.

We compared the results obtained with the BiLSTM against the results obtained with the CRF which was trained on the same representation of word embeddings. As mentioned in Subsect. 4.2, the evaluation measures were defined over two levels of prediction, namely sequence and token levels. Table 3 shows how our model performs against the baseline at predicting constraint sequences. The sloppy and strict sequence matching conditions provide an approximate upper and lower bound of the performance. In order to analyze the performance between these two extreme conditions of matching, we introduce a matching threshold. This threshold determines a minimum size of overlapping that a sloppy matching condition needs to satisfy for accurate predictions.

Table 3. Results of the sequence-level predictions by the means of precision, recall, and F1-score when using our model (the BiLSTM-based) and the baseline (CRF)

Matching		Model					
		BiLSTM			CRF		
Type	Threshold	P	R	F1	P	R	F1
Strict	N/A	0.58	0.56	0.57	0.30	0.29	0.30
Sloppy	Unconstrained	0.97	0.93	0.95	0.96	0.92	0.94
	25%	0.95	0.92	0.93	0.91	0.87	0.89
	50%	0.94	0.91	0.93	0.86	0.82	0.84
	75%	0.93	0.89	0.91	0.83	0.80	0.81

Table 4. Results of the token-level predictions by the means of precision, recall, and F1-score when using of our model (the BiLSTM-based) and the baseline (CRF)

Class	Model					
	BiLSTM			CRF		
	P	R	F1	P	R	F1
B-label	0.83	0.80	0.82	0.72	0.69	0.71
I-label	0.83	0.90	0.86	0.79	0.77	0.78
O-label	0.98	0.97	0.98	0.90	0.91	0.91

Several observations can be made from Table 3. First, concerning the strict matching criterion, our model significantly outperforms the baseline (0.57 vs. 0.30 F1-score). On the other hand, with the unconstrained sloppy matching, both models have shown comparable performance (0.95 and 0.94 F1-score). However, the results observed under the sloppy matching condition differ once we imposed constraints on the overlapping sequences. Two general observations are noted regarding the constrained sloppy matching; (1) The overall performance is proportional to the inverse of the overlapping threshold, i.e. the higher the threshold the lower the performance of both models; (2) The drop observed with our model is not as steep as with the baseline. Using our model, the drop in F1-score is about 4% as we change the settings from constrained to unconstrained matching with a matching threshold of 75%. However, a similar change causes the F1-score of the baseline to drop by about 13%. To justify these observations, we refer to the token-level predictions (i.e., Beginning, Inside, and Outside labels) as shown in Table 4. Considering, the relatively high performance shown by our model at the token-level predictions as well as at sequence predictions under the constrained sloppy matching condition, one can assume that in the vast majority of cases our model is capable to capture at least 75% of the span of text marked as a constraint. Hence, the relatively low scores shown in the case of the strict matching can be attributed to erroneous predictions of at most 25% of the constraint. In other words, our BiLSTM successfully detects most tokens ($\geq 75\%$) that are constraints with an F1 score of 0.91.

To answer RQ2, we employed the set of rules defined in the normalization phase, see Subsect. 3.3. For implementation, we used spaCy[5] for dependency parsing and practNLPTools[6] for semantic parsing. The rules achieved a precision of 0.93, recall of 0.89 and F1 score of 0.90 in extracting the exact conflicting values, forming constraints, from each policy sentence. Overall, the results indicate the soundness of the proposed rules. A closer inspection of the incorrect extractions shows that in the majority of cases erroneous results were due to errors in parsing. Enhancing the parsing effectiveness is most likely to improve the performance [5].

---

[5] https://spacy.io/.
[6] https://github.com/biplab-iitb/practNLPTools.

5   Related Work

Several studies have been focused on specifying constraints in an ABAC system. Jin et al. have proposed ABACα, a policy specification language, that can also specify constraints on attributes assignment [11]. In ABACα, the set of constraints are defined as the set of values a subject may have given that the subject has already been assigned a particular set of attribute values. Such definition of constraint is regarded as event dependent. Unlike ABACα, Attribute-based Constraint Specification Language (ABCL), presented by Bijon et al., can be used to represent different kinds of conflicting relations among attributes in a system [2,3]. The constraints in ABCL are event independent and are to be uniformly enforced regardless the event causing an attribute value to change. The enforcement of ABCL constraints and its performance is discussed in [2]. Jha et al. have introduced the problem of separation of duty (SoD) specification, verification, and enforcement in ABAC systems [10]. Helil et al. have defined ABAC constraints based on the potential relationships within and across the sets of subjects and objects [7]. Despite the importance of developing formal constraints specification languages, the manual effort involved in the actual process of defining constraints can be difficult, labor-intensive, and error-prone. Our work takes a step to facilitate this process, i.e., converting natural language constraints to machine-readable form.

The necessity to automate the manual analysis of natural language requirement documents has motivated several studies in developing automated tools to derive elements of access control rules directly from the raw text [1,18,19,22,23, 25]. Previous research efforts have mostly been restricted to three areas: (1) automatic identification of the access control policy sentences from natural language artifacts, (2) automatic extraction of the—subject, resource and action—triples from these sentences, and (3) automatic extraction of the attributes of these elements. Our current work goes beyond the previous efforts in this area by addressing the automated extraction of constraints from natural language policies. This contribution is essential to serve the ultimate goal of developing a comprehensive top-down ABAC policy engineering ecosystem.

6   Conclusion and Future Work

In this work, we designed a two-phase framework to extract ABAC constraints from unstructured NLACPs. We modeled the first phase of the framework, i.e., constraints identification, as sequential labeling problem leading to the choice of BiLSTM. We addressed the second phase, i.e., constraints normalization, using heuristics. The trained BiLSTM successfully detects at least 75% of tokens that are constraints in each policy sentence with F1 score of 0.91. The proposed normalization heuristics, on the other hand, achieved F1 score of 0.90 in extracting the conflicting values from each NLACP sentence. We also designed an annotation scheme that captures constraints expressions within NLACPs. This is an essential contribution to create an annotated corpus that allows natural language constraint analysis studies. The output of the current model is a set of conflicting values that form a constraint, i.e. the relation set R. As a future work, we will study the possibility to incorporate techniques proposed in [17] to integrate a formal analysis component -with natural language front-end- to the proposed framework in order to generate a formal representation of the constraints using the relation set R.

References

1. Alohaly, M., Takabi, H., Blanco, E.: A deep learning approach for extracting attributes of ABAC policies. In: Proceedings of the 23rd ACM Symposium on Access Control Models and Technologies (2018)
2. Bijon, K.Z., Krishman, R., Sandhu, R.: Constraints specication in attribute based access control. Science **2**(3), 131 (2013)
3. Bijon, K.Z., Krishnan, R., Sandhu, R.: Towards an attribute based constraints specification language. In: 2013 International Conference on Social Computing (SocialCom), pp. 108–113. IEEE (2013)
4. Chen, D., Manning, C.: A fast and accurate dependency parser using neural networks. In: Proceedings of the 2014 Conference on Empirical Methods in Natural Language Processing (EMNLP), pp. 740–750 (2014)
5. Fader, A., Soderland, S., Etzioni, O.: Identifying relations for open information extraction. In: Proceedings of the Conference on Empirical Methods in Natural Language Processing, pp. 1535–1545. Association for Computational Linguistics (2011)
6. Franzén, K., Eriksson, G., Olsson, F., Asker, L., Lidén, P., Cöster, J.: Protein names and how to find them. Int. J. Med. Inform. **67**(1–3), 49–61 (2002)
7. Helil, N., Rahman, K.: Attribute based access control constraint based on subject similarity. In: 2014 IEEE Workshop on Advanced Research and Technology in Industry Applications (WARTIA), pp. 226–229. IEEE (2014)
8. Hu, V.C., et al.: Guide to attribute based access control (ABAC) definition and considerations (draft). NIST special publication 800-162 (2013)
9. Huang, Z., Xu, W., Yu, K.: Bidirectional LSTM-CRF models for sequence tagging. arXiv preprint arXiv:1508.01991 (2015)
10. Jha, S., Sural, S., Atluri, V., Vaidya, J.: Specification and verification of separation of duty constraints in attribute-based access control. IEEE Trans. Inf. Forensics Secur. **13**(4), 897–911 (2018)
11. Jin, X., Krishnan, R., Sandhu, R.: A unified attribute-based access control model covering DAC, MAC and RBAC. In: Cuppens-Boulahia, N., Cuppens, F., Garcia-Alfaro, J. (eds.) DBSec 2012. LNCS, vol. 7371, pp. 41–55. Springer, Heidelberg (2012). https://doi.org/10.1007/978-3-642-31540-4_4
12. Johansson, R., Nugues, P.: Dependency-based semantic role labeling of PropBank. In: Proceedings of the Conference on Empirical Methods in Natural Language Processing, EMNLP 2008, pp. 69–78. Association for Computational Linguistics, Stroudsburg (2008). http://dl.acm.org/citation.cfm?id=1613715.1613726
13. Kang, T., Zhang, S., Xu, N., Wen, D., Zhang, X., Lei, J.: Detecting negation and scope in chinese clinical notes using character and word embedding. Comput. Methods Prog. Biomed. **140**, 53–59 (2017)
14. Kübler, S., McDonald, R., Nivre, J.: Dependency Parsing. Morgan & Claypool Publishers, San Rafael (2009)

15. Lafferty, J., McCallum, A., Pereira, F.C.: Conditional random fields: probabilistic models for segmenting and labeling sequence data (2001)
16. Mikolov, T., Chen, K., Corrado, G., Dean, J.: Efficient estimation of word representations in vector space. arXiv preprint arXiv:1301.3781 (2013)
17. Miyao, Y., Butler, A., Yoshimoto, K., Tsujii, J.: A modular architecture for the wide-coverage translation of natural language texts into predicate logic formulas. In: Proceedings of the 24th Pacific Asia Conference on Language, Information and Computation (2010)
18. Narouei, M., Khanpour, H., Takabi, H., Parde, N., Nielsen, R.: Towards a top-down policy engineering framework for attribute-based access control. In: Proceedings of the 22nd ACM on Symposium on Access Control Models and Technologies, pp. 103–114. ACM (2017)
19. Narouei, M., Takabi, H.: Towards an automatic top-down role engineering approach using natural language processing techniques. In: Proceedings of the 20th ACM Symposium on Access Control Models and Technologies, pp. 157–160. ACM (2015)
20. Pennington, J., Socher, R., Manning, C.: GloVe: global vectors for word representation. In: Proceedings of the 2014 Conference on Empirical Methods in Natural Language Processing (EMNLP), pp. 1532–1543 (2014)
21. Singh, M.P.: AHCSABAC: attribute value hierarchies and constraints specification in attribute-based access control. In: 2016 14th Annual Conference on Privacy, Security and Trust (PST), pp. 35–41. IEEE (2016)
22. Slankas, J., Williams, L.: Access control policy identification and extraction from project documentation. Acad. Sci. Eng. Sci. **2**(3), 145–159 (2013)
23. Slankas, J., Xiao, X., Williams, L., Xie, T.: Relation extraction for inferring access control rules from natural language artifacts. In: Proceedings of the 30th Annual Computer Security Applications Conference, pp. 366–375. ACM (2014)
24. Tjong Kim Sang, E.F., De Meulder, F.: Introduction to the CoNLL-2003 shared task: language-independent named entity recognition. In: Proceedings of the Seventh Conference on Natural Language Learning at HLT-NAACL 2003, vol. 4, pp. 142–147. Association for Computational Linguistics (2003)
25. Xiao, X., Paradkar, A., Thummalapenta, S., Xie, T.: Automated extraction of security policies from natural-language software documents. In: Proceedings of the ACM SIGSOFT 20th International Symposium on the Foundations of Software Engineering, p. 12. ACM (2012)

Removing Problems in Rule-Based Policies

Zheng Cheng[1], Jean-Claude Royer[2]([⊠]), and Massimo Tisi[2]

[1] ICAM, LS2N (UMR CNRS 6004), Nantes, France
zheng.cheng@icam.fr
[2] IMT Atlantique, LS2N (UMR CNRS 6004), Nantes, France
{jean-claude.royer,massimo.tisi}@imt-atlantique.fr

Abstract. Analyzing and fixing problems of complex rule-based policies, like inconsistencies and conflicts, is a well-known topic in security. In this paper, by leveraging previous work on enumerating all the problematic requests for a rule-based system, we define an operation on the policy that removes these problems. While the final fix remains a typically manual activity, removing conflicts allows the user to work on unambiguous policies, produced automatically. We prove the main properties of the problem removal operation on rule-based systems in first-order logic. We propose an optimized process to automatically perform problem removal by reducing time and size of the policy updates. Finally we apply it to an administrative role-based access control (ARBAC) policy and an attribute-based access control (ABAC) policy, to illustrate its use and performance.

Keywords: Conflict · Inconsistency · Policy · Problem · Removing · Rule

1 Introduction

Analyzing inconsistent and conflicting situations in security policies is an important area of research and many proposals exist. Several approaches are focused on detecting specific kinds of problems [2,7,9–11], while others are interested in fixing these problems [5,6,8,12,14]. We consider inconsistencies, or conflicts, or undefined requests, called *problems* here, as they lead to bugs or security leaks. Fixing these problems is difficult because of the policy size, the number of problems, their complexity and often the right fix needs human expertise. Our purpose in this paper is to suggest a new method in order to assist specifiers in fixing discovered problems in their policies. To make the conflicts explicit, we reuse a general and logical method defined in [2]. This method, in addition to revealing the problems, provides some information which can be exploited to cure those problems. In our current work we focus on *removing* some problems in the specification. Removing the problem means modifying the policy so that

© IFIP International Federation for Information Processing 2019
Published by Springer Nature Switzerland AG 2019
G. Dhillon et al. (Eds.): SEC 2019, IFIP AICT 562, pp. 120–133, 2019.
https://doi.org/10.1007/978-3-030-22312-0_9

the problem disappears and no new problems are added. Of course we need to preserve, as far as possible, the original behavior of the policy while minimizing the time and the size of the policy update. While the final fix (i.e. obtaining a system that gives the right reply for every request) remains a typically manual activity, removing conflicts allows the user to work on unambiguous policies, produced automatically. We also believe that this step is a useful basis for future work on assisted policy fixing. We first provide a naive approach to remove the problem but its time complexity becomes exponential if we try to minimize the size of the modifications. Exploiting the enumerative method of [2] we are able to provide an optimized version of this process.

Our first contribution is the formal approach to remove a problem while minimizing the rule modifications in a first-order rule-based logical context. The second contribution is an experiment on an ARBAC policy, illustrating how to apply the approach in practice. Regarding the performance of the problem removal process, we confirm our results on another case study, based on an XACML policy, and we demonstrate getting the same minimal modifications in dividing the global time by a factor of 4.

The content of this paper is structured as follows. Section 2 describes related work in the area of fixing conflicting problems. Section 3 provides the necessary background and a motivating ARBAC example. Section 4 describes the general process to remove problems. Section 5 provides the formal results regarding the optimization of the removing process. In Sect. 6 we evaluate our method on our initial use case and another ABAC use case. Lastly, in Sect. 7 we conclude and sketch future work.

2   Related Work

Our work is under the umbrella of automatic software/system repair, which aims at automatically finding a solution to software/system bugs without human intervention. There exists extensive work under this broad topic, and we refer to [8] for a review. In this section, we discuss some of related work on automatic repairing of bugs in rule-based systems. Researchers try to find automatic fixes for various kinds of bugs, e.g. redundancies [6], misconfigurations [5]. In this work we focus on conflicts and inconsistencies in rule-based systems. These problems may lead to runtime failures in the sense that sending a request to the system, may return several incompatible replies. This separates our work from efforts that address other kinds of bugs. Meta-rules are one of the most common ways to handle conflicts in rule-based systems. The general idea is that when conflicts occur, pre-defined meta-rules (e.g. first-applicable, prioritization [3]) will govern rule applications to have compatible replies. However, the problem persists to resolve potential conflicts in meta-rules. Hu et al. propose a grid-based visualization approach to identify dependency among conflicts, which aims to guide user in defining conflict-free meta-rules [6]. Son et al. [12] repair access-control policies in web applications. They first reverse engineer access control rules by examining user-defined annotations and static analysis. Then, they encapsulate domain specific knowledge into their tool to find and fix security-sensitive operations that are not protected by appropriate access-control logic. Wu focuses on detecting inconsistency bugs among invariant rules enforced on UML models [14]. The author presents a reduction from problem domain to MaxSMT, and then proposes a way to fix bugs by solving the set cover problem. Our approach is specific in targeting FOL rule systems and providing an automatic fix, without considering additional information, while minimizing the time to proceed and the size of the modifications.

3   Background

In this section, we introduce concepts/notations that will be consistently used in the rest of the paper. To facilitate our introduction, we illustrate on a variation of an ARBAC policy given by [13][1]. This is a middle size example with roles and a hierarchy of roles, role exclusivity constraints, permissions assignment and revocation. The original example contains 61 rules. In this work, we rewrite them in FOL, and modularize them into 4 modules, and parametrize appropriate rules with one integer for discrete time. The four modules are: roles, hierarchy and exclusion rules (11 rules), permissions (24 rules), assignment (13 rules) and assignment revocation (13 rules). In this section, we show only the role module for illustration purpose in Listing 1.1.

Listing 1.1. Rules of roles for an administrative RBAC policies

```
 1   And(Patient(T, X), PrimaryDoctor(T, X)) => False      %first rule
 2   And(Receptionist(T, X), Doctor(T, X)) => False
 3   And(Nurse(T, X), Doctor(T, X)) => False
 4   Nurse(T, X) => Employee(T, X)                          %4th rule
 5   Doctor(T, X) => Employee(T, X)
 6   Receptionist(T, X) => Employee(T, X)
 7   MedicalManager(T, X) => Employee(T, X)
 8   Manager(T, X) => Employee(T, X)
 9   Patient(T, X) => PatientWithTPC(T, X)
10   Doctor(T, X) => ReferredDoctor(T, X)
11   Doctor(T, X) => PrimaryDoctor(T, X)                    %last rule
```

A *rule* is a logical implication, taking the form of $D => C$, with D being the condition and C the conclusion of the rule, expressed in a logical language (in our case FOL). For example, line 4 specifies that at any given time T, if X is a nurse, it is also an employee. A *rule system* (R) is simply a conjunction of rules. *Requests* are FOL expressions. When they are sent to a rule system at runtime, they will be evaluated against all rules in that system to generate *replies* (which are also FOL expressions). For example, when a request Nurse(1, Jane) is sent to the system shown in Listing 1.1, Employee(1, Jane) is implied as a reply. A request is called *undefined request*, if it is satisfiable by itself, but unsatisfiable when in conjunction with R. The phenomenon caused by an undefined request is that when it is evaluated, R would give contradictory/unsatisfiable replies, therefore making the system unrealizable.

We previously propose in [2] an optimized method to enumerate and classify all undefined requests in a rule system. The method translates the original rule

---

[1] The original example with comments is available at http://www3.cs.stonybrook.edu/~stoller/ccs2007/.

system into an equivalent system made of exclusive rules. Each *exclusive rule* abstracts what kind of replies will be generated, provided that a certain set of rules in the original rule system is applied. We call *1-undefined request* a request which, in conjunction with one rule alone, is unsatisfiable. One result of our approach is that any undefined request is a union of 1-undefined requests associated to exclusive rules. The exclusive rules we generate are analyzed by the Z3 SMT solver[2]. Based on the result from the solver, we separate exclusive rules into two categories:

- Unsafe exclusive rules. These are exclusive rules that under request will always return *unsat* by the solver. They abstract undefined requests that are certain to cause conflicts.
- Not unsafe exclusive rules. These are exclusive rules that under request return *sat* or *unknown* by the solver. Therefore, undefined requests, that are abstracted by not unsafe exclusive rules, are uncertain to cause conflicts (conservatively, we also pick them up to rise developer's attention).

A special representation called *binary characteristic* links each (not) unsafe exclusive rule to the original rule system. Each binary characteristic is a list of values, where the position i represents a rule at the corresponding position in the original system. Values have enumeration type with three possibilities, i.e. $0/1/-1$, indicating that the condition of the rule is negatively/positively/not presented in the exclusive rule. We call a binary characteristic *complete* if it does not contain -1, and has the length equal to the total number of rules in the original system (*incomplete* otherwise).

Listing 1.2. The roles module analysis

```
1   ---------- UNSAFE --------------
2   [0,  0,  0,  1,  1,  1,  -1,  -1,  -1,  -1,  -1]
3     And(Not(Nurse(T, X)), Doctor(T, X),
4         Not(PrimaryDoctor(T, X)),
5         Not(Receptionist(T, X)), Patient(T, X)) => False
6   [1, -1, -1, -1, -1, -1, -1, -1, -1, -1, -1]
7     And(Patient(T, X), PrimaryDoctor(T, X)) => False
8
9   ... another 2 unsafe exclusive rules
10  ---------- NOT UNSAFE --------------
11  [0, 0, 0, 1, 0, 0, 1, -1, -1, -1, -1]
12    And(Nurse(T, X), Not(Doctor(T, X)), Patient(T, X), Not(PrimaryDoctor(T, X)))
13        => And(PatientWithTPC(T, X), Employee(T, X))
14
15  ... another 9 not unsafe exclusive rules
```

Applying the method given in [2] on the example shown in Listing 1.1, a total of 4 unsafe and 10 not unsafe exclusive rules are generated. Listing 1.2 shows a snippet of these rules. The shown rules are three typical kinds of exclusive rules, that in our experience, help in identifying problems in a rule system:

- Implicit unsafe rules, which are not contained in the original system, are the primary source of problems. They usually imply some overlapping condition

---

[2] The Z3 Theorem Prover. https://github.com/Z3Prover/z3.

that would cause conflicts. For example, from lines 2 to 5, the unsafe rule points out a problem, as several undefined requests. For instance in the original system Doctor(T, X) implies PrimaryDoctor(T, X) (last rule 11) which clashes with Patient(T, X) by rule 1. Interestingly, this problem is not mentioned under the radar of [13].

- Explicit unsafe rules (e.g. lines 6 and 7), are unsafe rules contained in the original system and whose conclusion is unsatisfiable.
- Not unsafe rules. As discussed before, they are uncertain sources of problems. It is a problem if the request implies the conjunction of the condition and the negation of the conclusion else it is an admissible request.

4   The Removing Process

Undefined requests detected by our optimized enumeration method may identify problems in the original rule system. However, fixing all identified undefined requests is not desirable, since it would result in a tautological and rather useless rule system: the system would not be able to infer new facts from the logical context. Also fixing a subset of the undefined requests is not trivial. It is difficult to guarantee that the iterative fixing process terminates, since it is hard to tell that there will be no new undefined requests after a fix.

Therefore, we propose an alternative solution, which is similar to the "quick fix" feature that appears in most of integrated develop environments. First, by analyzing the result of our enumeration method, the rule developers select a set of critical undefined requests to fix. Next, our solution performs a "quick fix" to remove the selected undefined requests. In the process, we strive to pinpoint a minimal set of rules in the original system for removing the selected undefined requests. By doing so, we minimize modifications that need to be made, and preserve the semantic of the original rule system as much as possible. Another important property of our solution is that it is effective. It means that regardless of choosing which undefined requests to fix first, applying our solution on each iteration will completely remove the selected undefined requests while not introducing new problems.

For example, let us consider the removal of the undefined requests characterized by the unsafe rule shown on lines 2 and 5 of Listing 1.2. Our approach takes its binary characteristic as input, and produces Listing 1.3 as output. The output states that one rule (i.e. the 11th rule) in the original system needs to be changed, and what it should be changed to (lines 2 and 5). Identified problems are existentially quantified (that explains the Exists quantifier) and composed by union with the selected rule conclusion.

Listing 1.3. Removing the Problem

```
1   Target rule: [11],
2   Suggested fix: Doctor(T, X) =>
3   Or(PrimaryDoctor(T, X), Exists([T, X], And(Not(Nurse(T, X)), Doctor(T, X),
4           Not(PrimaryDoctor(T, X)), Not(Receptionist(T, X)), Patient(T, X))))
```

Once the change has been made, applying our optimized enumeration method again will result in 10 not unsafe rules as before, but the selected unsafe rule has been removed and only the 3 explicit unsafe rules remain. At this stage, rule developers can choose once again undefined requests to remove, or stop if the rule system satisfied their expectation. In our final version of this module we simply forget this rule (the 11th) as it seems an error in the roles specification.

In what follows, we present an overview of our removing process (Sect. 4.1), and its main properties (Sect. 4.2).

4.1   Overview of Removing Process

Let U be the undefined request (represented by exclusive rules) that rule developer choose to fix. One way to remove U is by adding a new rule in the original rule system, which takes the shape of u => False. It explicitly alerts the rule developer on some undefined cases. However, this rule is redundant and the resulting system contains more and more rules when fixes iterate, which compromises understandability and maintainability. Another way is to globally restrict the set of input requests by adding a condition to all the defined requests. But this condition is far from readable in case we have several problems in the rule system. Therefore, for understandability and maintainability, we design a removing process that aims at automatically modifying a selected set of rules in the original system without the need to understand the real nature of the problem to fix. The principle of our removing process is to exclude U from the original rule system, i.e. $\neg U => R$ which is equivalent to

$$\forall * \bigwedge_{1 \leq i \leq n} (D_i => (C_i \vee U)) \tag{1}$$

All free variables in (1) are denoted by $*$, and i is the index of a rule in the rule system of size n. Moreover, U is the selected problem to fix, and represented by a FOL existential expression without free variables. Obviously modifying all the rules is not always required, and sometimes the modification of one rule alone $D => (C \vee U)$ could be sufficient. Thus we will consider how to do optimal modifications rather than modifying all the rules, that is to modify F, a subset of all the rules. We denote $R/F/U$ the system resulting by this modification.

4.2   Properties

We have illustrated how to remove selected undefined requests by pinpointing a minimal set of rules in the original system to modify. In this section, we discuss the effectiveness and semantics preservation of this approach.

Effectiveness. By effectiveness, we mean that removing the problem U does not add new problems. From now on, *all* represents the set of all the rules in R. We know that we have $R => R/F/U => R/all/U$, and $R/all/U = (R \vee U)$. It is easy to see that $R/F/U = R \vee (U \wedge R_{\neg F})$, where $R_{\neg F}$ is the subsystem of the rules in R which are not modified. We should note that if U is a problem at least one rule should be fixed and we know that fixing all the rules may be required (if U is 1-undefined for all the rules in R). If we fixed a problem U in R and get $R/F/U$, we expect that U is not a problem of $R/F/U$ and also this does not add new problems. But if U' is a problem after the fix we have $U' => \neg(R/F/U)$ then $U' => \neg R \wedge \neg U \wedge \neg R_{\neg F}$ which implies that $U' => \neg U$ and $U' => \neg R$. Thus it means that U' was not a new problem but an already existing one for R but not included in the fix U. Note also that we do not have $R/F/U => R$ to be valid (provided that U is satisfiable) meaning that we have strictly less problems after removing U. This shows that the removing process is effective whatever the removing ordering is.

Behaviour Preservation. What behaviours of the original rule system could rule developers expect to preserve after the removing process applied. Let a given request $req = (req \wedge \neg R) \vee (req \wedge R)$, applying it to R we get $req \wedge R$. But applied to $R/F/U$ we get $(req \wedge R) \vee (req \wedge U \wedge R_{\neg F})$. Thus preserving is equivalent to $req \wedge U \wedge R_{\neg F} => req \wedge R$ which is equivalent to $req \wedge U \wedge R_{\neg F}$ unsatisfiable. If req intersects both R and its negation we get $(req \wedge R) \vee (req \wedge U \wedge R_{\neg F})$ meaning that the new reply widens the original reply. Behaviour preservation is not possible for all requests.

Property 1 (Behaviour Preservation). Let req a satisfiable request thus $req =>$ $\neg U \vee \neg R_{\neg F}$ is equivalent to $req \wedge R = req \wedge R/F/U$.

The above property states that behaviour is strictly preserved after removing U if and only if the request satisfies $req => \neg U \vee \neg R_{\neg F}$. If $req => \neg U$ the behaviour is preserved for any selection F and if $req => R$ the behaviour is preserved for any problems and any selection. The next section explain how to make the process more efficient by pinpointing a minimal set of rules in the original system for removing the selected undefined requests (Sect. 5).

5 Finding the Minimal Selection

Let U be a problem and R a set of rules, our goal is to modify R in order to make the requests in U defined. The challenge is to do that efficiently and minimizing the modifications in the rule system. We will get a new system $R/F/U$ where the rules in F are modified in order to avoid U to be undefined. The fixing principle is either to add $\neg U$ in the selected rule conditions or to add U in the rule conclusions. Modifying conclusions is simpler since we have the same enumerative decomposition for R and $R/F/U$ only the conclusions are different.

Definition 1 (Correct Fix of a Rule System). *Let R, F, U be a closed and satisfiable sentence, F is not empty, $R/F/U$ is a correct fix for R with F and U if each rule in F has its conclusion enlarged with U and U is not a problem for $R/F/U$.*

A simple fact to observe is: If F is a correct fix then any G such that $F \subset G$ is also a correct fix. Then our challenge is to find a selected set F to fix, smaller than all the rules. Thus we need to show that if U is a problem for R it is defined for $R/F/U$. It means that a direct, called here *naive*, solution is to check this property with a SAT solver.

Definition 2 (Naive Check). *$R/F/U$ is a correct fix if and only if $U \wedge R_{\neg F}$ is satisfiable.*

This comes from the fact that U is a problem, $R/F/U = R \vee (U \wedge R_{\neg F})$ and the definition of a correct fix.

It is easy to see that if U is a problem for R with a set of conditions $D_{1 \leq i \leq n}$ then $U => \bigvee_{j \in J} \exists * D_j$, where J is a subset of $1 \leq i \leq n$. Exclusive rules as built by the enumerative method have some interesting properties, particularly because $U => (\forall * D_j)$ means that only one rule (the jth) applies. A *single* problem is associated to a complete binary characteristic, while a *complex* problem has an incomplete binary characteristic. From [2] we know that any undefined request U satisfies $U => \exists * (\bigwedge_{i \in I_1} D_i \bigwedge_{j \in I_0} \neg D_j)$, where I_1 (respectively I_0) is the set of positive (respectively negative) rules in the binary characteristic.

Property 2 (Application to Exclusive Rules). Let R an exclusive rule system if $U => \exists * D_j$ then $R \wedge U$ is equivalent to $(U \wedge (\forall * D_j \wedge C_j)) \vee (U \wedge (\exists * D_j \wedge \exists * \neg D_j)) \wedge R$.

In this paper we omit the full proofs, they can be found in the full version of the paper on our repository https://github.com/atlanmod/ACP-FIX. The proof relies on the fact that the universally quantified part triggers only one exclusive rule. We show with Property 3 that any problem found by the enumerative method can be split into disjoint parts called, respectively, *universal* and *existential* parts.

Property 3 (Universal and Existential Parts). Let U a satisfiable problem such that $U => \exists * (\bigwedge_{i \in I_1} D_i \bigwedge_{j \in I_0} \neg D_j)$ then $U = U \wedge (\forall * (\bigwedge_{i \in I_1} D_i \bigwedge_{j \in I_0} \neg D_j) \vee U \wedge (\exists * (\bigvee_{i \in I_1} \neg D_i \bigvee_{j \in I_0} D_j)$.

This property results from the partition of U related to the universally quantified part and its negation. Exploiting the information given by the enumerative method we expect to optimize the definedness checking for problems found by this method. We analyze now two cases: single or complex problem in order to expect to optimize the naive approach.

Property 4 (Definedness of Single Problem). Let R a rule system and a single problem $U => \exists * (\bigwedge_{i \in I_1} D_i \bigwedge_{j \in I_0} \neg D_j)$ with a complete binary characteristic, if $U \wedge \forall * (\bigwedge_{i \in I_1} D_i \bigwedge_{j \in I_0} \neg D_j \bigwedge_{i \in I_1} C_i)$ is satisfiable then U is defined for R.

From R we can build an equivalent exclusive system using the enumerative method and thus we use Property 2. We consider the universal part of the problem, that is $U \wedge \forall * \bigwedge_{i \in I_1} D_i \bigwedge_{j \in I_0} \neg D_j$. With these conditions only one rule applies and others do not apply, they lead to the universal part and then the result of $R \wedge U$ comes from a single enumerative rule for R and gives $U \wedge \forall * (\bigwedge_{i \in I_1} D_i \bigwedge_{j \in I_0} \neg D_j \bigwedge_{i \in I_1} C_i)$. We now consider the enumerative process but for $R/F/U$ since we need to prove that U is defined for it. Computing the enumerative process for $R/F/U$ gives new rules of the form:

$\forall * ((\bigwedge_{i \in I_1} D_i \bigwedge_{j \in I_0} \neg D_j) => ((\bigwedge_{i \in I_1 \wedge \neg F} C_i) \bigwedge_{i \in I_1 \wedge F} (C_i \vee U)))$. We start by analyzing the case of a single problem with a complete binary characteristic.

Property 5 (Removing Criterion for Single Problem). Let U a single problem with positive rules I_1 thus $R/F/U$ is a correct fix if either $I_1 \subset F$ or $I_1 \cap F \neq \emptyset$ and $U \wedge \forall * (\bigwedge_{i \in I_1} D_i \bigwedge_{j \in I_0} \neg D_j \bigwedge_{i \in I_1 \cap \neg F} C_i)$ is satisfiable.

In this case there is a unique enumerative rule which applies and we use Property 4 for $R/F/U$. This is only a sufficient condition as we only check the universal part of the problem.

In case of a complex problem U with an incomplete binary characteristic we can obtain a set of complete binary characteristics adding digits not already in the incomplete binary characteristic. A completion $G_1 \cup G_0$ is a subset of $\{1..n\} \backslash (I_1 \cup I_0)$ with positive and negative rules.

Property 6 (Removing Criterion for a Complex Problem). Let U a complex problem, $R/F/U$ is a correct fix if $\neg F \subset I_0$ or $F \cap \neg I_0$ and $U \wedge \forall * (\bigwedge_{i \in I_1} D_i \bigwedge_{j \in I_0} \neg D_j \bigwedge_{i \in I_1 \cap \neg F} C_i) \bigwedge_{g \in \neg I_1 \cap \neg I_0 \cap \neg F} ((\forall * D_g \forall * C_g) \vee \forall * \neg D_g)$ is satisfiable.

This criterion generalizes the previous one for single problem.

In the previous cases we defined a sufficient condition to remove a single or a complex problem. From the previous criterion and the decomposition into a universal and an existential part we have: If U is a problem for R then U is defined for $R/F/U$ if and only if the criterion for complex problem is satisfied or if $U \wedge (\exists * (\bigvee_{i \in I_1} \neg D_i \bigvee_{j \in I_0} D_j) \wedge R_{\neg F})$ is satisfied.

Property 7 (CNS for complex problem). If U is a complex problem, $R/F/U$ is a correct fix if and only if the universal or the existential part is defined for $R/F/U$.

5.1   Looking for Minimal Size

This subsection discusses how to find a set of rules to modify but with a minimal size. We can define a top-down and a bottom-up process to find a minimal solution. Both ways have a worst complexity which is exponential in the number of satisfiability checks. But the bottom up approach is preferable since it stops once the solution is found. Instead the top down approach, once the solution of size m is found, must prove the minimality of it by checking all the smaller combinations of size $m - 1$. The naive approach consists in modifying a subset of rules and checking the satisfiability of the new system in conjunction with the problem to remove. Minimal core satisfiability techniques cannot be used here (for instance [14]), since they do not respect the structure of the rule system.

Using our criteria we optimize this search. We know that the set of all rules is a solution but in case of a single or complex problem it is also true if we take F as all the positive rules in the binary and its completion. Indeed, if we are looking for minimal solutions it is sufficient to look inside these positive rules. The reason is that if the criterion is satisfied with $I \cap F$ the part of F not in I does not matter and can be forgotten. Given a problem we defined a `lookup_complex` algorithm which looks for a minimal set of rules. It simply starts with the least possible solution (that is a single rule) and checks the criterion an all the combinations until reaching a minimal solution.

There are two critical points in the time performances of our two solutions: the number of rule combinations to test and the size of the expression to check for satisfiability. Both these aspects have an exponential nature in general. Exploiting the binary information the `lookup_complex` algorithm looks for less combinations than the naive algorithm. Regarding the satisfiability checking we expect to gain but the size of the formula is not a reliable indicator here. The informal reason lies in the form of the universal formula which is closed to CNF (Conjunctive Normal Form) which is at the heart of most of the solvers. To justify it we consider a problem associated to an unsafe rule. We also assume that our rule system contains only free variables and rules with a conjunction of predicates as condition and a disjunction of predicates as conclusion. If $K = 1$ the maximal number of predicates in a condition or a conclusion, the universal part can be seen as a 2-SAT CNF which satisfiability time is polynomial. If $K \geq 2$ we get CNF in the NP complete case but our optimisation relies on the transition phase phenomenon [1]. Analysing the CNF transformation we get a $2 * K$-SAT CNF and we estimate the maximal number of clauses $M \leq 2 * K * n$ while the total number of literals in the clauses is $N \geq (2 * K + n - 1)$. Thus the ratio $\alpha = M/N$ is below the threshold $2^{2*K} * ln(2) - 2 * K$, (as soon as $K \geq 2$), the area where the universal part is probably satisfiable in a small amount of time.

6   Application Examples

The purpose of this section is to show that our removing approach succeeds on middle-size examples. We will focus on removing problems coming from unsafe rules in these examples.

Our first specification is compound of the four previous modules introduced in Sect. 3. The permissions module is rather straightforward, assignment and revocation need to manage discrete time changes. In the assignment of permissions we choose to set the effect at next time. One example is `And(Doctor(T, X),` `Doctor(T, Y), assign(T, X, Y)) => ReferredDoctor(T+1, Y)`. In this specification the effect of a permission assignment is done at `T+1` which is a simple solution avoiding clashes with the roles module. The revocation module has similar rules to the assignment of permissions. However, we need more complex conditions because before to revoke an assignment it should have been previously done. The corresponding example for revocation of the above rule is `And(Doctor(T, X), revoke(T, X, Y),`

`(P < T), assign(P, X, Y), Not(assign(T, X, Y)))` => `Not(ReferredDoctor(T+1, Y)))`. An assign and a revocation are not possible at the same time instant because of inconsistency. We already analyzed the roles module and it was easy to process the three new ones in isolation since they have no unsafe rule. One interesting fact is that their composition does not generate new unsafe rules, indeed we get the three explicit unsafe rules coming from the roles module (see Sect. 4).

6.1   A Second Specification

An alternative solution for the specification of the assignment module is to write rules without changing the time instant in the conclusion. In this new specification our example above becomes: `And(Doctor(T, X), Doctor(T, Y), assign(T, X, Y))` => `ReferredDoctor(T, Y)`. It generates unexpected conflicts we will solve now. Our analysis shows that we get 91 not unsafe rules and three unsafe rules in nearly 8 s. Thus using our `lookup_complex` procedure we find that these problems are all removed by modifying the rule: [5]. The enumerative computation of the new system shows that it has no more unsafe rule. Now these modifications could produce new interactions with the other modules. In fact only the roles module has new unsafe rules with the assignment module, indeed there are 3 new unsafe rules. These unsafe rules are coming from the negation of the 11th rule in assignment and the `lookup_complex` shows that the 4th rule is the minimal fix for all these problems. Fixing these three problems we compute the enumerative solution for the 4 modules together and we do not get new unexpected unsafe rules. The result was computed in nearly 5200 s and generates 20817 not unsafe rules and the three explicit unsafe rules from the roles module. This example shows that we can select some problems and remove them from the specification while minimizing the impact on the rule system.

Table 1. Measures for two policies

Usecase	Naive algorithm		Lookup algorithm		Additional measures		
	NS	NT	LS	LT	PR	DS	TF
Healthcare policy	1	1.01 s	1	0.2 s	10.1	0	509%
ContinueA policy	1.53	115 s	1.53	31 s	3.9	0	794%

We compare the `naive` and our `lookup_complex` algorithms and compute several measures which are summarized in the Table 1. We consider 123 unsafe problems occurring before the final fix in the composition of the four modules. Note that in this setting our example is not simply variable-free because we fix two rules adding some complex existential expressions. We compute[3] the following measures in Table 1: for the naive approach the mean of minimal size (NS), mean

---

[3] These results were computed with 10 runs when it was sensible in time, that is all cases except three (amongst 530) for the ContinueA policy.

of time (NT), the same for the lookup method with LS, LT and in addition the mean of positive rules in each problem (PR), the maximum of differences between size of the selection (DS) and the mean of the ratio: naive time divided by lookup time (factor time TF). But this example is specific on one point: the problems are not so numerous and related to some specific rules in the assignment module. Thus most of the problems (but the first three) are related to the 4th rule of the assignment module.

6.2   The ContinueA Example

To consolidate our results we consider the ContinueA policy[4] we already analyzed in [2] and which was the study of several previous work [4,6]. This policy has 47 rules, which are pure first-order with at most two parameters. The original example is in XACML which forces the conflict resolution using combining algorithms. To stress our algorithms we do not consider ordering or meta-rules but a pure logical version of the rules. The result is that we have a great amount of problems amongst them 530 unsafe rules while the number of not unsafe rules is 302 (computed in 97 s). We process all the unsafe problems that is 530, see Table 1. The following observations confirm what was observed on the health-care example, except that now we have many more problems to analyze. First we observed that the minimal set of fixing rules is generally low (between 1 and 5 rules) and this shows that finding it is relevant to minimize the modifications in the rule system. Another point is that due to the combinatorial explosion it is really costly to go up to more than 4 rules (see Table 2). The second point is that the lookup algorithm does not deviate from the naive one regarding the size of the minimal set. We do not get exactly the same selection set in 33% of the cases, due to the different ordering in the search for minimal, but the minimal sizes are always the same. Regarding the time to proceed, the lookup outperforms the naive one by a factor between 60% and 5000% with a median of nearly 800%. For this example we also compute the distribution per selection size, and the mean time for each algorithms.

Table 2. Selection distribution

Selection size	Frequency	Naive mean time	Lookup mean time	Time factor
1	63%	0.33 s	0.04 s	800%
2	27%	6.6 s	1.15 s	573%
3	8%	124 s	24 s	517%
4	3%	1573 s	281 s	560%
5	0.5%	88890	3433	256%

---

[4] http://cs.brown.edu/research/plt/software/margrave/versions/01-01/examples/.

6.3 Discussion

Regarding the healthcare example, we defined two versions which have finally only three explicit unsafe rules and we remove only few problems. For the ContinueA example removing all the unsafe problems can be done modifying all the 47 rules with an increase in size of $24910 * US$, where US is the median size of the problems. Using the minimal selection of rules the increase in size is $796*US$. The naive algorithm needs nearly 17 h to compute the minimal selections while the lookup takes 4.6 h. Our experiments also confirm that our time improvement is twofold: the restricted space to search for a minimal selection and checking first the universal part. We do not detail this here but the picture appears in our repository (https://github.com/atlanmod/ACP-FIX) as well as the source of the prototype, the examples and some measures.

It is not relevant to expect to remove all the problems. Furthermore we should also cope with the set of real requests which will decrease the amount of such undefined requests. Nevertheless an assistance should be provided to identify what are the critical problems. This is a tricky issue. The presented technique is also correct for 1-undefined problems arising in not unsafe rules. We also process some of these problems: for the 320 problems in ContinueA we get a mean for $TF = 120\%$ while with 3000 problems (nearly 10% of the problems) of the healthcare we get a mean of 800%. Thus our technique is generally useful for any kinds of problems and furthermore the reader may note that our examples after fixing are not longer simply variable-free. This is the case when we fixed the role and assign modules of the healthcare example since we add existentially quantified expressions. However, the performances were similar and we need more experiments and analysis to precisely understand the applicability of the proposed method.

Fixing a problem means to associate to U a single reply rather than inconsistent replies. This is similar to removing the problem but in addition we need to choose a reply which in general cannot be automatic. In this case the fixing principle is to change the conclusion of a rule $(D => C)$ in $(D => (C \vee (U \wedge OK))$ where OK stands for the correct reply to U. We did not yet investigate it but our current work is a good basis to solve this more complex problem.

7 Conclusion

Automatically removing problems in a policy is important to sanitize it. Sometimes there are too many problems and they are difficult to understand at least for non experts of the system. Getting simplified problems can help in solving them, however it is a complex and costly issue. Our work demonstrates that, under the conditions of the satisfiability decision and the time to proceed, we can automatically remove a selection of problems. Furthermore, we are able to minimize the size of the modifications as well as improving the time to proceed. We demonstrate it on two policies of middle size: an ARBAC and an ABAC.

This work leaves open many questions, first is about checking the existential part while getting a minimal size close to the exact minimal size. However, it seems tricky because most of our attempts to relax the existential part lead to an unsatisfiable expression. Second, we can benefit by more case studies from related work (e.g. [6,12,14]) to statistically justify that we can get a low minimum in real cases, or we can synergize with related works. The main problem is how to faithfully encode the complete case studies (e.g. encoding explicit/implicit UML semantics plus invariants as rules in our system). The third track is to explore the benefit of checking the universal part only.

References

1. Achlioptas, D., Naor, A., Peres, Y.: Rigorous location of phase transitions in hard optimization problems. Nature **435**, 759–764 (2005)
2. Cheng, Z., Royer, J.-C., Tisi, M.: Efficiently characterizing the undefined requests of a rule-based system. In: Furia, C.A., Winter, K. (eds.) IFM 2018. LNCS, vol. 11023, pp. 69–88. Springer, Cham (2018). https://doi.org/10.1007/978-3-319-98938-9_5
3. Cuppens, F., Cuppens-Boulahia, N., Garcia-Alfaro, J., Moataz, T., Rimasson, X.: Handling stateful firewall anomalies. In: Gritzalis, D., Furnell, S., Theoharidou, M. (eds.) SEC 2012. IAICT, vol. 376, pp. 174–186. Springer, Heidelberg (2012). https://doi.org/10.1007/978-3-642-30436-1_15
4. Fisler, K., Krishnamurthi, S., Meyerovich, L.A., Tschantz, M.C.: Verification and change-impact analysis of access-control policies. In: International Conference on Software Engineering (2005)
5. Garcia-Alfaro, J., Cuppens, F., Cuppens-Boulahia, N., Martinez, S., Cabot, J.: Management of stateful firewall misconfiguration. Comput. Secur. **39**, 64–85 (2013)
6. Hu, H., Ahn, G.J., Kulkarni, K.: Discovery and resolution of anomalies in web access control policies. IEEE Trans. Dependable Secure Comput. **10**(6), 341–354 (2013). https://doi.org/10.1109/TDSC.2013.18
7. Jha, S., Li, N., Tripunitara, M., Wang, Q., Winsborough, W.H.: Towards formal verification of role-based access control policies. IEEE Trans. Dependable Secure Comput. **5**(4), 242–255 (2008)
8. Monperrus, M.: Automatic software repair: a bibliography. ACM Comput. Surv. **51**(1), 17:1–17:24 (2018). https://doi.org/10.1145/3105906
9. Montangero, C., Reiff-Marganiec, S., Semini, L.: Logic-based conflict detection for distributed policies. Fundamantae Informatica **89**(4), 511–538 (2008)
10. Neri, M.A., Guarnieri, M., Magri, E., Mutti, S., Paraboschi, S.: Conflict detection in security policies using semantic web technology. In: Satellite Telecommunications (ESTEL), pp. 1–6. IEEE (2012). https://doi.org/10.1109/ESTEL.2012.6400092
11. Ni, Q., et al.: Privacy-aware role-based access control. ACM Trans. Inf. Syst. Secur. **13**(3), 24:1–24:31 (2010). https://doi.org/10.1145/1805974.1805980
12. Son, S., McKinley, K.S., Shmatikov, V.: Fix Me Up: repairing access-control bugs in web applications. In: 20th Annual Network and Distributed System Security Symposium. Usenix, San Diego (2013)
13. Stoller, S.D., Yang, P., Ramakrishnan, C.R., Gofman, M.I.: Efficient policy analysis for administrative role based access control. In: Proceedings of the 2007 ACM Conference on Computer and Communications Security, CCS 2007, Alexandria, Virginia, USA, 28–31 October 2007, pp. 445–455 (2007)
14. Wu, H.: Finding achievable features and constraint conflicts for inconsistent meta-models. In: Anjorin, A., Espinoza, H. (eds.) ECMFA 2017. LNCS, vol. 10376, pp. 179–196. Springer, Cham (2017). https://doi.org/10.1007/978-3-319-61482-3_11

Is This Really You? An Empirical Study on Risk-Based Authentication Applied in the Wild

Stephan Wiefling[1]([✉])(iD), Luigi Lo Iacono[1](iD), and Markus Dürmuth[2]

[1] TH Köln - University of Applied Sciences, Cologne, Germany
{stephan.wiefling,luigi.lo_iacono}@th-koeln.de
[2] Ruhr University Bochum, Bochum, Germany
markus.duermuth@rub.de

Abstract. Risk-based authentication (RBA) is an adaptive security measure to strengthen password-based authentication. RBA monitors additional implicit features during password entry such as device or geolocation information, and requests additional authentication factors if a certain risk level is detected. RBA is recommended by the NIST digital identity guidelines, is used by several large online services, and offers protection against security risks such as password database leaks, credential stuffing, insecure passwords and large-scale guessing attacks. Despite its relevance, the procedures used by RBA-instrumented online services are currently not disclosed. Consequently, there is little scientific research about RBA, slowing down progress and deeper understanding, making it harder for end users to understand the security provided by the services they use and trust, and hindering the widespread adoption of RBA.

In this paper, with a series of studies on eight popular online services, we (i) analyze which features and combinations/classifiers are used and are useful in practical instances, (ii) develop a framework and a methodology to measure RBA in the wild, and (iii) survey and discuss the differences in the user interface for RBA. Following this, our work provides a first deeper understanding of practical RBA deployments and helps fostering further research in this direction.

1 Introduction

Weaknesses in password-based authentication have been known for a long time [21]. They range from weak and easy to guess passwords [4,29] or password re-use [9] to being susceptible to phishing attacks. Still, passwords are the predominant authentication mechanism deployed by online services today [6,23]. To increase the users' security, service operators should implement additional measures. *Two-factor authentication (2FA)* [22] is one widely offered measure that improves account security, but is rather unpopular (e.g. in January 2018,

© IFIP International Federation for Information Processing 2019
Published by Springer Nature Switzerland AG 2019
G. Dhillon et al. (Eds.): SEC 2019, IFIP AICT 562, pp. 134–148, 2019.
https://doi.org/10.1007/978-3-030-22312-0_10

less than 10% of active Google accounts used 2FA [19]). *Risk-based authentication (RBA)* [11] is an approach that increases security with minimal impact on user interaction, and thus has the potential to provide secure authentication with good usability. It is among the approaches suggested by the NIST digital identity guidelines to mitigate online guessing attacks [14].

Risk-Based Authentication (RBA). RBA is typically used in addition to passwords or other forms of user authentication. It is designed to protect against a rather strong attacker that either knows the correct credentials (i.e., username/password pair) or can guess correct credentials with a low number of guesses. Examples include *credential stuffing attacks* [30] where an attacker tries credentials leaked from another service, *phishing attackers*, or *online guessing attacks* [29]. During password entry RBA monitors and records additional features that are contextually available. In principle, a number of various distinct features can be taken into account (see Table 1), including the *IP address* and derived features such as *geolocation* or *country*, and the *user agent* [5,11]. Some features are better suited for risk assessment than others: The IP address, e.g., could be rated as "more important" than the user agent string since spoofing an IP address is considered as more difficult than the latter [3].

From these features a *risk score* is calculated. It is then typically classified into three buckets (low, medium and high risk) [11,16,20]. Depending on the risk score and its classification, a variety of actions can be performed by the service. When a risk score exceeds e.g. the low threshold and falls into the medium risk category, the service typically requests additional authentication factors from the user (e.g. verification of email address or phone number [11,17,24]), requires to solve a CAPTCHA [24], or informs the user about suspicious activities [13]. If the risk score is deemed high, the service can decide to block access altogether, but this event is rare, as it will not allow legitimate users mistakenly classified as a high risk to recover. The thresholds of when a user becomes suspicious have to be carefully chosen for each individual RBA use case scenario.

Contribution. We investigate how RBA is used on eight high-traffic online services (Amazon, Facebook, GOG.com, Google, iCloud, LinkedIn, Steam and Twitch). We created 28 virtual identities and 224 user accounts for this purpose. During a period of 3.5 months we conducted studies to determine (an approximation to) a set of features that contributes to the risk score computation, and studied the influence of these features. We also captured and analyzed the deployed additional authentication factors. Our studies revealed serious vulnerabilities emphasizing the need for an open discussion on RBA in science.

To achieve reliable and repeatable results, we developed an automated browser testing framework and simulated human-like user behavior with individual activities on each of the online services. The framework contains enhanced technical camouflage measures to be indistinguishable from human users. The developed testing framework[1] can be used to analyze black boxed services for

---

[1] Provided as open source software at https://github.com/DASCologne/HOSIT.

RBA features. Our work is intended to support both research and development. Researchers benefit from an increased transparency on the current practice of RBA deployment. Also, they obtain a test methodology and tooling for running replication or follow-up studies. Developers obtain guided insights on how to best create or improve own RBA implementations. The same is true for administrators aiming at integrating RBA as an additional line of defense in their online services. This all contributes to an open scientific discussion on RBA, ultimately leading to a comprehensively understood security measure, leaving no room for obscurities. We hope that public research on RBA will enable a broader adoption of RBA and thus protect a larger user base, while currently only larger online services are capable to offer RBA techniques (beyond very basic and inaccurate service).

Outline. The rest of the paper is organized as follows. Section 2 reviews related work. Section 3 describes the developed automated testing framework, created identities and prerequisites for the studies. The study setup and obtained results are described in Sect. 4. We discuss findings and limitations in Sect. 5 and conclude with the main contributions and an outlook on future work in Sect. 6.

2   Related Work

The features and authentication factors deployed by RBA-instrumented online services are currently either not disclosed or just briefly mentioned by the respective companies [17–19]. This lack hinders any scientific debate and rigorous analysis to facilitate the effective and open use of RBA. These debates and analyses are even more important today since RBA is recommended by NIST [14] and therefore becoming a requirement for many IT security professionals.

Most of the RBA-related research is focused on evaluating the reliability and robustness of certain features. A RBA method based on mouse and keyboard dynamics was developed and tested by Traore et al. [27]. Judging from the observed equal error rate, they concluded that this method is not suitable for RBA inside the login process. Hurkala and Hurkala [16] published a software architecture of a RBA system. The features *IP address, login time, availability of cookie, device profiling* and *failed login attempts* are implemented in the RBA system. The limitations and effectiveness of these features were not estimated. Freeman et al. [11] presented the, to the best of our knowledge, first publicly known RBA algorithm using *IP address* and *user agent* as features. Steinegger et al. [26] presented another RBA implementation, with *browser fingerprint, failed login attempts* and *IP based geolocation* as features. Alaca and van Oorschot [3] classified and rated 29 distinct methods for device fingerprinting regarding possible *"distinguishing info"*. They rated *IP address* and *geolocation* as *"high"*. Daud et al. [10] introduced an adaptive authentication method applying HTML5 canvas fingerprinting. The effectiveness of this method is unclear due to the lack of testing with participants. Herley and Schechter [15] presented a method for authentication servers to distinguish attacks from legitimate traffic. They rated the *password used for a failed attempt* as a strong feature to identify attacks.

Petsas et al. [22] estimated the quantity of Google user accounts with enabled 2FA functionality. They used headless browser automation with enhancements for user simulation. Their methodology, using browser automation and observing reactions, is roughly similar to ours. However, due to the complex nature of RBA and novel browser automation detection methods [28], a considerably higher amount of effort was necessary in our studies.

3   Black Box Testing RBA

In this section we introduce the developed approach for black box testing RBA implementations in the wild. The basic methodology is to create accounts on the inspected online services and to observe the behavior when accessing the service using these accounts for a variety of scenarios. This seemingly simple procedure is complicated by a number of factors: (i) The account's login history may influence the risk score. Thus, testing multiple scenarios with the same account may produce unreliable results. (ii) Automated testing is likely influencing the outcome, as one of the tasks of RBA is specifically to detect bots. (iii) The list of features that potentially may be used by online services to determine the risk score is vast, and simply testing all combinations is next to impossible. (iv) Depending on the service's implementation of RBA, the feedback can be coarse-grained, i.e., giving mostly binary information (RBA triggered/not triggered), while other online services provide more fine-grained information.

Our approach considers these issues and mitigates their effects on the results. We created a larger number of virtual identities and spent several weeks to train them on legitimate behavior. The data collection uses an extensively patched version of Chromium and a careful planning to protect against detection.

3.1   Creation of Identities

We created 28 identities for our studies. User accounts for all eight inspected online services were created with each identity. We used a random identity generator for identity creation. Each identity consisted of first and last name, birthday, gender (50% male, 50% female), job title (function, company) as well as typing speed. Each identity owns an individual IP address (geolocation: TH Köln) and a personal computer (virtual machine running Ubuntu Linux 16.04 LTS). We conducted a one month pilot phase with one identity in order to optimize our identity creation, training and testing automation. Afterwards, we started the automated training and testing with the remaining 27 identities. The account creation for Facebook required some extra care, as RBA is not activated per se for all accounts [17]. We manually conducted extra training to these accounts (e.g. friend requests) prior to the studies. Resulting of the higher effort, 14 Facebook accounts (5 male, 9 female) were created. Six accounts (4 male, 2 female) were suspended during training because of "suspicious" activities. Since female accounts had higher success rates in terms of accepted friend requests or messages, we preferred them in Facebook account creation. Thus, in total we created

224 accounts of which 210 remained available for training and 204 for inspecting the targeted online services.

3.2   Training of Identities

Each online service was trained with individual user activities for each identity in a 3.5 month period between December 2017 and March 2018. Each identity executed 20 user sessions lasting between 1.5 and 2 h within a training period of 2 or 4 weeks. The start of the browsing sessions varied randomly between two time spans (9:00–9:30 AM, 1:00–1:30 PM) to mitigate possible automation detection by the online services. For further mitigation, the identities were created iteratively in small batches of three to four identities per week.

We developed individual automated user activities for each online service. Activities include the login process, actions on the online services at logged in state (*user action*) and the logout process. In the login process, our user opens the targeted online service in a new browser tab, enters its login credentials and accesses this service. We considered typical user activities for the user actions, e.g. scrolling in the news feed or browsing on the online service. These actions included randomness and fine-grained variations to avoid being spotted as a "scripted human". Also, the user behavior differed between genders. For the logout, our user logs out of the online service and closes the tab.

We simulated browsing activities on other websites in separate tabs, as online services may track this browsing behavior [7]. Users visited a search engine and looked for current events in local media. They followed some of the links in the search results and "read" the website's content by scrolling and waiting.

These activities were conducted inside browsing sessions. Each session was initiated with an empty browser history including cookies and local cache. The cookies were retained inside each browsing session. Afterwards, the testing sequence of online services was shuffled to a random order. We did this to prevent that our user logs into online services at the same time throughout the study.

3.3   Implementation of RBA Inspection System

The implemented RBA inspection system is based on the browser Chromium 64.0.3253.3. For browsing automation, the library Puppeteer 0.13.0 is used. The obtained observations during the test phase are stored in a MongoDB log.

Chromium was operated in a custom *headful mode* (browser is launched with visible graphical user interface inside a virtual window session). We used the headful mode to avoid detection of our automated browsing. When Chromium is executed in *headless mode*, which is specifically designed for browsing automation, a number of differences in Chromium's behavior allow websites to detect the automation mode [28]. In fact, during pilot testing we experienced situations in which inspected online services treated a browser in headless mode differently.

Furthermore, we modified the Chromium source code to minimize possible detection of our automated RBA inspection system.

We implemented the user automation framework using Puppeteer, a library to control Chromium. We found that several of the provided automation functions can be detected by online services. The constant delay in the standard Puppeteer key typing function is used to detect automated input. We therefore modified and enhanced several Puppeteer library functions to mimic human behavior more closely: (i) We added randomized delays between pressing and releasing key buttons as well as consecutive button presses. (ii) We adjusted the default mouse input behavior of clicking on the exact center of a specified element by selecting a random click point in the center quarter of the element. Moreover, the default time between pressing and releasing the mouse button of zero was replaced with a more realistic randomized click time. (iii) We implemented a scrolling function to imitate human-like reading of website contents.

We integrated external services providing CAPTCHA solving capabilities in order to allow our RBA inspection system to operate fully automated.

Table 1. Comparison of possible RBA features (bold: selected for the studies)

Feature	RBA references (except [3])	Distinguishing info [3]
IP address[#]	$[2, 8, 11, 12, 16, 26]$	High
User agent string	$[11, 16, 25]$	High*
Language	$[8, 11, 16]$	High*
Display resolution	$[10, 25]$	High*
Login time	$[8, 11, 12, 16, 25]$	Low[+]
Evercookies	[16]	Very high
Canvas fingerprinting	$[10, 20, 26]$	Medium
Mouse and keystroke dynamics	[27]	- (*Low for scroll wheel fingerprinting*)
Failed login attempts	[16, 26]	-
WebRTC	-	Medium
Counting hosts behind NAT	-	Low
Ad blocker detection	-	Very low

[#] Includes IP based *geolocation*.
* Refers to *major software and hardware details*.
[+] Refers to *system time and clock drift*. Alaca and van Oorschot did not consider the login time. Hurkala and Hurkala [16] estimated a medium risk level for unusual login times.

3.4 Inspection of RBA Features

A wide variety of features can be used for RBA deployments, ranging from browser provided information to network information [3, 27]. To reduce complexity, we selected five features based on the number of mentions in literature and the evaluations in [3] in terms of highest *"distinguishing info"* (see Table 1).

We selected the features *IP address, user agent string, language, login time* and *display resolution* for our investigations.

Canvas fingerprinting and evercookies provide a high level of information [1, 3, 10]. Canvas fingerprinting can be seen as a more robust and fine-grained version of user agent strings. Evercookies can uniquely identify a device. Since both features are considered as harder to fake, they add a high level of trust, possibly bypassing RBA security mechanisms. Since we aimed to test the "uncertain" area in terms of RBA risk scores, we did not consider both for our studies.

Prior to the study design, we estimated possible risk score results for specific variations of feature values. We used these estimations to design the final studies. Since no public information on the analyzed RBA implementations was known, we considered three publications [8,11,16] as a baseline for the estimation. We made use of the maximum possible range of ratings. However, since IP addresses are considered as more spoofing resistant than the other features [3], we expect this feature to be weighted highest inside the black box RBA implementations.

We assume that the *IP address* risk score increases with both geographical distance towards the usual values and changes in IP address and internet service provider (ISP). Since users are more likely moving in their current region, we expect the risk score to be *medium* at a maximum inside the same country. We assume changes in continents to be more unusual, so we expect a *high* risk score in that case. We rated the risk score for IP addresses of the anonymization service *Tor* as *unknown* for two reasons: (i) Tor exit nodes (and Tor users) can be identified through a public list. Thus, one publication [16] estimated a high risk score for Tor. (ii) Facebook explicitly supports Tor. Hence, lower risk scores can also be possible. We subdivided the *user agent string* into *browser*, *operating system* (OS) and *version*. We expect users to switch browsers more likely than the OS, which is why we weighted browser changes lower than those in the OS. For the remaining three features, we assume that changes in one or more parameters will increase the score equally.

4   Studies

In this section, we describe the setup and results of the studies we conducted to evaluate the eight analyzed online services for their RBA behavior. We conducted two studies. In the first one, we tested how the online services reacted to six different variations of IP addresses to reduce the number of required test conditions for the second study (see Sect. 4.1). In the second and main study we then determined which of the five investigated features (see Table 1) play a role in RBA decision-making (see Sect. 4.2). We tested all possible combinations of these features for each online service and observed the results. We did this to determine whether a certain feature was included in the online service's feature set and to ascertain how a particular feature was weighted in the online service's RBA decision-making. Finally, we also did several activities on user accounts so that online services might offer diverse selections of additional authentication factors. We did this to capture as many additional authentication factors applied by the targeted online services as possible (see Sect. 4.3). An extended version of our results including all captured dialogs can be found online [31].

4.1   Study 1: Determining IP Feature Thresholds

The feature space that can be used for RBA is huge, and even with the restrictions put forth in Sect. 3.4 the search space is still too large for the type of study we envision. Even the particularly important IP address feature has a wide range of possible values. Possibly interesting variations range from dynamic IPs (same ISP, same geolocation) or different access points (work, home, mobile) at similar locations, to national or international travelling or Tor (see Table 2). Thus, in a first step we treated the IP space separately and tried to find thresholds for the individual online services that are close to the decision boundary of the decision procedure. This will simplify the subsequent experiments and reduce the number of required probes.

Methodology. In this first study varied the IP address only. We equipped seven of the trained identities with new IP addresses (Table 2). Probe 0 uses the identical IP from which the online services were trained before. Probe 1 and probe 2 are located in close vicinity of the training IP (same city, physical distance less than 1 km), where probe 1 is from the same ISP (a university) and probe 2 is from a different ISP. Probes 3 to 5 used IPs with an increasing distance from the training origin. We used VPN tunnels through Amazon Web Services (AWS) instances for these probes. Probe 6 used the Tor network, with an IP of the exit node that is potentially known by service providers and sometimes treated differently. Logins at all online services were conducted with the new IP address and reactions of the online services were recorded.

Results. The obtained results are depicted in Table 3. We see that the thresholds seem to be at IP variation probe 4 (Google, Amazon, LinkedIn) and probe 1 (GOG.com). Facebook, Steam, Twitch and iCloud did not request additional authentication factors, if only the IP address was varied. A CAPTCHA inside the Steam login form was visible in probe 6 (Tor). A reCAPTCHA on the Twitch login form was not displayed in probe 2 (Netcologne) while being visible vice versa. These might rather be signs for blacklisting (Steam) or whitelisting (Twitch) than for RBA. Google sent an email containing a security alert on two occasions before reaching the threshold of asking for additional authentication factors.

Based on the results, we extracted three IP settings for use in the subsequent experiments. These were selected for each online service separately, reflecting the determined thresholds. We set probe 0 (TH Köln) for GOG.com, probe 3 (Frankfurt) for Google, Amazon and LinkedIn as well as probe 5 (Oregon) for Facebook, Steam, Twitch and iCloud. We did not use Tor in subsequent studies, due to its unpredictable nature (frequent variations of IP addresses) which could produce unreliable results. Varying the ISP to AWS (probe 3) inside the same country did not result in requesting additional authentication factors. Hence, we assume that using AWS IP addresses did not affect the reliability of our results.

4.2 Study 2: Examining RBA Usage

In the second and main study, we determined which features play a role in the overall RBA decision-making and under which circumstances the inspected online services request additional authentication factors.

Table 2. Setup of study 1 to determine the RBA triggering threshold for the IP feature

	IP	ISP	Geolocation	Description
Probe 0	Fixed	TH Köln	Cologne, Germany	Same IP as used during training
Probe 1	Fresh	TH Köln	Cologne, Germany	Fresh IP in the same building
Probe 2	Fresh	Netcologne	Cologne, Germany	Different provider in the same city
Probe 3	Fresh	AWS	Frankfurt, Germany	Same country, different provider
Probe 4	Fresh	AWS	Paris, France	Same continent, different provider
Probe 5	Fresh	AWS	Oregon, USA	Different continent
Probe 6	Fresh	Random	random (Tor exit node)	Tor exit node at random location

Table 3. Results of study 1 showing the determined RBA triggering thresholds for the IP feature (bold lines).

IP variation	Identity	Facebook	Google	Amazon	LinkedIn	GOG.com	Steam	Twitch	iCloud
probe 0 (TH Köln, fixed)	*All identities*	-	-	-	-	-	-	-	-
probe 1 (TH Köln, fresh)	IDA, IDAA[+]	-	-	-	-	A	-	-	-
probe 2 (Netcologne)	IDB	-	S	-	-	A	-	Ø	-
probe 3 (Frankfurt)	IDC	-	S	-	-	A	-	-	-
probe 4 (Paris)	IDD	-	A	A	A	A	-	-	-
probe 5 (Oregon)	IDE	-	A	A	A	A	-	-	-
probe 6 (Tor)	IDF	-	A	A	A	A	O	-	-

A: Additional authentication factors requested   O: CAPTCHA displayed before login
S: Security alert submitted (via email)   Ø: reCAPTCHA not displayed before login
- : No RBA triggered   [+]: Facebook login was conducted with this identity

Methodology. We tested all 31 possible combinations of the five parameters *IP address, user agent string, language, time parameters* and *display resolution* for triggering RBA. Each trained account conducted one or two login attempts with different parameter combinations. The *IP address* was chosen one step beneath the determined RBA triggering threshold. The remaining parameters were chosen to represent the highest possible risk estimation as defined in Sect. 3.4 (see Table 4). We chose a far distance country with a different national language than in the training country as the testing country. Based on the online services' behavior of all 31 parameter combinations, we are able to derive possible feature set parameters.

Results. *Google* sent a security alert via email when either of the features *IP address, user agent* or *resolution* changed (see Table 5). Changes in one of the features *language* and *time* didn't result in a warning instead. In contrast to that, we have seen before that strong variations of the IP address result in a request for additional authentication factors (see Table 3). When modifying two features, all combinations resulted in a security warning, except for the combination of *language* and *time*. Modifying three features resulted at least in a security warning, and the combination of *IP address, user agent,* and *time parameters* led to an additional authentication factor requested. Concluding all results, our derived Google feature set contains *IP address* (highest weighting), *time parameters* (lower weighted than IP), *user agent* and *resolution*.

LinkedIn's RBA was triggered with combinations of *IP address* and at least one of the other parameters (see Table 6). Thus, LinkedIn's feature set comprises *IP address, user agent, language, time parameters* and *resolution*. The IP address seems to be higher weighted since it triggered RBA in the prior study alone.

Table 4. Setup of study 2 showing the probed features. We tested all possible combinations, i.e., $2^5 - 1 = 31$ variations per online service.

	Neutral/Training	Testing
IP address	As in training	As determined in Sect. 4.1
User agent	Chrome/Linux	Firefox/Windows 10
Languages	de-DE, de, en-US, en	es-MX, es, en-US, en
Time Timezone	UTC+1 (Europe/Berlin)	UTC-6 (Mexico/General)
Login times [UTC+1]	9:00 AM–2:30 PM	0:00–1:00 AM
Display resolution	1366 × 768	1280 × 1024

Table 5. Results of study 2 for Google modifying a *single feature* (left), *two features* (middle), and *more than two features* (right).

(A: Additional authentication factors requested, - : No RBA triggered, S: Security alert, C: Critical security alert)

	Result
IP address	S
User agent	S
Language	-
Time	-
Resolution	S

	IP	UA	L	T	R
IP address		S	S	S	S
User agent	S		S	S	S
Language	S	S		-	S
Time	S	S	-		S
Resolution	S	S	S	S	

IP	UA	L	T	R	Result
X		X	X		S
	X	X	X		S
		X	X	X	S
X	X		X		A/C
X	X	X	X		A/C
X	X		X	X	A/C
X	X	X	X	X	A/C

Table 6. Results of study 2 for LinkedIn modifying a *single feature* (left) and *two features* (right).

(A: Additional authentication factors requested, - : No RBA triggered)

	Result
IP address	-
User agent	-
Language	-
Time	-
Resolution	-

	IP	UA	L	T	R
IP address		A		A	A	A
User agent	A		-		-	-
Language	A	-			-	-
Time	A	-		-		-
Resolution	A	-		-	-	

Facebook seems to have RBA deactivated by default. We could not trigger RBA on accounts having at least 50 connections to other accounts (friends). However, we could trigger RBA on two female accounts having both 40–50 friends and a high interaction rate based on received friendship requests and messages from other users. Due to the possible dissimilarities between the test accounts

(RBA enabled or disabled), we cannot deduce the exact feature set here. However, our results show that Facebook requested additional authentication factors when at least *IP address, user agent* and *resolution* were changed.

On *Amazon* and *GOG.com* we could not trigger RBA with more or other parameters than the IP address. Thus, their derived feature sets contain only the *IP address* of our probed features.

The remaining online services *Steam, Twitch* and *iCloud* did not show any reaction in both studies. Possible reasons for this behavior could be: (i) RBA was not implemented or not activated by the user behavior. (ii) Other features than the five tested were rated as more important. (iii) An internal warning was triggered informing operational staff about suspicious behavior.

4.3   Study 3: Analyzing Additional Authentication Factors

With RBA being triggered, additional authentication factors are requested by the respective online service. Depending on internal account settings, online services might vary the set of requested additional authentication factors. Overviews of neither the additional authentication factors nor the corresponding RBA user interfaces in current practice were published in literature to date. For this reason, we tried to capture as many variations as possible. In order to achieve this, we added a mobile phone number, a smartphone or tablet as a second device and did additional user actions (e.g. writing a private message with phone number included). We triggered RBA on desktop and mobile devices with all possible combinations and monitored the demanded authentication factors (see Table 7).

Table 7. Captured additional authentication factors

Service	Requested authentication factors
Facebook	Approve login on another computer*
	Identify photos of friends*
	Asking friends for help*
	Verification code (text message)
Google	Enter the city you usually sign in from
	Verification code (email, text message, app, phone call)
	Press confirmation button on second device (tablet, smartphone)
LinkedIn	Verification code (email)*
Amazon	Verification code (email*, text message)
GOG.com	Verification code (email)*

(*: Authentication factor was offered in all tested parameter variations)

5   Discussion

According to our findings, all tested RBA-instrumented online services used the *IP address* in their feature sets. Most online services also used additional features as *user agent* or *display resolution*. All tested online services offered verification codes as an additional authentication factor. The test results confirmed our hypothesis that online services rated the *IP address* higher than other parameters.

Facebook's verification code feature leaked the full phone number. We consider this as a bad practice and a threat for privacy. In so doing, phone numbers of users can be obtained. Also, attackers can call the number and gain access to the verification code by social engineering. We are convinced that such a RBA solution will not mitigate incentives for credential stuffing or online password guessing attacks. Thanks to the prompt reaction by Facebook, this vulnerability is now fixed: We contacted Facebook about the phone number leak on September 4th, 2018. Facebook resolved the issue on September 6th, 2018. Since this issue seemingly remained undiscovered by Facebook before our disclosure, this underlines the demand for more research on RBA to improve its overall security.

5.1   Derived RBA Models Applied in Practice

Based on our findings, we are able to derive three distinct types of conceptual RBA models. Note that due to the abstract nature of these models, they do not provide implementation details.

The **Single-Feature Model** relies on a single feature only. The password authentication process is extended to search for an exact match of the IP address in the IP address history of the user. If there is no such match, additional authentication steps are requested. We assume that GOG.com adopted this model. This model is easy to implement, since only one feature has to be stored and evaluated. Thus, a minimum of sensitive data has to be collected and stored. However, this approach entails potential usability problems. Since IP addresses might change frequently in time [3], this can result in frequent re-authentication. Hence, we do not consider this as a sensible RBA solution for practical use.

The **Multi-Features Model** extends the single-feature model. It derives additional features from the IP address. These are evaluated together with additional features in a scoring model, which compares the current feature values with the authentication history. Depending on the resulting risk score, multiple types of actions are performed (e.g. sending security alerts or requesting additional authentication factors). According to our observations this model was adopted by Google and—in slightly more simplified form without security alerts—by Amazon and LinkedIn. This model has the potential to increase usability compared to the single-feature model since additional authentication factors can be requested less frequently. However, attackers are possibly able to learn about the RBA implementation based on detailed information delivered in security alerts.

The **VIP Model** protects only special users. Depending on the user's status (e.g. important or not important), RBA is active or inactive. We assume that

Facebook used this model. This procedure will make it harder for attackers to gain information about the used RBA implementation. However, if such a mechanism is known, attackers are able to find out whether an account is considered as important by the online service (which is the case when RBA is triggered). Also, this model puts some users at risk since it does not protect all users.

5.2 Limitations

We were able obtain a high amount of information with the described studies. However, the RBA behavior could only be determined from visible reactions disclosed by the online services. Hence, we can only estimate internal weightings for features. It is still possible that the real weightings might vary in detail. In addition, RBA is required to be activated anytime for determining feature sets accurately. It is still possible that online services (additionally) use other features which were not tested in the studies (e.g. canvas fingerprinting).

Although we took a lot of care of not being detectable as an automated user, we cannot fully exclude that the inspected online services identified our identities as non-humans. Judging some of the hints we obtained during our pilot phase, we are strongly convinced, though, that our investigations remained under respective detecting thresholds.

5.3 Ethical Considerations

It is commonly found that tools and techniques used for security analysis are "dual use", i.e., can be used for illegitimate purposes as well. We believe our work is justified, as the expected security gain (from broader adoption of RBA) outweighs the expected security implications. Furthermore, we designed our study to keep the potential impact on the server infrastructure minimal. Finally, we followed the principle of responsible disclosure.

6   Conclusion

RBA is becoming more and more important to strengthen password-based authentication without affecting the user interface at the same time. As RBA is still in its infancy, it is of paramount importance that RBA approaches and implementations are rigorously analyzed following common scientific policies. Unfortunately, almost all early adopters of RBA restrain their approaches and experiences, preventing the required scientific dialogue and the widespread adoption. To close this information gap, we developed distinct studies enabling to verify whether a particular online service adopted RBA. Moreover, we were able to determine the underlying feature sets and requested authentication factors.

We can confirm the general trend in RBA of using the IP address as a high weighted indicator to determine risks of login attempts. Some services also used additional lower weighted indicators (e.g. user agent). Furthermore, verification codes are currently the unwritten standard for additional RBA authentication factors. Our research disclosed potential vulnerabilities and usability problems on specific RBA implementations (one vulnerability was fixed after we contacted the company in charge). Since RBA usually evaluates sensitive data, there is need for more open research on this technology to mitigate such potential risks.

Acknowledgments. This research was supported by the research training group "Human Centered Systems Security" (NERD.NRW) sponsored by the state of North-Rhine Westphalia.

References

1. Acar, G., Eubank, C., Englehardt, S., Juarez, M., Narayanan, A., Diaz, C.: The web never forgets: persistent tracking mechanisms in the wild. In: CCS 2014, pp. 674–689. ACM (2014)
2. Akhtar, N., Haq, F.: Real time online banking fraud detection using location information. In: Das, V.V., Thankachan, N. (eds.) CIIT 2011. CCIS, vol. 250, pp. 770–772. Springer, Heidelberg (2011). https://doi.org/10.1007/978-3-642-25734-6_136
3. Alaca, F., van Oorschot, P.C.: Device fingerprinting for augmenting web authentication. In: Proceedings of ACSAC 2016, pp. 289–301. ACM (2016)
4. Bonneau, J.: The science of guessing: analyzing an anonymized corpus of 70 million passwords. In: 2012 IEEE Security & Privacy, pp. 538–552. IEEE, May 2012
5. Bonneau, J., Felten, E.W., Mittal, P., Narayanan, A.: Privacy concerns of implicit secondary factors for web authentication. In: WAY Workshop (2014)
6. Bonneau, J., Herley, C., van Oorschot, P.C., Stajano, F.: Passwords and the evolution of imperfect authentication. Commun. ACM **58**(7), 78–87 (2015)
7. Bujlow, T., Carela-Espanol, V., Lee, B.R., Barlet-Ros, P.: A survey on web tracking: mechanisms, implications, and defenses. Proc. IEEE **105**(8), 1476–1510 (2017)
8. Cser, A., Maler, E.: The Forrester Wave: Risk-Based Authentication, Q1 (2012)
9. Das, A., Bonneau, J., Caesar, M., Borisov, N., Wang, X.: The tangled web of password reuse. In: NDSS 2014, San Diego, vol. 14, pp. 23–26, February 2014
10. Daud, N.I., Haron, G.R., Othman, S.S.S.: Adaptive authentication: implementing random canvas fingerprinting as user attributes factor. In: ISCAIE, pp. 152–156. IEEE (2017)
11. Freeman, D., Jain, S., Dürmuth, M., Biggio, B., Giacinto, G.: Who are you? A statistical approach to measuring user authenticity. In: NDSS 2016, February 2016
12. Golan, L., Orad, A., Bennett, N.: System and method for risk based authentication. US Patent 8,572,391, October 2013
13. Google: Notifying Android users natively when devices are added to their account (2016). https://gsuiteupdates.googleblog.com/2016/08/notifying-android-users-natively-when.html
14. Grassi, P.A., et al.: Digital identity guidelines. Technical Report NIST SP 800-63b (2017)
15. Herley, C., Schechter, S.: Distinguishing attacks from legitimate authentication traffic at scale. In: NDSS 2019, San Diego (2019)
16. Hurkala, A., Hurkala, J.: Architecture of context-risk-aware authentication system for web environments. In: Proceedings of ICIEIS 2014, Lodz, Poland, September 2014

17. Iaroshevych, O.: Improving second factor authentication challenges to help protect Facebook account owners. In: SOUPS 2017, Santa Clara, CA, USA, July 2017
18. Johansson, J., Canavor, D., Hitchcock, D.: Risk-based authentication duration. US Patent 8,683,597, March 2014
19. Milka, G.: Anatomy of account takeover. In: Enigma 2018. USENIX, January 2018
20. Molloy, I., Dickens, L., Morisset, C., Cheng, P.C., Lobo, J., Russo, A.: Risk-based security decisions under uncertainty. In: CODASPY 2012, pp. 157–168. ACM (2012)
21. Morris, R., Thompson, K.: Password security. Commun. ACM **22**(11), 594–597 (1979)
22. Petsas, T., Tsirantonakis, G., Athanasopoulos, E., Ioannidis, S.: Two-factor authentication: is the world ready? In: EuroSec 2015, pp. 4:1–4:7. ACM (2015)
23. Quermann, N., Harbach, M., Dürmuth, M.: The state of user authentication in the wild. In: Who are you? Adventures in Authentication Workshop 2018, August 2018
24. Shepard, L., Chen, W., Perry, T., Popov, L.: Using social information for authenticating a user session, December 2014
25. Spooren, J., Preuveneers, D., Joosen, W.: Mobile device fingerprinting considered harmful for risk-based authentication. In: EuroSec 2015, pp. 6:1–6:6. ACM (2015)
26. Steinegger, R.H., Deckers, D., Giessler, P., Abeck, S.: Risk-based authenticator for web applications. In: Proceedings of EuroPlop 2016, pp. 16:1–16:11. ACM (2016)
27. Traore, I., Woungang, I., Obaidat, M.S., Nakkabi, Y., Lai, I.: Combining mouse and keystroke dynamics biometrics for risk-based authentication in web environments. In: Proceedings of ICDH 2012, pp. 138–145. IEEE, November 2012
28. Vastel, A.: Detecting Chrome headless (2018). https://antoinevastel.com/bot %20detection/2018/01/17/detect-chrome-headless-v2.html
29. Wang, D., Zhang, Z., Wang, P., Yan, J., Huang, X.: Targeted online password guessing: an underestimated threat. In: CCS 2016, pp. 1242–1254. ACM (2016)
30. Wang, X., Kohno, T., Blakley, B.: Polymorphism as a defense for automated attack of websites. In: Boureanu, I., Owesarski, P., Vaudenay, S. (eds.) ACNS 2014. LNCS, vol. 8479, pp. 513–530. Springer, Cham (2014). https://doi.org/10.1007/978-3-319-07536-5_30
31. Wiefling, S., Lo Iacono, L., Dürmuth, M.: Risk-Based Authentication (2019). https://riskbasedauthentication.org

Organizational and Behavioral Security

Differentially Private Generative Adversarial Networks for Time Series, Continuous, and Discrete Open Data

Lorenzo Frigerio[1], Anderson Santana de Oliveira[2(✉)], Laurent Gomez[2], and Patrick Duverger[3]

[1] Polytech Nice, Biot, France
[2] SAP Labs France, Mougins, France
anderson.santana.de.oliveira@sap.com
[3] Ville d'Antibes, Antibes, France

Abstract. Open data plays a fundamental role in the 21st century by stimulating economic growth and by enabling more transparent and inclusive societies. However, it is always difficult to create new high-quality datasets with the required privacy guarantees for many use cases. In this paper, we developed a differential privacy framework for privacy preserving data publishing using Generative Adversarial Networks. It can be easily adapted to different use cases, from the generation of time-series, to continuous, and discrete data. We demonstrate the efficiency of our approach on real datasets from the French public administration and classic benchmark datasets. Our results maintain both the original distribution of the features and the correlations among them, at the same time providing a good level of privacy.

Keywords: Differential privacy · Generative Adversarial Networks

1 Introduction

The digital revolution has changed societies and democracies across the globe, making personal data an extremely valuable asset. In such a context, protecting individual privacy is a key need, especially when dealing with sensitive information, such as political preferences. At the same time, the demand for public administration transparency has introduced guidelines and laws in some countries to release open datasets.

To limit personal data breaches, privacy-preserving data publishing techniques can be employed. This approach aims at adding the noise directly to the data, not only to the result of a query (like in interactive settings). The result is a completely new dataset, where analysts can perform an infinite number of requests without increasing the privacy costs, nor disclosing private information. Meanwhile, it is difficult to preserve the utility of the data.

© IFIP International Federation for Information Processing 2019
Published by Springer Nature Switzerland AG 2019
G. Dhillon et al. (Eds.): SEC 2019, IFIP AICT 562, pp. 151–164, 2019.
https://doi.org/10.1007/978-3-030-22312-0_11

A strong standard privacy guarantee widely accepted by the research community is differential privacy. It ensures that each individual participating in a database does not disclose any additional information by participating in it. Traditionally, many approaches tried to reach differential privacy by adding noise to the data in order to protect personal information [5,8,12], however they have never been able to provide satisfying results on real semantically-rich data; most of the implementations were limited to very specific purposes such as histogram queries or counting queries [18]. Generative models represent the most promising approach in this field. Interesting results have been obtained through Generative Adversarial Networks (GANs) [9]. These models are able to generate new samples coming from a given distribution. The advantage of generative models is that the noise to guarantee privacy is not added directly to the data, causing a significant loss of information, but it is added inside the latent space, reducing the overall information loss, but guaranteeing meanwhile privacy.

This paper extends the notion of dp-GAN, an anonymized GAN with a differential privacy mechanism, to handle continuous, categorical and time-series data. It introduces an optimization called clipping decay that improves the overall performances. This new expansion shapes the noise addition during the training. This allows to obtain a better data utility at the same privacy cost. A set of analysis on real scenarios evidence the flexibility and applicability of our approach, which is supported by an evaluation of the membership inference attack accuracy, proving the positive effects of differential privacy. We provide experimental results on real industrial datasets from the French public administration and over well-known publicly available datasets to allow for benchmarking.

The remainder of the paper is organized as follows: Sect. 2 provides the theoretical background for the paper; In Sect. 3, presents our framework for anonymization, together with the mathematical proofs of differential privacy. Section 4 provides a set of experiments on diverse use cases to highlight the flexibility and effectiveness of the approach. Section 5 discusses related work; and finally Sect. 6 concludes the paper.

2   Preliminaries

This section brings some important background for the paper.

2.1   Generative Adversarial Networks.

GANs (Generative Adversarial Networks) are one of the most popular type of generative models, being already defined as the most interesting idea in the last 10 years in machine learning[1], moreover, a lot of attention has been given to the development of new variations [3,11,14]. Given an initial dataset, a GAN is able to mimic its data distribution, for that, a GAN employs two different networks: a generator and a discriminator. The architecture of the two networks

---

[1] "GAN and the variations that are now being proposed is the most interesting idea in the last 10 years in ML, in my opinion", Yann LeCun.

is separate from the definition of GAN; depending on the application, different network configurations can be used. The role of the generator is to map random noise into new data instances capturing the original data distribution. On the opposite side, the discriminator tries to distinguish the generated samples from the real ones estimating the probability that a sample comes from the training data rather than the generator. In this way, after each iteration, the generator becomes better at generating realistic samples, while the discriminator becomes increasingly able to tell apart the original data from the generated ones. Since the two networks play against each other, the two losses will not converge to a minimum like in a normal training process but this minmax game has its solution in the Nash equilibrium. Nevertheless, the vanilla GAN [9] suffers from several issues that make it hardly usable especially for discrete data. A new definition of the loss function, Wasserstein generative adversarial network (WGAN) [2] and Improved Training of Wasserstein GANs [10] partially solved this problem. We are going to use this latest loss function to train our dp-GAN models.

2.2   Differential Privacy

The state of the art anonymization technique is differential privacy. This concept ensures that approximately nothing can be learned about an individual whether she participates or not in a database. Differential Privacy defines a constraint on the processing of data so that the output of two adjacent databases is approximately the same. More formally: A randomized algorithm M gives (ϵ, δ)-differential privacy if, for all databases d and d', differing on at most one element and all $S \in Range(M)$,

$$Pr[M(d) \in S] \leq exp(\epsilon) \times Pr[M(d') \in S] + \delta. \tag{1}$$

This condition encapsulates the crucial notion of indistinguishability of the results of a database manipulation by introducing the so-called privacy budget ϵ. It represents the confidence that a record was involved in a manipulation of the database. Note that the smaller ϵ is, the more private the output of the mechanism. According to [6] the optimal value of δ is less than the inverse of any polynomial in the size of the database. Any function M that satisfies the Differential Privacy condition can be used to generate data that guarantees the privacy of the individuals in the database. In the non-interactive setting, a mechanism M is a function that maps a dataset in another one. The definition states that the probability of obtaining the same output dataset from M is similar, using either d or d' as input for the mechanism. Composability is an interesting property of Differential Privacy. If M and M' are ϵ and ϵ'-differential private respectively, their composition $M \circ M'$ is $(\epsilon + \epsilon')$-differentially private [5]. This property allows to craft a variety of mechanisms and combinations of such mechanisms to achieve differential privacy in innovative ways.

Concerning deep learning, Abadi et al. [1] developed a method to train a deep learning network involving differential privacy. This method requires the addition of a random noise, drawn from a normal distribution, to the computed gradients, to obfuscate the influence that an input data can have on the final model.

As for any anonymization methods, one must assess the likelihood of membership inference attacks. This kind of attack evaluates how much a model behaves differently when an input sample is part of the training set rather than the validation set. Given a machine learning model, a membership inference attack uses the trained model to determine if a record was part of the training set or not. In the case of generative models such as the one of GAN, a high attack accuracy means that the network has been able to model only the probability distribution of the training set and not the one of the entire population. This kind of attack has been proven to be effective especially when overfitting is relevant [17].

3   The Framework

In our framework proposed we assume a trusted curator interested in releasing a new open dataset with privacy guarantees to the users present in it. Outside the trusted boundary, an analyst can use the generator model, result of our algorithm, to perform an indefinite number of queries over the data the generator produces. Such outputs can be eventually released as open data. Even by combining the generated data with other external information, without ever having access to the original training data, the analyst would not be able to violate the privacy of the information, thanks to the mathematical properties of differential privacy.

The dp-GAN model is constituted by two networks, a generator and a discriminator, that can be modelled based on the application domain. We adopted Long Short Term Memories (LSTM) inside the generator to model streaming data and multilayer perceptron (MLP) to model discrete data. In addition, to manage discrete data, we also used a trick that does not influence the training algorithm but it changes the architecture of the generator network. Specifically, an output is created for each possible value that a variable can assume and a softmax layer is added for each variable. The result of the softmax layer becomes the input of the discriminator network. Indeed, each output represents the probability of each variable instance; the discriminator compares these probabilities with the one-hot encoding of the real dataset. On the contrary, the output nodes associated with continuous variables are kept unchanged.

At the end of the training, the generator network can be publicly released; in this way, the analyst can generate new datasets as needed. Moreover, since the generator only maps noise into new data the process is really fast and data can be generated on the fly when a new analysis is required.

We used the differentially private Stochastic Gradient Descent (dp-SGD) proposed by [1] to train the discriminator network and the Adam optimizer to train the generator. The dp-GAN implementation relies on a traditional training in which the gradients computed for the discriminator are altered. This due to the fact that we want to limit the influence that each sample has on the model. On the contrary, the training of the generator remains unaltered; indeed, this network bases its training only on the loss of the discriminator without accessing directly the data.

The approach is independent from the chosen loss and therefore can be applied to the vanilla GAN implementation [9] but also to the improved WGAN one. The dp-SGD works as follows: once the gradients are calculated, it clips them by a threshold C and alter them by the addition of a random noise with variance proportional to C. Each time an iteration is performed, the privacy cost increases and the objective is to find a good balance between data utility and privacy costs.

Our implementation is an extension to the improved WGAN framework combining it with the dp-SGD. Therefore, the loss functions are calculated as in a normal WGAN implementation, except that the computed gradients are altered to guarantee privacy. Moreover, for the first time up to our knowledge, the dp-GAN concept is adapted to handle discrete data. Algorithm 1 describes our training procedure.

Algorithm 1. Algorithm for training a GAN in a differentially private manner

---

Input: Samples from x_1 to x_N, group size L, number of samples N, clipping parameter C, noise scale σ, privacy target ϵ, number of iterations of the discriminator per each iteration of the generator $Ndisc$, batch size b, Wasserstein distance $\mathcal{L}$, learning rate η, number of discriminator' s parameters m, clipping decay C_{decay}.

Output: differentially private Generator G

Initialize weights randomly both for the Generator $\theta_{G(0)}$ and the discriminator $\theta_{D(0)}$

Convert discrete variables into their One-Hot encodings

while (While privacy cost $\leq \epsilon$) **do**

    for $t = 0$ to $Ndisc$ **do**

        for $j = 0$ to b **do**

            sample L_t with sample probability $L/N = q$

            For each x_i in L_t, compute $g_t(x_i) \leftarrow \nabla_\theta \mathcal{L}(\theta_t, x_i)$ ▷ Compute gradient

            $g_t(x_i) \leftarrow g_t(x_i)/max(1, \|g_t(x_i)\| /C)$ ▷ Clip gradient

            $g_t \leftarrow \frac{1}{L}(\sum\limits_{i=0}^{L} g_t(x_i) + N(0, (\sigma * C)^2 I))$ ▷ Add noise

            $\theta_{D(t+1)} \leftarrow \theta_{D(t)} - \eta * g_t$ ▷ Gradient descent

        end for

    end for

    $C *= C_{decay}$ ▷ Clipping decay

    Update the overall privacy cost ϵ ▷ Moment accountant

    Sample m values $z_i \sim$ Random noise   ▷ Sample random noise

    $\theta_{G(t+1)} \leftarrow Adam(\nabla_\theta \frac{1}{m} \sum\limits_{i=0}^{m} -D(G(z_i)))$ ▷ Update Generator

end while

return G

3.1 Clipping Decay

The role of the clipping parameter is to limit the influence that a single sample can have on the computed gradients and, consequently, on the model. Indeed, this parameter does not influence the amount of privacy used. A big clipping parameter allows big gradients to be preserved at the cost of a noise addition with a proportionally high variance. On the contrary, a small clipping parameter limits the range of values of the gradients, but it keeps the variance of the noise small. The bigger the clipping parameter the bigger the gradients' variance. Similarly to what it is done with the learning rate, it is possible to introduce a clipping parameter decay. In this way, the gradients not only tend to descend over time to better reach a minimum but, in addition, they mimic the descending trend of the gradients allowing to clip the correct amount at each step. In fact, when the model tends to converge to the solution, the gradients decrease. Therefore, the noise may hide the gradients if its variance is kept constant. By reducing the clipping parameter over time, it is possible to reduce the variance in the noise in parallel with the decrease of the gradients, thus improving the convergence of the model. This without influencing the overall privacy costs that are not altered by the clipping parameter but only by the amount of noise added.

3.2   Moment Accountant

A key component of the dp-GAN is the moment accountant. It is a method that allows to compute the overall privacy costs by calculating the cost of a single iteration of the algorithm and cumulating it with the other iterations. Indeed, thanks to the composability property of differential privacy it is possible to cumulate the privacy costs of each step to compute the overall privacy cost. Given a correct value of σ and thanks to weights clipping and the addition of noise, Algorithm 1 is $(O(\epsilon, \delta))$-DP with respect to the lot. Since each lot is sampled with probability $q = L/N$, each iteration is $(O(q\epsilon, q\delta))$-DP. In the formula, q represents the sampling probability (the number of samples inside a lot divided by the total number of samples present in the dataset). The clipping decay optimization has no influence on the moment accountant. Indeed, it alterates only the clipping parameter and not the variance of the noise that is the variable that influences the cost of a single iteration by changing the value of ϵ. Each time a new iteration is performed the privacy costs increase. However, thanks to the definition of moment accountant, these costs do not increase linearly. Indeed, by cumulating the privacy costs for each iteration, an overall level of $(O(q\epsilon\sqrt{T}), \delta)$-DP is achieved where T represents the number of steps (the number of epochs divided by q).

The moment accountant is based on the assumption that the composition of Gaussian mechanisms is being used. Assessing that a mechanism M is $(O(\epsilon, \delta))$-DP is equivalent to a certain tail bound on $M's$ privacy loss random variable. The moment accountant keeps track of a bound on the moments of the privacy loss random variable defined as:

$$c(o; M, aux, d, d') = log\frac{Pr[M(aux, d) = o]}{Pr[M(aux, d') = o]} \qquad (2)$$

In (2) d and d' represent two neighbouring databases, M the mechanism used, aux an auxiliary input and o an outcome. What we are computing are the log moments of the privacy loss random variable that can be cumulated linearly. In order to bound this variable, since the approach is the sequential application of the same privacy mechanism we can define the λth moment $\alpha M(\lambda, aux, d, d')$ as the log of the moment generating function evaluated at the value λ :

$$M(\lambda, aux, d, d') = logE_{o \sim M(aux,d)}[exp(\lambda c(o, M, aux, d, d'))]. \tag{3}$$

And consequently we can bind all possible $\alpha M(\lambda, aux, d, d')$. We define

$$\alpha M(\lambda) = max_{aux,d,d'} \alpha M(\lambda, aux, d, d') \tag{4}$$

Theorem 1. *Using the definition (4) then $\alpha M(\lambda)$ has the following characteristics: given a set of k consecutive mechanisms, for each λ:*

$$\alpha_M(\lambda) \le \sum_{i=1}^{k} \alpha M_i(\lambda)$$

for any $\epsilon > 0$, a mechanism M is (ϵ, δ)-differentially private for

$$\delta = min_\lambda exp(\alpha_M(\lambda) - \lambda * \epsilon)$$

Proof (Proof of Theorem 1). A detailed proof of Theorem 1 can be found in [1].

Theorem 2. *Algorithm 1 is $(O(q\epsilon\sqrt{T}), \delta)$-differentially private for appropriately^2 chosen settings of the noise scale and the clipping threshold.*

Proof (Proof of Theorem 2). By Theorem 1, it suffices to compute, or bound, $\alpha M_i(\lambda)$ at each step and sum them to bound the moments of the mechanism overall. Then, starting from the tail bound we can come back to the (ϵ, δ)-differential privacy guarantee. The last challenge missing is to bind the values $\alpha M_t(\lambda)$ for every single step. Let μ_0 denote the Probability Density Function (PDF) of $N(0, \sigma^2)$, and μ_1 denote the PDF of $N(1, \sigma^2)$. Let μ be the mixture of two Gaussians $\mu = (1 - q)\mu_0 + q\mu_1$. Then we need to compute $\alpha(\lambda) = log(max(E_1, E_2))$ where

$$E_1 = E_z[(\mu_0(z)/\mu(z))^\lambda] \tag{5}$$
$$E_2 = E_z[(\mu(z)/\mu_0(z))^\lambda] \tag{6}$$

In the implementation of the moment accountant, we carry out numerical integration to compute $\alpha(\lambda)$. In addition, we can show the asymptotic bound

$$\alpha(\lambda) \le q^2\lambda(\lambda + 1)/(1 - q)\sigma^2 + O(q^3/\sigma^3)$$

This inequation together with Theorem 1 implies Theorem 2. □

---

^{2} The appropriate values for the noise scale and for the threshold will depend on the desired privacy cost and on the size of the dataset.

4   Experiments

In this section, we evaluate empirically our framework. The experiments are designed to assess the quality of the generated data, measure the privacy of the generated models and understand how differential privacy influences the output dataset. Moreover, we evaluate the solidity of the different models against membership inference attacks. Since it is notoriously arduous to assess the results of a GAN, we decided to combine qualitative and quantitative analysis to obtain a reliable evaluation. Qualitative analysis allows us to graphically verify the quality of the results and to observe the effects of differential privacy; while quantitative analysis provides a most accurate evaluation of the results; in particular, we measured some distance metrics to compare the generated data with the real data. Finally, our process included evaluating our model on a classification problem. This highlights the high utility of the data even when anonymization is used. For all experiments, when differential privacy is used, δ is supposed to be less than 10^{-5}, a value that is generally considered safe [1] because it implies that the definition of differential privacy is true with a probability of 99.999%. Indeed δ is the probability that a single record is safe and not spoiled. We kept the value of δ fixed to be able to evaluate the privacy of a mechanism with a single value ϵ that summarizes in a clearer manner the privacy guarantees.

In the different settings we applied only minor changes to the dp-GAN architecture, since we proved that it adapts well to each of them. In particular, in every case the discriminator is composed of a deep fully connected network. On the other hand, the architecture of the generator is adapted to the different datasets used. To generate time-series we used an LSTM which output becomes the input of the discriminator. On the contrary, in the case of discrete datasets we used a fully connected network which outputs the probability distribution for each value that a variable can assume. The interested reader can find an exhaustive explanation of the experiments, including additional datasets in the following GIT repository: https://github.com/Lory94/dp-GAN.

4.1   Synthetic Dataset

In order to provide a first evaluation of the performances of the dp-GAN and understand the effects of differential privacy, we conducted a first experiment on a synthetic dataset. The dataset is constituted by samples coming from six 2D-gaussian distributions with the same variance, but with different centers. The quality of the results using dp-GAN is similar in both marginals and joint distributions.

Figure 1 plots the kernel density estimation of the data to visualize the bivariate distribution. As expected, differential privacy introduces a small noise in the results, thus increasing the variance of the six gaussian distributions, while at the same time replicating faithfully the original distribution.

We use the Wasserstein distance to measure the distance between two distributions, to ascertain the quality of the GAN models. Figure 2 plots the distance values for the non-anonymized GAN, a dp-GAN, and a dp-GAN using clipping

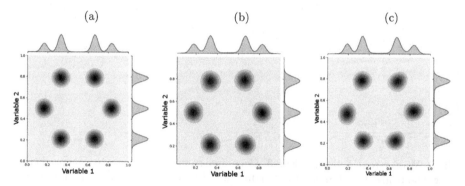

Fig. 1. Kernel estimation for: (a) the original points, (b) WGAN, and (c) dp-GAN

decay. Both the dp-GAN models have $\epsilon = 8$. The different measures tend to converge to similar results, especially when clipping decay is applied, demonstrating the high quality of the results. Indeed, clipping decay allows the Wasserstein distance to drop to values comparable to those of the non-anonymized version in the second half of the graph. The main difference resides in the higher number of epochs necessary to reach the convergence, due to the noise addition.

4.2 Time-Series Data

To test our implementation on a real dataset we decided to use a set of data coming from the IoT system of the City of Antibes, in France. This dataset is private because it contains sensitive information about the water consumption and water pipeline maintenance, obtained directly from sensors in each neighborhood. The purpose is to support public administration in releasing highly relevant open data, while hiding specific events in the time-series, and preventing individual re-identification. With minor adjustments, the solution can constitute a valid framework applicable to other purposes, such as electricity consumption and waste management.

The dataset is an extract of one month of measurements, where each sample is a time-series containing 96 values (one every 15 min). Each sample is labelled with the name of the neighborhood. The data has been normalized before the training. The goal is to generate a new time-series that contains the same number of records and the same distribution as the original dataset, while providing differential privacy guarantees, that is, each sample does not influence whatever analyses more than a certain threshold. In this way, anomalous situations such as maintenance works, a failure in a water pipe or an unexpected water usage by a person living in a certain area are protected and kept private.

Figure 3 compares real samples and generated ones, using non-anonymized-GAN and a dp-GAN with $\epsilon = 6$. We plot the sensor values for a generated sample and the closest sample coming from the original data, in terms of the dynamic time warping distance. The distribution of the original time series is

(a)   (b)

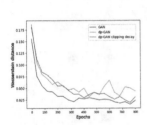
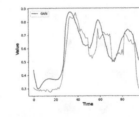
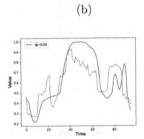

Fig. 2. Wasserstein distance using anonymized and non anonymized GANs

Fig. 3. Generated sample from a non-anonymized GAN (a) and dp-GAN (b), in blue. In orange, the closest sample present in the dataset, in terms of dynamic time warping (Color figure online)

kept, but in the dp-GAN samples, the curves tend to be smoother, hiding some of the variability of the original data.

For time-series data, the quality assessment for GANs represents a challenge. While for images the inception score [13] has become the standard measure to evaluate the performance of a GAN, there is no counterpart for the assessment of time series. We believe that this represents an interesting area of research for the future.

4.3   Discrete Data

We analyzed the performances of our model on the UCI adult dataset: this dataset is an extract of the US census and contains information about working adults, distributed across 14 features. A classification task on it is a reliable benchmark, because of its widespread use in several studies. Records are classified depending on whether the individual earns more or less than $50k$ dollars each year.

To use accuracy as an evaluation metric, we decided to sample the training and test data in such a way that both classes would be balanced. We built a random forest classifier on the dataset generated by the dp-GAN. We evaluated the accuracy on the test set and compared it with the one of the model built on the real non-anonymized dataset. If the dp-GAN model behaves correctly, all the correlations between the different features should be preserved. Therefore, the final accuracy should be similar to what was achieved by using the real training set. We also tracked the privacy costs to verify that the generated data were correctly anonymized. Finally, we examined how much membership inference attacks can influence our model and compared it to a non-anonymized GAN model.

Table 1 evaluates the degradation of the performances when differential privacy is adopted. The target accuracy of 77.2% was reached by using the real training set to train the random forest classifier. As expected, a non-anonymized GAN is able to produce high quality data: its accuracy loss is very low, 0.5%,

Table 1. Classification accuracy for training sets generated by different models

Method	Epsilon	Accuracy
Real dataset	Infinite	77.2%
GAN	Infinite	76.7%
dp-GAN	3	73.7%
dp-GAN clipping decay	3	75.3%
dp-GAN	7	75.0%
dp-GAN clipping decay	7	76.0%

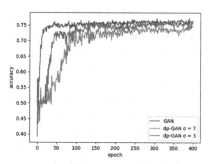

Fig. 4. Classification accuracy average for 5 runs for different noise values

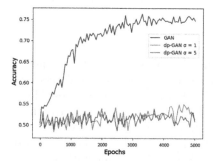

Fig. 5. Membership inference attack accuracy for non-anonymized GAN, dp-GAN with $\sigma = 1$ and $\sigma = 5$

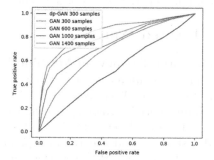

Fig. 6. ROC curves for membership inference attacks for GAN and dp-GAN using generated samples of different size

compared to the target. Interestingly, even when we adopted the dp-GAN framework the accuracy remained high. Using $\epsilon = 7$ and clipping decay, we obtained results similar to the ones without anonymization. In addition, Table. 1 points out the positive effect of clipping decay, that it is able to increase the accuracy of about 1%.

Figure 4 highlights the effects on the classification accuracy when dp-GAN is adopted using different amounts of noise. The main effect is to slow down the training process, but not to significantly impact accuracy. Indeed, the added noise requires more epochs to reach convergence, which is amplified by σ. In most of the use cases, this is a minor drawback, considering the classification accuracy. Moreover, the dp-GAN is trained once; then the generator can be released to produce samples on demand.

Figure 5 shows the analysis of the accuracy of membership inference attacks on the model using different training procedures with different levels of privacy guarantees. This analysis has been done at different epochs of the training process. The accuracy of the model increases over time, however this makes the model more subject to membership inference attacks. As can be seen from Fig. 6, training the model with no anonymization rapidly increases the accuracy of attacks: this highlights the problems that still afflict many generative models that cannot effectively generalize the training data. On the contrary, by increasing the privacy level, the accuracy of the attacks tends to remain close to 50%. This is obtained at the costs of losing about 1% of accuracy during the final classification.

The size of the dataset is another important factor that influences significantly the results, since GANs need a good amount of data to generalize effectively. Figure 6 confirms the results obtained in [17], but at the same time it shows how differential privacy works well even when the dataset is small. Indeed, the dp-GAN provides random accuracy towards membership inference attacks independently from the size of the dataset. It is interesting to notice that since the dataset is small, the level of privacy ϵ is big compared to what it is commonly used; however, the effects of differential privacy can be still perceived clearly.

5   Related Work

Differential Privacy on Machine Learning Models. In [16] it is proposed an innovative approach for training a machine learning model in a differentially private manner. On the contrary of the dp-SGD, they proved that it is possible to reach differential privacy by transferring knowledge from some models to others in a noisy way. A set of models, called teachers, are trained on the real dataset and a student model learns in a private way what the teachers have grasped during the training. However, it is still unclear how this implementation can be extended to a non-interactive setting. [19] developed a dp-GAN based on the dp-SGD providing some optimizations in order to improve performances focusing on the generation of images. In contrast, our work highlights that dp-GAN can be adapted to a variety of different use cases and in particular we developed a variation dedicated to discrete data. In addition, we provide, also, an overview of the effects that differential privacy has on membership inference attacks; [17] pointed out how severe the risk of this kind of attack in a general machine learning model can be. We have confirmed the issue while highlighting that the noise introduced by differential privacy reduces overfitting and consequently the accuracy of membership inference attacks.

Generative Adversarial Networks on Discrete Data. [20] developed Seq-GAN, an approach dedicated to the generation of sequences of discrete data. SeqGAN is based on a reinforcement learning training where the reward signal is produced by the discriminator. However, it is not clear how this approach can be extended to include differential privacy. On the contrary, [15] uses Cramer GANs to combine discrete and continuous data. These recent works did not address data privacy concerns.

Differential Privacy Without Deep Learning. Interesting results have been also obtained through other types of generative models. In the context of non-interactive anonymization, [21] developed a differentially private method for releasing high-dimensional data through the usage of Bayesian networks. This kind of network is able to learn the correlations present in the dataset and generate new samples. In particular, the distribution of the dataset is synthetized through a set of low dimensional marginals where noise is injected and from which it is possible to generate the new samples. However, the approach suffers from an extremely high complexity, thus being unpractical to anonymize large datasets. We have also analysed the literature to prove that the amount of privacy that dp-GAN guarantees is comparable to the one of the other most common implementations. Although there is no specific value for which ϵ is considered safe, we obtained most often lower privacy costs compared to [4,7], which are the two most relevant works dealing with real-life datasets. Similar privacy costs have been used in the most recent literature in the differential privacy field [1,16].

6 Conclusion

In this paper we extended the notion of dp-GANs to privacy-preserving data publishing of continuous, time-series and discrete data. We introduced clipping decay to preserve data utility while providing data privacy: it can be used for any differentially private gradient descent on any neural network to improve learning accuracy. We have shown how our implementation is resistant to membership inference attacks, being suitable for open data releases. In the future, we will work on the reduction of privacy costs and investigate the potential benefits of transfer learning to data anonymization.

References

1. Abadi, M., et al.: Deep learning with differential privacy. In: 23rd ACM Conference on Computer and Communications Security (ACM CCS), pp. 308–318 (2016). arXiv:1607.00133
2. Arjovsky, M., Chintala, S., Bottou, L.: Wasserstein generative adversarial networks. In: Precup, D., Teh, Y.W. (eds.) Proceedings of the 34th International Conference on Machine Learning. Proceedings of Machine Learning Research, 06–11 August 2017, vol. 70, pp. 214–223. PMLR, International Convention Centre, Sydney. http://proceedings.mlr.press/v70/arjovsky17a.html
3. Chen, X., Duan, Y., Houthooft, R., Schulman, J., Sutskever, I., Abbeel, P.: Info-GAN: interpretable representation learning by information maximizing generative adversarial Nets. In: Lee, D.D., Sugiyama, M., Luxburg, U.V., Guyon, I., Garnett, R. (eds.) Advances in Neural Information Processing Systems 29, pp. 2172–2180. Curran Associates, Inc. (2016)
4. Differential Privacy Team - Apple: Learning with privacy at scale. Apple Mach. Learn. J. **1**, 1–25 (2017)
5. Dwork, C.: Differential Privacy: A Survey of Results. In: Agrawal, M., Du, D., Duan, Z., Li, A. (eds.) TAMC 2008. LNCS, vol. 4978, pp. 1–19. Springer, Heidelberg (2008). https://doi.org/10.1007/978-3-540-79228-4_1
6. Dwork, C., Roth, A.: The algorithmic foundations of differential privacy. Now Foundations and Trends (2014). https://doi.org/10.1561/0400000042. https://ieeexplore.ieee.org/xpl/articleDetails.jsp?arnumber=8187424

7. Erlingsson, Ú., Pihur, V., Korolova, A.: RAPPOR: randomized aggregatable privacy-preserving ordinal response. In: Proceedings of the 2014 ACM SIGSAC conference on computer and communications security, pp. 1054–1067. ACM (2014)
8. Geng, Q., Viswanath, P.: The optimal noise-adding mechanism in differential privacy. IEEE Trans. Inf. Theory **62**, 925–951 (2016)
9. Goodfellow, I., et al.: Generative adversarial nets. In: Ghahramani, Z., Welling, M., Cortes, C., Lawrence, N.D., Weinberger, K.Q. (eds.) Advances in Neural Information Processing Systems 27, pp. 2672–2680. Curran Associates, Inc. (2014). http://papers.nips.cc/paper/5423-generative-adversarial-nets.pdf
10. Gulrajani, I., Ahmed, F., Arjovsky, M., Dumoulin, V., Courville, A.C.: Improved training of Wasserstein GANs. In: Guyon, I., et al. (eds.) Advances in Neural Information Processing Systems 30, pp. 5767–5777. Curran Associates, Inc. (2017), http://papers.nips.cc/paper/7159-improved-training-of-wasserstein-gans.pdf
11. Juefei-Xu, F., Boddeti, V.N., Savvides, M.: Gang of GANs: generative adversarial networks with maximum margin ranking. CoRR arXiv:abs/1704.04865 (2017)
12. Kellaris, G., Papadopoulos, S., Xiao, X., Papadias, D.: Differentially private event sequences over infinite streams. Proc. VLDB Endow. **7**(12), 1155–1166 (2014). https://doi.org/10.14778/2732977.2732989
13. Lucic, M., Kurach, K., Michalski, M., Gelly, S., Bousquet, O.: Are GANs created equal? a large-scale study. In: Advances in Neural Information Processing Systems, pp. 697–706 (2018)
14. Mirza, M., Osindero, S.: Conditional generative adversarial Nets. CoRR arXiv:abs/1411.1784 (2014)
15. Mottini, A., Lheritier, A., Acuna-Agost, R.: Airline passenger name record generation using generative adversarial networks. CoRR arXiv:abs/1807.06657 (2018)
16. Papernot, N., Abadi, M., Erlingsson, Ú., Goodfellow, I.J., Talwar, K.: Semi-supervised knowledge transfer for deep learning from private training data. CoRR arXiv:abs/1610.05755 (2016),
17. Shokri, R., Stronati, M., Song, C., Shmatikov, V.: Membership inference attacks against machine learning models. In: 2017 IEEE Symposium on Security and Privacy (SP), pp. 3–18, May 2017. https://doi.org/10.1109/SP.2017.41
18. Xiao, X., Wang, G., Gehrke, J.: Differential privacy via wavelet transforms. IEEE Trans. Knowl. Data Eng. **23**(8), 1200–1214 (2011). https://doi.org/10.1109/TKDE.2010.247
19. Xie, L., Lin, K., Wang, S., Wang, F., Zhou, J.: Differentially private generative adversarial network. CoRR arXiv:abs/1802.06739 (2018)
20. Yu, L., Zhang, W., Wang, J., Yu, Y.: SeqGAN: sequence generative adversarial Nets with policy gradient. CoRR arXiv:abs/1609.05473 (2016). http://dblp.uni-trier.de/db/journals/corr/corr1609.html=YuZWY16
21. Zhang, J., Cormode, G., Procopiuc, C.M., Srivastava, D., Xiao, X.: Privbayes: private data release via Bayesian networks. ACM Trans. Database Syst. **42**(4), 25:1–25:41 (2017). https://doi.org/10.1145/3134428

ESARA: A Framework for Enterprise Smartphone Apps Risk Assessment

Majid Hatamian[1]([✉]), Sebastian Pape[1,2][iD], and Kai Rannenberg[1]

[1] Chair of Mobile Business & Multilateral Security,
Goethe University Frankfurt, Frankfurt, Germany
{majid.hatamian,sebastian.pape,kai.rannenberg}@m-chair.de
[2] Chair of Information Systems, University of Regensburg,
Regensburg, Germany

Abstract. Protecting enterprise's confidential data and infrastructure against adversaries and unauthorized accesses has been always challenging. This gets even more critical when it comes to smartphones due to their mobile nature which enables them to have access to a wide range of sensitive information that can be misused. The crucial questions here are: How the employees can make sure the smartphone apps that they use are trustworthy? How can the enterprises check and validate the trustworthiness of apps being used within the enterprise network? What about the security and privacy aspects? Are the confidential information such as passwords, important documents, etc. are treated safely? Are the employees' installed apps monitoring/spying the enterprise environment? To answer these questions, we propose *Enterprise Smartphone Apps Risk Assessment (ESARA)* as a novel framework to support and enable enterprises to analyze and quantify the potential privacy and security risks associated with their employees' installed apps. Given an app, *ESARA* first conducts various analyses to characterize its vulnerabilities. Afterwards, it examines the app's behavior and overall privacy and security perceptions associated with it by applying natural language processing and machine learning techniques. The experimental results using app behavior and perception analyses indicate that: (1) *ESARA* is able to examine apps' behavior for potential invasive activities; and (2) the analyzed privacy and security perceptions by *ESARA* usually reveal interesting information corresponding to apps' behavior achieved with high accuracy.

Keywords: Smartphone · App · Security · Privacy · Risk · Enterprise

1 Introduction

The amount of available apps for smartphones seems to be almost endless. The developers range from spare time developers to large companies. However, none

The original version of this chapter was revised: An acknowledgement was added. The correction to this chapter is available at https://doi.org/10.1007/978-3-030-22312-0_27

G. Dhillon et al. (Eds.): SEC 2019, IFIP AICT 562, pp. 165–179, 2019.
https://doi.org/10.1007/978-3-030-22312-0_12

of the app stores offers a dedicated security or privacy score for those apps [19]. This does not only challenge individuals but also companies. However, smartphones are often used for personal matters and official business. *Bring your own device (BYOD)* is an attractive employee IT ownership model that enables employees to bring and use their personal devices in enterprises. Such a model provides more flexibility and productivity for the employees, but may impose some serious privacy and security risks. This way not an administrator decides about the installation of apps but the user. Similar problems arise if users are allowed to install apps on the devices provided by the company. The problem raises when enterprise's confidential data is endangered as smartphones now being used to access enterprise email, calendars, apps and data. As a result, enterprises are facing the tricky task of protecting valuable data from threats such as data leakage and malware. As a consequence, it is quite challenging for enterprises to balance both their employees' needs and their security concerns. But even if the employees are not allowed to decide by themselves, then the decision would have to be made by the IT department. As a consequence, enterprises would have to provide black lists that contain apps that are not allowed to be used, or white lists that contain apps allowed for use. Grey lists may be established to list apps, where no decision was made. In any case either the IT department needs to make decisions which app belongs to which list or the employees need to make their own decisions, whether a specific app is to be used. Decisions will be made as a trade off between the necessity of the app for business purposes and the risk with regard to enterprise assets.

Our Work: In this paper, we propose *Enterprise Smartphone Apps Risk Assessment (ESARA)* as a novel framework aimed at supporting enterprises to protect their data against adversaries and unauthorized accesses. Our framework eases the process of privacy and security risk assessment for the use of smartphone apps. To achieve this goal, we propose two concepts regarding the privacy and security assessment of smartphone apps namely app *Behavior Analyzer (BA)* and app *Perception Analyzer (PA)*. We develop these two concepts along with two essential requirements namely *vulnerability checker* and *malware checker* that are cooperated with each other aiming at supporting enterprises to discriminate privacy and security misbehaviors. To the best of our knowledge, we are the first proposing the combination of these concepts and requirements that are jointly working with each other. Through experiments and implementations, we investigate how efficient and reliable the newly proposed concepts are.

Outline: The rest of this paper is organized as follows. Section 2 reviews the existing works in the area of smartphone app privacy and security preservation for general and enterprise use cases. In Sect. 3 the respective components and architecture of *ESARA* framework are presented. Section 4 elaborates on the main results obtained from the evaluation of different components of *ESARA* and highlights the key insights. Finally, we present the main conclusions in Sect. 5.

2   Related Work

In this section, we provide an overview of the relevant related work in the area of privacy and security enhancement in smartphone ecosystems and enterprise environments.

Agarwall and Hall [14] propose an approach called *ProtectMyPrivacy (PMP)* for iOS devices to detect access to private information at run-time and protect users by substituting anonymized data to be sent instead of sensitive information. Enck et al. [17] proposed *TaintDroid* for real-time tracking of information flows of smartphone apps. By focusing on personal resources, the system can reveal the manipulation or transfer of sensitive data and thus analyze the app's behavior. The monitoring procedure is based on identifying privacy-related information sources and labeling associated data. Moreover, other impacted data are tracked and identified before being transferred outside the system. The evaluation on 20 popular apps showed data leakage, e.g. phone identifier, location information and phone number being transferred to remote advertising servers. The *Apex* mechanism [24] is a name for an additional component for Android which enables users to selectively allow, deny or limit access to specific permissions requested by apps. Beresford et al. [15] propose an approach called *Mockdroid* to substitute private data with mock data when they are asked to be accessed by installed apps. *TISSA* [27] is another component for Android that enables user to choose a list of untrusted apps, and based on this list it provides mock data in place of private data at run-time. Appicaptor [12] is a framework that helps enterprises for app risk management. The goal of Appicaptor is to detect the potential privacy and security risks associated with mobile app by benefiting from static analysis of app binaries. Based on app's behavior, a ranking list is provided to classify apps into white and black lists. BizzTrust [9] is another framework that suggested the use of restricted and open areas on the employee's smartphone. Both approaches are mainly focused on security risks resulted from malicious apps.

We believe that one efficient solution should not only be focused on security behavior of mobile apps, but also privacy behavior. Importantly, consideration of users' perception about the behavior of apps plays an important role to have a more comprehensive solution. These are interesting works, but neither of them focuses on a comprehensive solution that fulfills the essential requirements of enterprise environment. In our work, we propose a solution that enhances the existing works and revamps the current enterprise app risk assessment models.

3   ESARA Framework

3.1   Goal and Requirements

Our framework makes use of different approaches from literature and combines them with our app behavior analyzer and our app perception analyzer to get a more realistic and holistic picture of installed apps. E.g., a malware checker does not detect data leakages or vulnerabilities in an app and a vulnerability scanner does not detect malicious behavior. Requirements for the development of *ESARA* were: (1) Reusing existing approaches; (2) Limiting the effort needed (since there is a large number of apps); (3) Scalability in the way that it should be easy to rely on external services and allowing several companies to share a same infrastructure; (4) Independence from app markets since even after several years none of them offers a decent security or privacy score; (5) Involving employees for feedback when using an app; (6) Involving employees for decisions.

3.2 Architecture Design

Figure 1a shows an overview of the proposed architecture for *ESARA*. As can be seen, *ESARA* consists of three main modules: employee's smartphone, server and enterprise IT department. On the employee's device an app is running which analyzes the behavior of a certain installed app and ultimately communicates the results to the employee. It also stores the employee's security and privacy perception and receives results regarding the perception analysis and risk assessment from the other two modules. The server or an outsourced service is supposed to check apps for vulnerabilities and malicious activities by running a malware and vulnerability scanner, therefore, it does not collect any data from employees. This server/service is also responsible to analyze employees' and other users' perception about security and privacy behavior of apps. If security policies are put in place, black, white and gray lists can also be stored here. The enterprise IT department takes the final decisions about which app is to place on which list – either manually or automatically by defining certain rule sets.

3.3 Components

Malware Checker. The impact of infection by a malware can be huge ranging from enterprise's infrastructure to the entire network. This component ensures the protection of enterprise's confidential data against malware. Therefore, we should not neglect the importance of this aspect of mobile apps while designing the *ESARA*'s architecture. As deploying a malware checker on resource constrained smartphones can be challenging [16], we propose the use of malware checker within the cloud as it has more computational resources. Therefore, this component is running on the server side. Checks need to be repeated with each update of the app or update of the malware checker's signature file.

Vulnerability Checker. Vulnerabilities are exploited by hackers to gain access to the device's or enterprise's resources. Statistics and observations showed that mobile platforms are among the most vulnerable operating systems in 2017 [13]. An observation by NowSecure [10] demonstrated that 25% of mobile apps have at least one high risk security vulnerability. Also, the latest security report published by Arxan [11] showed that 59% of the analyzed Android finance apps contained three OWASP mobile top 10 risks [3]. Surprisingly, all the analyzed

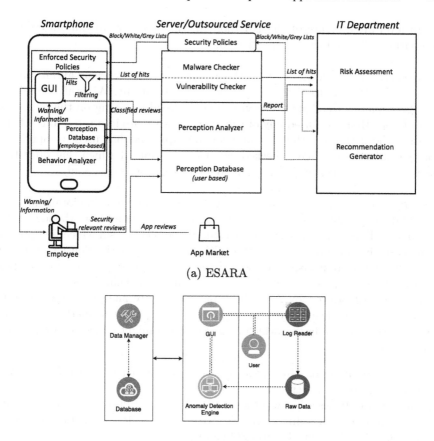

(a) ESARA

(b) Behavior analyzer

Fig. 1. High level overviews

iOS apps had at least 3 top risks. Due to such shocking statistics, an in-depth vulnerability analysis is required to investigate the potential vulnerabilities imposed by the employees' installed apps. Therefore, we also consider the importance of vulnerability analysis in *ESARA*'s architecture. Similar to the malware checker, this component is also running on the server side. There is an availability of a diverse number of vulnerability checkers both for Android and iOS that can be exploited based on the requirements [2, 4–7, 22].

Behavior Analyzer. *Behavior Analyzer (BA)* is an extension of our previous work [18,20] and a monitoring tool that analyzes the behavior of employee's installed apps. In contrast to run-time monitoring, where one could conclude what an employee was doing, we analyze the apps' behavior only by looking at the apps' permission requests. This way, the employees' privacy will be respected, while on the other hand security intrusive apps can be identified. Figure 1b shows a high level architecture of the *BA* tool. In what follows, we elaborate on the core parts of *BA* and their respective role.

Log Reader. The log reader collects the logs from `AppOpsCommand` and it sends a timer to the `PermissionUsageLogger` service periodically. When it is received, the logger queries the AppOps service that is already running on the phone for a list of apps that have used any of the operations we are interested in tracking. We then check through that list and for any app that has used an operation more recently than we have checked, we store the time at which that operation was used. These timestamps are then counted to get a usage count.

Anomaly Detection Engine. This component is supposed to behaviorally analyze the installed apps by getting help from the results obtained from the log reader component. This is done according to a rule-based mechanism which is supposed to increase the functionality and flexibility of our approach. Consequently, we have defined a set of invasive behavior detection rules that are aimed to analyze the behavior of employees' installed apps. We initially defined a set of sensitive permissions (introduced by Android[1]) and we mainly analyze the accesses to these resources. While implementing the *BA* tool, we paid special attention to the following elements to discern which resource access might be legitimate (needed by a certain app):

- Device's Orientation: This gives us information about the orientation of the device, e.g., if the screen is down or up;
- Screen State: It describes whether the device's screen is on or off at a certain time. As long as a scan is running, we register a `Receiver` for the events `ACTION_SCREEN_ON` and `ACTION_SCREEN_OFF`;
- Proximity Sensor: The screen state alone, however, is not meaningful enough, as it may happen that the screen is indeed off but certain personal resources may still be accessed (e.g., when talking on the phone, the screen turns off when the phone is approaching the ear, but access to `RECORD_AUDIO` is justified at this time). Therefore, we read the proximity sensor to indicate whether an object is within a defined range of the mobile phone;
- App State: We also consider the app state (at the time of access to a certain resource) as an important element while monitoring the apps' behavior. We distinguish the following app states: `SYSTEM_APP`, `PRE_INSTALLED_APP`, `INACTIVE`, `BACKGROUND` and `FOREGROUND`.

The *BA* tool operates like a watchdog and it only checks whether sensitive device resources are accessed. To protect the employees' privacy, the *BA* does not have the right/capability to access the sensitive data itself or track/monitor employee's activities. Furthermore, privacy controls are given to the employees to selectively choose the information that they want to share with the IT department. In particular, the IT department does not learn about all apps on the employees' device but only about those where the employees submit a report.

---

[1] https://developer.android.com/guide/topics/permissions/requesting.html.

For the sake of user interface design and risk indicator communication to the employees, we designed user interfaces for *BA* as shown in Fig. 2.

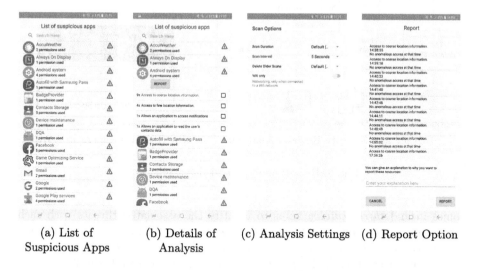

(a) List of Suspicious Apps   (b) Details of Analysis   (c) Analysis Settings   (d) Report Option

Fig. 2. The proposed GUI for the BA tool

Perception Analyzer. The *BA* tool enables employees to write (optional) reviews regarding each privacy and security invasive activity that they observe. The main goal of *Perception Analyzer (PA)* as an extension of our previous work [21] is to mine these bunch of reviews to investigate how much privacy and security relevant claims/statements can be extracted that can be ultimately used for the risk assessment component. These self-written reports are sent to the IT security department of the enterprise as well. The main idea is to not only rely on individual's report, but also to consider a high level overview of apps' real behavior. This would enable the enterprise to improve the fairness of their decisions. Additionally, we enriched the reviews of the employees with reviews from app markets (e.g. Google Play). The reviews in app markets are in general more concerned about features and performance of the apps and only little of them contain comments about security or privacy. However, since for some of the apps there are tons of reviews, even a low percentage of reviews dealing with security and privacy can be helpful. Therefore, we used a machine learning approach to find the relevant reviews. Figure 3a shows the proposed architecture for *PA*.

The *app reviews* are first pre-processed in *text pre-processing* component using typical natural language processing (NLP) techniques (e.g. tokenization, stemming and removing stop words). Further, we propose the use of *sentiment analysis* techniques to find both positive and negative reviews that talk about privacy and security aspects of apps. Afterwards, the machine learning model

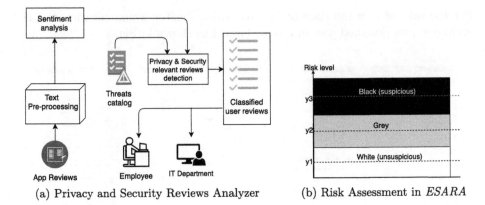

(a) Privacy and Security Reviews Analyzer   (b) Risk Assessment in *ESARA*

Fig. 3. Review analyzer and risk assessment

comes in and *threat catalog* helps to identify the associated threats with each user review by getting help from *privacy and security relevant reviews detection*. Finally the *classified reviews* are communicated to the *IT department* for risk assessment procedure. As it is obvious, *BA* is supposed to tell the IT security department how good/bad is a certain app in terms of privacy and security aspects based on its behavior in reality, and *PA* is aimed at providing a fair comparison by considering a consensus from employees and crowdsources. We detect not only a privacy and security relevant user review, but also determine the threat hidden in it. To this end, we take the most relevant threats in the context of smartphone ecosystems introduced in [21] into account. These threats are used as the input for the supervised classification algorithm as described in Table 1.

Table 1. Identified threats

#	Threat	Description
T1	Tracking & Spyware	Allows an attacker to access or infer personal data to use it for marketing purposes, such as profiling or targeted ads
T2	Phishing	An attacker collects user credentials (e.g. passwords and credit card numbers) by means of fake apps or messages that seem genuine
T3	Unauthorized charges	The hidden and unauthorized charges through registration to a premium service AND/OR installation a certain app
T4	Unintended data disclosure	Users are not always aware of all the functionality of smartphone apps. Even if they have given explicit consent, users may be unaware that an app collects and publishes personal data
T5	Targeted ads	Refers to unwanted ads and push notifications
T6	Spam	Threat of receiving unsolicited, undesired or illegal messages. Spam is considered an invasion of privacy. The receipt of spam can also be considered a violation of our right to determine for ourselves when, how, and to what extent information about us is used
T7	General	Comprises all the threats that are not categorized into other categories, e.g. permission hungry apps, general security concerns, etc.

Risk Assessment. Risk assessment examines the potential privacy and security risks associated with each employee's installed app. Therefore, it is highly dependent on the results obtained from *BA*, *PA*, *malware checker* and *vulnerability checker*. In this paper, we assume three different risk levels, including black (seems suspicious), grey (requires more investigation) and white (seems unsuspicious) as shown by Fig. 3. If a certain app does not successfully pass the investigations done by malware and vulnerability checkers, then it is automatically ranked as black and the outcome will be communicated to the employee. Otherwise, the risk assessment considers the real behavior and overall perception results in order to provide the recommendation generator with sufficient decision making information. It is worth mentioning that our main focus is on the grey risk level.

Recommendation Generator. Recommendation generator gets the input from the risk assessment component. It helps the IT security departments to better classify apps as allowed or not allowed (e.g. blacklists and whitelists). Thus, it ranks similar functionality apps, i.e. those apps that have similar functionality (e.g. weather forecasting apps, navigation apps, etc.) are assigned ranks based on the analysis done by risk assessment. Moreover, it maintains a history of privacy and security behavior records (analyses) based on apps' versions, meaning that once a certain installed app is updated, a trend containing the behavior measurements related to the current and older versions will be issued. Therefore, the IT security department can follow and analyze the trend analyses done by *ESARA* from version to version. This would enable them to analyze the behavior of current versions and compare it with older versions. This is a substantial impact of *ESARA* that is based on the fact that there is no guarantee for privacy and security friendly apps to behave nicely in the future.

4   Evaluation

Since the efficiency of malware and vulnerability scanner is a topic of its own, we do not discuss it here. However, we discuss the results of the app behavior analyzer and the app perception analyzer. For the overall evaluation, we will discuss which component covers which kind of risk.

4.1   App Behavior Analysis Results

To evaluate the applicability of *BA* and its importance in the overall performance of *ESARA*, we demonstrate some initial results regarding the behavior analysis done by *BA*. To make our scope as narrow as possible and to have a fair analysis, we mainly focused on Android apps and chose one general purpose app category. To this end, we found *Health & Fitness* as the most interesting option that is widely used by people and has raised serious privacy and security concerns [23, 26]. Therefore, we selected the top 20 apps in the *Health & Fitness* category and started the case study. We purchased six Android smartphones and installed all the 20 apps on each of them. We then used *BA* to analyze the behavior of the aforementioned apps. While we were implementing the case study, the *BA* tool was running in the background the whole time (i.e. it was monitoring apps' behavior). We ran the apps once and let them to be executed in the background. Thus, we never interacted with the mobile devices during the experiment period. This is mainly because mobile apps are task-specific and expected to only access resources when needed for their functionality. When the employee is using an app, it is harder to infer whether the app needs to have access to a certain sensitive resource as this requires to know what exactly the employee is doing which may violate his/her privacy. But when the app is not used, it is easy to detect non-security friendly sensitive resource accesses, e.g. access to enterprise's confidential data (e.g. employees' calendar, contacts, ...), since most of them will be unsolicited. Afterwards, we collected and analyzed the data generated by *BA*. In total, nine sensitive resources were accessed by the apps. The results of the analysis for each app and resources are shown in Table 2. The numbers in each cell show the number of times that each app accessed a certain resource.

Table 2. Resource access behavior pattern extracted by *BA*.

Resources	1	2	3	4	5	6	7	8	9	10	11	12	13	14	15	16	17	18	19	20
STORAGE	561	106	143	106	175	103	124	196	702	186	53	394	137	87	156	747	95	331	184	1376
CAMERA	0	0	0	0	0	0	0	25	53	86	0	0	0	0	0	0	0	15	0	14
READ_SMS	0	0	0	0	0	0	0	0	17	0	0	0	0	0	0	0	0	0	0	0
READ_CONTACTS	0	62	0	0	0	0	0	53	31	653	0	0	34	0	0	0	0	0	0	142
LOCATION	0	32	0	0	0	0	985	183	650	403	217	0	116	96	412	3780	0	670	566	1526
PHONE_STATE	0	0	0	35	0	0	284	0	534	87	0	0	0	0	0	0	0	0	0	0
MICROPHONE	0	0	0	0	0	0	0	0	0	0	0	0	0	552	0	0	34	0	0	0
GET_ACCOUNTS	0	126	0	0	0	0	93	407	279	363	0	0	0	0	0	0	0	0	101	455
BODY_SENSOR	0	1	0	0	0	0	0	0	0	0	0	0	0	0	0	0	0	0	0	0

The accesses to READ_STORAGE are not surprising because the smartphones were not completely turned off and all apps could read and write files placed on the external storage (e.g. cache files). However, five apps accessed CAMERA (apps 8, 9, 10, 18 and 20). These accesses are not privacy-friendly, since the user does not know that the app currently accesses the camera. Furthermore, READ_CONTACTS was accessed by six apps. In general, such accesses to the contacts should not be done by apps. In our case the apps are health-based, where it is not clear why they need access to the user's contacts. PHONE_STATE is an interesting data resource since the respective information is highly sensitive. This permission enables an invasive party to gain access to sensitive resources such as phone number, cellular network information, outgoing call information, etc. The only relevant reason to access this permission is to stop the app when there is an ongoing call, however, we did not use SIM card on the devices, therefore, there is no reason of such resource access. This also happened to other sensitive resources such as MICROPHONE, READ_SMS, LOCATION, etc. We also observed that many of these resource accesses happened when: (1) the devices were in horizontal orientation, (2) the devices' screen was off and (3) the proximity sensor indicated that there is no nearby object.

4.2 App Perception Analysis Results

To validate the capability of our novel perception analyzer, we collected a dataset consisting of 75,601 user reviews corresponding to these 20 health-based apps using the scraper in [1]. Three experts went manually through the data and labeled them. We then used $CountVectorizer$ and $TfidfTransformer$ packages in scikit-learn [25] for the feature extraction phase. We then split the data set into training and testing data (70% for training and 30% for testing). Using scikit-learn we exploited several classification algorithms such as *Support Vector Machines (SVMs), Random Forest, Logistic Regression (LR)*, etc. We observed *LR* outperforms others, therefore, we only show the results for *LR*. We used recall, precision and F-score metrics to evaluate the performance of the classifier. The values of these metrics show how well the classifier's results correspond to the annotated results. Table 3 shows the values for the aforementioned metrics corresponding to each identified threat. The observation is that the overall recall and precision values are of 88.95% and of 91.16%, respectively. Moreover, the values obtained for F-score show the good performance of our approach.

Table 3. Performance measures of the classification algorithm

Classes	Recall	Precision	F-score
Tracking & Spyware	0.7549	0.8311	0.8214
Phishing	0.8588	0.8653	0.8601
Unauthorized charges	0.7912	0.9583	0.8296
Unintended data disclosure	0.9010	0.9765	0.9218
Targeted ads	0.9374	0.9971	0.9663
Spam	0.9374	0.9514	0.9388
General	0.7576	0.8639	0.8492
Overall	0.8895	0.9116	0.9059

Table 4 shows some examples regarding the strength of perception analyzer in distinguishing different types of user reviews with different sentiments and relevant threat (shown by T). The obtained results clearly confirm the applicability and the positive influence of perception analyzer in the overall risk assessment done by *ESARA*.

4.3 Risk Coverage

As *ESARA* is a privacy and security risk assessment tool for mobile apps in enterprises, it is of particular importance to check its coverage of the most prevalent

Table 4. An example of classified user reviews

#	Sample user review	T
1	*You don't need to spy on my activities outside of this app. they don't care about their customers, they want to ruin the device with horrible bloatware spyware*	T1
2	*Im still getting warnings that my phone is infected with virus after i update and scan again. If its not going to work why download it. I have very limited memory to use. No need to download stupid apps that dont work*	T2
3	*Cheating Y the hell.. u cut my 50 rupees for nothing.. i just enter my card details and u cut my money without asking me.. i want it back*	T3
4	*SHit!Takes control of device.. why my photo is there??!!*	T4
5	*Ads are terrible Sorry but the ads are comparing to the website really irritating.*	T5
6	*Had this problem about these Annoying full screen PoP-ups!*	T6
7	*Dangerous! requires unnecessary access to sensitive permissions! Uninstalled*	T7

mobile app risks. We took Veracode [8] as one of the well-established references that categorizes the top 10 mobile app risks (considering the top 10 risks introduced by OWASP [3]) and we investigate the robustness of *ESARA* in assessment and detection of each individual risk. In Table 5 we clarify which component of *ESARA* may detect which identified risk. Thanks to the novel combination of *BA*, *PA*, malware and vulnerability checkers, the *ESARA*'s components totally (shown by ✓) or partially (shown by (✓)) cover all the risks. We observed that each risk is at least covered by two components (except UI impersonation which is one the most complex risk scenarios in terms of identification and mitigation).

Table 5. Coverage of Veracode top 10 mobile app [8] risks by *ESARA*.

No.	Risk	Malware checker	Vuln. checker	Behavior analyzer	Perception analyzer
1	Activity monitoring and data retrieval	✓	–	(✓)	(✓)
2	Unauthorized dialing, SMS, and payments	✓	–	✓	(✓)
3	Unauthorized network connectivity	(✓)	–	–	(✓)
4	UI Impersonation	–	–	–	(✓)
5	System modification	✓	–	–	✓
6	Logic or Time bomb	✓	(✓)	–	–
7	Sensitive data leakage	(✓)	✓	✓	✓
8	Unsafe sensitive data storage	–	✓	–	(✓)
9	Unsafe sensitive data transmission	–	✓	–	✓
10	Hardcoded password/keys	✓	✓	–	–

4.4   Discussion and Limitations

We could address all the requirements we defined in Sect. 3.1 and cover the top 10 mobile app risks with at least two components (except for UI impersonation). The results from the evaluation of the app behavior analyzer and the app perception analyzer are very promising. However, there is a limitation of our work. We have not tested our framework in a real company environment, yet. We only did user studies in a laboratory environment. Therefore, it remains to respectively show that employees would like the idea of getting support for the decisions about which apps they want to install and therefore actively make use of the potentials provided by *ESARA*.

5   Conclusion and Future Work

Smartphones have become ubiquitous within enterprise environments. At the same time, with the increased interest and not only the adoption of *BYOD*, employees heavily rely on apps, sometimes also used for personal purposes, that have access to enterprise confidential data as well. As a result, security and privacy have become a big challenge in enterprises. In this paper, we proposed *ESARA* as a novel framework to analyze and quantify the potential privacy and security risks associated with employees' smartphone apps within an enterprise environment. After an in-depth analysis of the most relevant works in the literature, we proposed an approach that leverages a four-pillar mechanism, including malware checker, vulnerability checker, behavior analyzer and perception analyzer. The combination of these mechanisms that are jointly working together supports and enables enterprises to profoundly examine the privacy and security aspects of their employees' installed apps. Since malware and vulnerability checkers are well researched, we only evaluated the performance of the two newer components, the behavior analyzer (BA) and the perception analyzer (PA). We practically showed the applicability of using behavior and perception analyses to have a more fine-grained app risk assessment and our results confirmed that these two factors play a critical role in the overall quantification of app security and privacy risks. *ESARA* opens opportunities for further innovative solutions for risk assessment of mobile apps within enterprise environments, including easing the quantification of apps trustworthiness degree.

In our future work, we will further enhance the performance of the perception analysis component by providing more training and testing data. Additionally, user studies are planned to determine the employees' and IT departments' acceptance of our approach. Also, a comprehensive analysis in an enterprise environment to validate the whole framework in a real world scenario is planned in the future.

Acknowledgment. This research was supported by the European Union's Horizon 2020 Research and Innovation program under the Marie Skłodowska-Curie "Privacy&Us" project (GA No. 675730).

References

1. Google play scraper. https://github.com/facundoolano/google-play-scraper/
2. Mobile application security scanner. https://www.ostorlab.co/
3. Mobile top 10 2016-top 10. https://www.owasp.org/index.php/mobile_top_10_ 2016-top_10/
4. Nviso. apkscan. https://apkscan.nviso.be/
5. Quick android review kit. https://github.com/linkedin/qark
6. Quixxi integrated app management system. https://quixxisecurity.com/
7. Sanddroid - an automatic android application analysis system. http://sanddroid. xjtu.edu.cn
8. Veracode mobile app top 10. http://www.veracode.com/directory/mobileapp-top-10/
9. Protection of sensitive data and services (2012). https://www.sit.fraunhofer.de/ en/bizztrust/
10. NowSecure mobile security report (2016). https://www.nowsecure.com/blog/ 2016/02/11/2016-nowsecure-mobile-security-report-now-available/
11. Arxan's 5th annual state of application security report (2016). https://www. arxan.com/press-releases/arxans-5th-annual-state-of-application-security-report-reveals-disparity-between-mobile-app-security-perception-and-reality
12. Framework for app security tests (2016). https://www.sit.fraunhofer.de/en/ appicaptor/
13. Most vulnerable os of the year 2017 (2017). https://www.cybrnow.com/10-most-vulnerable-os-of-2017/
14. Agarwal, Y., Hall, M.: Protectmyprivacy: detecting and mitigating privacy leaks on IOs devices using crowdsourcing. In: Proceedings of MobiSys, pp. 97–110 (2013)
15. Beresford, A., Rice, A., Sohan, N.: Mockdroid: trading privacy for application functionality on smartphones. In: The Proceedings of the 12th Workshop on Mobile Computing Systems and Applications, Phoenix, Arizona, USA, pp. 49–54 (2011)
16. Chandramohan, M., Tan, H.B.K.: Detection of mobile malware in the wild. Computer **45**(9), 65–71 (2012). https://doi.org/10.1109/MC.2012.36
17. Enck, W., et al.: Taintdroid: an information-flow tracking system for realtime privacy monitoring on smartphones. In: The Proceedings of the the 9th ACM USENIX Conference on Operating Systems Design and Implementation, Vancouver, BC, Canada, pp. 393–407 (2010)
18. Hatamian, M., Serna, J., Rannenberg, K., Igler, B.: Fair: fuzzy alarming index rule for privacy analysis in smartphone apps. In: The Proceedings of the 14th International Conference on Trust and Privacy in Digital Business (TrustBus), Lyon, France, pp. 3–18 (2017)
19. Hatamian, M., Serna-Olvera, J.: Beacon alarming: Informed decision-making supporter and privacy risk analyser in smartphone applications. In: Proceedings of the 35th IEEE International Conference on Consumer Electronics (ICCE), USA (2017)
20. Hatamian, M., Kitkowska, A., Korunovska, J., Kirrane, S.: "It's Shocking!": analysing the impact and reactions to the A3: Android Apps behaviour analyser. In: Kerschbaum, F., Paraboschi, S. (eds.) DBSec 2018. LNCS, vol. 10980, pp. 198–215. Springer, Cham (2018). https://doi.org/10.1007/978-3-319-95729-6_13
21. Hatamian, M., Serna, J., Rannenberg, K.: Revealing the unrevealed: mining smartphone users privacy perception on app markets. Comput. Secur. (2019). https://doi.org/10.1016/j.cose.2019.02.010. http://www.sciencedirect.com/science/ article/pii/S0167404818313051

22. Maggi, F., Valdi, A., Zanero, S.: Andrototal: a flexible, scalable toolbox and service for testing mobile malware detectors. In: Proceedings of the 3rd ACM Workshop on Security and Privacy in Smartphones and Mobile Devices, pp. 49–54 (2013)
23. Martínez-Pérez, B., De La Torre-Díez, I., López-Coronado, M.: Privacy and security in mobile health apps: a review and recommendations. J. Med. Syst. **39**(1), 1–8 (2015)
24. Nauman, M., Khan, S., Zhang, X.: Apex: extending android permission model and enforcement with user-defined runtime constraints. In: Proceedings of the 5th ACM Symposium on Information, Computer and Communications Security, pp. 328–332 (2010)
25. Pedregosa, F., et al.: Scikit-learn: machine learning in python. J. Mach. Learn. Res. **12**, 2825–2830 (2011)
26. Plachkinova, M., Andres, S., Chatterjee, S.: A taxonomy of mhealth apps - security and privacy concerns. In: 2015 48th HICSS, pp. 3187–3196, January 2015
27. Zhou, Y., Zhang, X., Jiang, X., Freech, V.W.: Taming information-stealing smartphone applications (on android). In: the Proceedings of the 4th International Conference on Trust and Trustworthy Computing, Pittsburgh, PA, USA, pp. 39–107 (2011)

SocialAuth: Designing Touch Behavioral Smartphone User Authentication Based on Social Networking Applications

Weizhi Meng[1(✉)], Wenjuan Li[1,2], Lijun Jiang[3], and Jianying Zhou[4]

[1] Department of Applied Mathematics and Computer Science,
Technical University of Denmark, Lyngby, Denmark
weme@dtu.dk
[2] Department of Computer Science, City University of Hong Kong,
Kowloon, Hong Kong
[3] CyberTree Research Institute, Hong Kong, Hong Kong
[4] Singapore University of Technology and Design, Singapore, Singapore
jianying_zhou@sutd.edu.sg

Abstract. Modern smartphones expressed an exponential growth and have become a personal assistant in people's daily lives, i.e., keeping connected with peers. Users are willing to store their personal data even sensitive information on the phones, making these devices an attractive target for cyber-criminals. Due to the limitations of traditional authentication methods like Personal Identification Number (PIN), research has been moved to the design of touch behavioral authentication on smartphones. However, how to design a robust behavioral authentication in a long-term period remains a challenge due to behavioral inconsistency. In this work, we advocate that touch gestures could become more consistent when users interact with specific applications. In this work, we focus on social networking applications and design a touch behavioral authentication scheme called *SocialAuth*. In the evaluation, we conduct a user study with 50 participants and demonstrate that touch behavioral deviation under our scheme could be significantly decreased and kept relatively stable even after a long-term period, i.e., a single SVM classifier could achieve an average error rate of about 3.1% and 3.7% before and after two weeks, respectively.

Keywords: Behavioral user authentication · Touch gestures · Usable security · Smartphone security · Social networking · Machine learning

1 Introduction

Due to the capabilities and convenience, smartphones have been widely adopted by individuals. International Data Corporation (IDC) reported that up to 344.3 million smartphones have been shipped around the world in the first quarter of

© IFIP International Federation for Information Processing 2019
Published by Springer Nature Switzerland AG 2019
G. Dhillon et al. (Eds.): SEC 2019, IFIP AICT 562, pp. 180–193, 2019.
https://doi.org/10.1007/978-3-030-22312-0_13

2017, which achieved a growth rate of around 3.4% over the last year [4]. These devices have become a personal assistant, i.e., working as a social connection and work facilitator. A survey showed that nearly 40% of respondents play with their phones for three hours or more each day [1]. As modern smartphones can work like a mini-computer, users are willing to store personal data and complete sensitive tasks on the phones [7], such as personal photos, credit card information, transactions, etc. For example, 62% of phone users in Denmark were using their phone for viewing bank account and online payment [3].

As compared with PCs or laptops, smartphones are becoming a more private device (i.e., few people would like to share their phones) [8]. For profit purposes, cyber-criminals are always trying to exploit the stored data on smartphones. As long as having a victim's phone, cyber-criminals can launch various attacks, i.e., they can steal the identity of phone users and conduct impersonation attacks to threaten the whole networks, especially online social networks. As a result, designing appropriate user authentication mechanisms becomes very important to protect phones from unauthorized access.

Most smartphones adopt traditional password-based authentication mechanisms like PINs. However, this kind of authentication is known to be insecure, i.e., passwords are easily to be stolen via "shoulder surfing" [17], smudge attacks [2] and phone charging attacks (e.g., JFC attack [12]). To address this problem, research has been focused on behavioral authentication, which uses measurements from human actions to re-authenticate a user. Behavioral authentication is believed to complement the existing authentication mechanisms. Generally, behavioral authentication needs to build a normal profile at first and then detect an anomaly by identifying any great deviations between the current profile and the pre-defined normal profile. For instance, Frank *et al.* [6] proposed a behavioral authentication scheme with 30 features, which achieved a median equal error rate of nearly 4% using an SVM classifier.

Contributions. Up to now, there are many touch behavioral authentication schemes available in the literature, but how to design a behavioral authentication scheme for a long-term period still remains a challenge. Previous work ever showed that users' touch behavioral would become more stable after more trials [11]. Motivated by this observation, we advocate that the deviation of users' touch behavior would be reduced when they played some specific tasks. In this work, we focus on social networking applications due to their frequent usage by phone users [1], and design a touch gesture-based authentication scheme called *SocialAuth*. Our contributions in this work can be summarized as below.

- We revise and design a touch gesture-based authentication scheme with 22 features to authenticate a phone user, when they are playing a social networking application. As compared with some conventional tasks, i.e., inputting a PIN code, social networking applications allow users to perform more diverse touch gestures, like touch movement and multi-touch.
- To investigate the scheme performance, we performed a user study with a total of 50 Android phone users, who were required to use the phones in the same way as they would do in their daily lives. We mainly consider two

situations for data analysis. For the first situation, our scheme analyzes all touch gestures during the phone usage, while for the second situation, our scheme only considers the touch gestures when the users were playing with social networking applications.

- Experimental results with five popular classifiers demonstrated that the deviation of users' touch actions could be reduced when phone users were interacting with social networking applications, where an SVM could achieve a better average error rate of approximately 3.1% than other classifiers. Our study also verified the authentication performance after two weeks (as long-term period), and it is found that the SVM classifier could still reach an average error rate of nearly 3.7%.

Road Map. The reminder of this paper is organized as follows. Section 2 introduces related studies on touch behavioral authentication on mobile devices. We describe the authentication scheme, touch features, data collection and session identification in Sect. 3. In Sect. 4, we present a user study with 50 participants and analyze the scheme performance like authentication accuracy and long-term performance. We conclude our work in Sect. 5.

2   Related Work

Thanks to the rapid development of smartphones, touchscreens are becoming quite common and popular. Touch dynamics has thus received more attention worldwide. Feng et al. [5] designed a touchscreen-based authentication system called *FAST*, in which users utilize a digital sensor glove for authentication. Their approach could achieve a false acceptance rate (FAR) of 4.66% and a false rejection rate (FRR) of 0.13% using a random forest classifier. Meanwhile, Meng et al. [9] developed a behavioral authentication scheme with 21 features and performed a study with 20 participants. An average error rate of nearly 3% was reported by means of a PSO-RBFN classifier. Then, Frank et al. [6] developed *Touchalytics*, a touch behavioral authentication scheme with a total of 30 touch features. In the study, their system showed a median equal error rate of nearly 4%. Based on the observations obtained in their study, they claimed that *Touchalytics* could only be deployed as an optional rather than a stand-alone authentication mechanism. Later, Sae-Bae et al. [14] focused on multi-touch behavior and proposed to authenticate a user based on up to 22 multi-touch gestures, which could be extracted from both hand and finger actions.

Recent studies started combining behavioral authentication with other biometrics. For instance, Smith-Creasey and Rajarajan [15] described an authentication scheme by combining face and touch gestures based on a dataset with 50 users, and reported an equal error rate of 3.77% with a stacked classifier. Shahzad et al. [16] proposed an authentication scheme based on users' particular behavior when they perform a touch gesture and a signature. Nguyen et al. [13] proposed an authentication scheme called *DRAW-A-PIN*, which required users to draw a PIN on touchscreen instead of typing. Their system particularly employed a Content Analyzer and a Drawing Behaviour Analyzer to identify imposters. Meng *et al.* [11] proposed *TMGuard*, a touch movement-based authentication scheme with a combination of Android unlock patterns. Their study with 75 participants demonstrated that the security of Android unlock patterns can be enhanced without degrading its usability, and that users' touch behavior can become relatively stable after more trials.

3 Touch Gesture-Based User Authentication

3.1 Authentication Architecture

To secure a smartphone from unauthorized access, an ideal touch behavioral authentication scheme has to continuously monitor the behaviors and make an alert (or lock the phone) when any anomalies are detected. The high-level architecture of touch gesture-based authentication system is presented in Fig. 1.

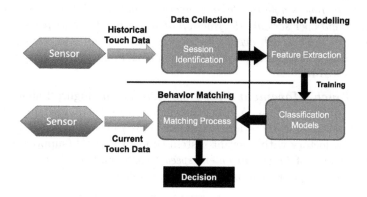

Fig. 1. The architecture of touch gesture-based authentication system.

A behavioral authentication system often contains three major phases: data collection, behavior modelling and behavior matching. The purpose of the first phase is to gather behavioral data from screen sensors and store them based on the particular session identification method. The second phase then refines the raw data and extracts features to build a normal behavioral profile for legitimate users. Various machine learning algorithms can be applied here. These two phases can help prepare the system for detecting behavioral anomalies. The last phase takes the current behavioral data from sensors and makes a decision by conducting a comparison with the pre-defined normal profile.

3.2 Touch Gesture Types and Features

Modern smartphones can provide a wide range of touch gestures, such as tap, swipe left or right, swipe up and down, and so on. Generally, these gestures on touchscreen can be categorized into the following types:

- **Single-Touch (ST):** this touch event starts with a touch-press down, and ends with a touch-press up without any touch movement in-between, like single-finger tap.
- **Touch-Movement (TM):** this touch event starts with a touch-press down, followed by a touch movement, and ends by a touch-press up, like swipe up and down.
- **Multi-Touch (MT):** this touch input starts with two or more simultaneous and distinct touch-press down events at different coordinates of a touchscreen, either with or without any touch movement before a touch press up event, like zoom, pinch and rotate.

To facilitate the comparison, in this work, we adopt and revise a touch behavioral authentication scheme on smartphones with up to 22 features, based on the work by Meng *et al.* [9]. These features can also be extracted when users interact with social networking applications, including *average touch movement speed per direction* (eight directions), *the fraction of touch movements per direction* (eight directions), *average single-touch time, average multi-touch time, the fraction of touch movements per session, the fraction of single-touch events per session,* and *the fraction of multi-touch events per session.* We further add one extra touch feature, namely touch pressure into the scheme, as many studies have proven its effectiveness [6, 14].

Average Touch Movement Speed per Direction. Figure 2 shows how to define each direction; thus, a touch movement can be divided into different features. If we assume there are two points $(x1, y1)$ and $(x2, y2)$ in a touch movement' trajectory with relevant system time $S1$ and $S2$ (suppose $S1 < S2$). Then the features of *touch movement speed* (*TMS*) and *touch movement angle* between these two points can be calculated as follows:

$$TMS = \frac{\sqrt{(x2 - x1)^2 + (y2 - y1)^2}}{S2 - S1}$$

$$Touch\ movement\ angle: \quad \theta = \arctan\frac{y2 - y1}{x2 - x1}, \theta \in [0, 360^\circ]$$

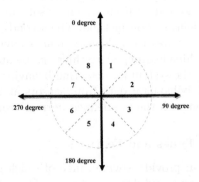

Fig. 2. Different directions for a touch action.

Let *ATMS* denote *average touch movement speed*. It is easy to calculate each feature based on the angles, i.e., *ATMS2* describes an average touch movement speed in direction 2, and *ATMS5* describes this feature in direction 5.

Fraction of Touch Movements per Direction. Intuitively, users may perform a touch movement more often in some certain directions. Therefore, the fraction of touch movements per direction varies among users and can be used for user authentication.

Average Single-Touch and Multi-touch Time. Single-touch and multi-touch are two types of touch gestures when users interact with their phones. Let *AST* denote *average single-touch time* and *MTT* denote *average multi-touch time*. The touch duration would be different between a single-touch and a multi-touch action.

Fraction of Touch Action Events. It is observed that users could have their own habit when interacting with the phone. For instance, some users would like to use single-touch more often than multi-touch, while some may prefer using multi-touch actions more, i.e., during web browsing. As a result, the fraction of touch events can be used for authenticating users. Three relevant features can be derived: *the fraction of touch movements per session* (denoted *FTM*), *the fraction of single-touch events per session* (denoted *FSTE*), and *the fraction of multi-touch events per session* (denoted *FMTE*).

Touch Pressure. With the development of modern smartphones, sensors are becoming more accurate and sensitive. In this case, average touch pressure (denoted *ATP*) has become one of the promising features for validating users. It is worth noting that all these features would be validated in Sect. 4.

3.3 Data Collection

Similar to [9, 10], we employ an Android phone - Google/HTC *Nexus One* for data collection, which has a capacitive touchscreen of 480×800 px. This type of phone is selected because its OS can be replaced with a modified OS version. In this work, we updated the phone with a modified Android OS version 2.2 based on *CyanogenMod*[1]. The changes were mostly on its application framework layer by inserting system level command to record raw data from the touchscreen, such as the timing of touch inputs, the coordinates x and y, and the touch pressure and various gestures like single-touch, multi-touch and touch movement.[2] A separate logcat application was installed to help extract and record the captured data from the phone.

A sample of collected raw data from the phone is depicted in Table 1. Each record contains five major items: *input type, x-coordinate, y-coordinate, touch*

---

[1] http://www.cyanogenmod.com/.

[2] We inserted *Slog.v* command to two java source files (*InputDevice.java* and *KeyInputQueue.java*) regarding the *Application framework layer*, and then recompiled the whole source codes of *Froyo* operating system to generate our demanded experimental platform.

Table 1. A sample of raw data collected from touchscreen on the Android platform.

Input type	X-coordinate	Y-coordinate	Touch pressure	Time (ms)
Press down	478.5686	658.6726	0.090196080	1870785
Press move	473.5593	660.5503	0.101960786	1870807
Press move	471.2780	660.9001	0.101960786	1870814
Press move	468.7645	662.0188	0.125686300	1870852
Press move	470.5872	660.5211	0.125686300	1870898
Press move	472.8723	658.5432	0.125686300	1870910
Press up	470.6778	660.6223	0.125686300	1870933

pressure, and *system time (S-time)*. The system time is relevant to the last start-up of the phone and is managed by the phone itself, while the duration of each touch gesture can be computed by measuring the difference in system-time between touch press down and up. As a complementary item to the system time, the deployed logcat application can record regular timing information (e.g., 06–29 22:08:48.080) for later potential data verification. This kind of data collection does not need any special hardware on phone's side. It is worth noting that additional information can be collected by updating certain parts of the Android application framework.

3.4 Session Identification

To build a behavioral profile, session identification is an important factor that could affect authentication performance. The purpose of session identification is to help decide the length of a session. To ensure the collection of enough touch gestures, in this work, we adopted an event-based session identification includes a total of 120 touch gestures in each session [10]. A session ends if the number of touch gestures reached the pre-define value and then a new session starts. For implementation, session start and end can be easily determined by checking the raw data record.

4   User Study

4.1   Study Methodology

In the study, we recruited a total of 50 regular Android phone users (including 26 female and 24 male), who were aged from 18 to 61 years. Participants have a diverse background including students, senior citizens, researchers and business people. Table 2 details the background information of participants.

During the study, each participant was provided with an Android phone (a Google/HTC Nexus One) equipped with our modified OS version. The main purpose is to ensure that all data were collected under the same settings. Before the study, we described our research objective to all participants, introduced how

Table 2. Background of participants in the user study.

Occupation	Male	Female	Age	Male	Female
Students	14	16	18–30	14	16
Business people	2	3	31–40	5	5
Researchers	7	5	40–50	2	3
Senior citizen	1	2	Above	3	2

to perform data collection, and explained what kind of data would be collected, i.e., we emphasized that no personal data would be collected during the study. Further, we seek approval from each participant for gathering and analyzing the data, before they started the experiment.

More specifically, all participants were required to use the Android phones freely as the same way they would use the phones in their daily lives. By considering the limitations of a lab study, we allowed participants to do the actual data collection out of the lab, motivating them to have enough time to get familiar with the phone. They could decide when to start the collection process, according to our provided manual with detailed steps and explanations.

In this study, we mainly consider two situations for data analysis. For the first situation ($S1$), our scheme analyzes all recorded touch gestures when participants use the phones, whereas for the second situation ($S2$), our scheme only considers the touch gestures when participants play with any social networking applications. Each participant was required to complete 15 sessions for each situation (each session contains 120 touch gesture events) within 3 days. As a result, we could collect up to 1500 sessions of raw data, that is, 750 sessions for each situation. All participants could get a $20 gift card.

4.2 Machine Learning Classifiers and Metrics

As a study, we mainly employed five popular machine learning classifiers in the comparison: namely, Decision tree (J48), Naive Bayes, Radial Basis Function Network (RBFN), Back Propagation Neural Network (BPNN) and Support Vector Machine (SVM). To avoid any unexpected implementation bias, we extracted the above classifiers from WEKA [18] (using default settings), which is an opensource collection of machine learning algorithms.

Intuitively, a machine learning classifier is expected to achieve high classification accuracy. However, it is not easy in practice due to the behavioral dynamics. There is a need to balance false acceptance rate and false rejection rate in real-world applications. In practice, a desirable user authentication system is expected to achieve both a low FAR and FRR.

- False Acceptance Rate (FAR): indicates the probability that an impostor is categorized as a legitimate user.
- False Rejection Rate (FRR): indicates the probability that a legitimate user is classified as an intruder.

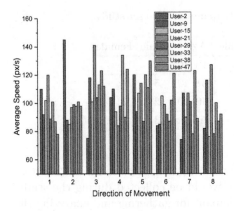

Fig. 3. The average touch movement speed per direction for eight different users.

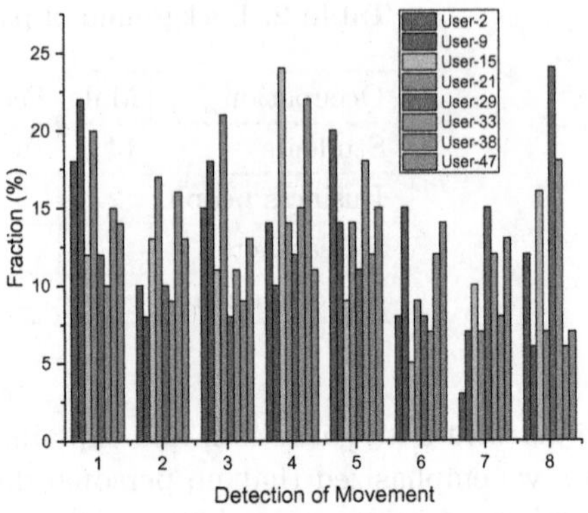

Fig. 4. The fraction of touch movements per direction for eight different users.

4.3 Result Analysis

As stated above, our authentication scheme is comprised of 22 touch features such as *ATMS1, ATMS2, ATMS3, ATMS4, ATMS5, ATMS6, ATMS7, ATMS8, FTM1, FTM2, FTM3, FTM4, FTM5, FTM6, FTM7, FTM8, AST, MTT, FTM, FSTE, FMTE* and *ATP*. In this part, we analyze the collected data regarding the effectiveness of features, touch behavioral deviation between two groups, authentication accuracy, and long-term performance after two weeks.

The Effectiveness of Features. Based on the collected 1500 sessions of touch gesture events, we calculate the touch features for each participant and randomly present 1/3 participants (about *eight* individuals) to validate the effectiveness of each feature in distinguishing users.

Figure 3 describes the average touch movement speed for different directions. It is found that the distributions varied with different users. For example, User-2 performed a higher speed in direction 1, 2 and 4; User-9 performed a higher speed in direction 3, 4 and 8; User-15 performed a higher speed in direction 1, 3, 5 and 6; User-21 achieved a higher speed in direction 1, 3, 5 and 8; and User-29 achieved a higher speed in direction 2, 3 and 7. The results prove that the use of *ATMS* per direction could help distinguish different phone users.

The fraction of touch movements for different directions is shown in Fig. 4. It is observed that User-2 conducted relatively more touch movements in direction 1, 3 and 5; User-9 performed more touch movements in direction 1, 3, 5, and 6; User-15 achieved a higher rate in direction 1, 2, 4, and 8; User-21 performed more touch movements in direction 1, 2 and 3; and User-29 had a higher rate in direction 1, 4, 7 and 8. The results validate that *FTM* in different directions can be used to characterize a user's touch behavior.

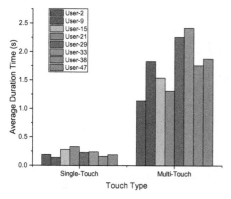
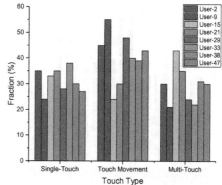

Fig. 5. The average duration time regarding single-touch and multi-touch for eight different users.

Fig. 6. The fraction of single-touch, touch movement and multi-touch for eight different users.

Figure 5 presents the average duration time regarding single-touch and multi-touch action. It is visible that time consumption could vary with different users. For single-touch action, User-21 consumed more time than the others while User-9 could finish the gesture with the minimum time among these users. For multi-touch action, User-33 and User-2 required the longest and the shortest time to finish the action. Based on our data, there is no direct relationship identified between single-touch and multi-touch. In this case, these features can be used to distinguish different users.

Figure 6 describes the fraction of single-touch, touch movement and multi-touch for eight users. It is found that User-2, User-21 and User-33 performed more single-touch actions than others. For touch movement, User-9 achieved a much higher rate than others, whereas User-15 achieved a higher rate than others regarding multi-touch. These results prove that these features can be used to model phone users' touch habits. It is similar to the feature of average touch pressure (ATP), it is found that User-9 achieved the biggest touch pressure of 2.012, while User-33 had the smallest touch pressure of 0.8892. The values of ATP for other users mainly ranged from 1 to 2.

Overall, our data analysis validates that our adopted 22 features could be effective in distinguishing phone users. This observation is in-line with the results in many previous studies like [6,9].

Touch Behavioral Deviation. Under *S1*, we considered all touch behavioral events when participants used the phone, while under *S2*, we only considered the touch gestures when they were using social networking applications. A total of four social networking applications were selected in the study: WeChat, Facebook, Twitter and Instagram. Our major purpose is to investigate the touch behavioral deviation between the two situations. Intuitively, a smaller deviation is desirable, indicating that users' touch actions are more stable. Figure 7

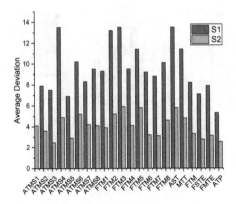 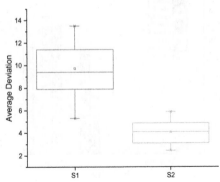

Fig. 7. The average behavioral deviation regarding all features under two situations.

Fig. 8. The distribution of average deviation under two situations.

depicts the average behavioral deviation regarding all features and Fig. 8 shows the distribution of behavioral deviation under two situations.

It is visible that the average deviation for each touch feature under *S2* is much smaller than that under *S1*. For example, participants under *S1* made a deviation above 10 for ATMS1, ATMS4, ATMS6, FTM2, FTM3, FTM5, FTM8, AST and MTT, while the corresponding deviation was only ranged from 4.1 to 5.2 under *S2*. Figure 8 indicates that the deviations made under *S2* are mostly half or less than those made under *S1*. Intuitively, a higher deviation means that participants' touch gestures are more unstable, which may increase the difficulty of behavioral modelling. In contrast, a smaller deviation makes it easier to build a robust touch behavioral authentication scheme.

Further, we informally interviewed all the participants about their habits of phone usage. Based on their feedback, most participants reflected that their touch behavior would be quite dynamic when they freely used the phone without a task, whereas their touch actions would become focused when they were using a particular application, like social networking application. The feedback validated the observation that users' touch actions could become relatively stable under certain scenarios.

Authentication Accuracy. To investigate the authentication performance, we applied 18 sessions (up to 60% of the total sessions) as training data to help each classifier build a touch behavioral profile for each participant. Then we used the remaining sessions for testing. The test was run in 10-fold mode provided by the WEKA platform. The false acceptance rate (*FAR*), false rejection rate (*FRR*), and average error rate (*AER*) are presented in Table 3.

It is found that under *S1*, the single classifier of SVM could reach a better error rate than other classifiers, i.e., SVM achieved an AER of 6.02% while the others could only reach a rate of nearly 10%. Under *S2*, it is visible that

Table 3. Authentication performance for different classifiers under two situations.

	J48	NBayes	RBFN	BPNN	SVM
S1					
FAR (%)	22.55	18.66	9.72	9.12	5.22
FRR (%)	23.78	20.73	10.45	10.34	6.82
AER (%)	23.17	19.70	10.09	9.73	6.02
S2					
FAR (%)	15.13	11.56	6.88	6.42	2.89
FRR (%)	16.55	13.23	7.11	7.88	3.24
AER (%)	15.84	12.40	7.00	7.15	3.07

the performance was much better than that under *S1*. For example, SVM still achieved the best performance among single classifiers, but could offer an AER of 3.07% under *S2* vs. 6.02% under *S1*. For other classifiers like J48 and NBayes, their AER could be reduced by around 7% under *S2*.

Overall, these results demonstrate that with a smaller deviation, it is easier for a classifier to model phone users' touch behavior and to provide desirable authentication accuracy. In addition, users' touch behavior can become relatively stable under our scheme of *SocialAuth*, when they play with certain phone applications like a social networking application, as compared to the situation by considering all touches during the phone usage.

Long-Term Authentication. In the study, up to 16 participants (seven males) chosen to attend our task on long-term authentication, in which they could keep using our provided phone and returned to our lab after two weeks. They then required to complete 5 sessions for each *S1* and *S2* within two days. After the experiment, they could get a $30 gift card.

Our goal is to investigate the behavioral deviation after two weeks. Similarly, Figs. 9 and Fig. 10 shows the average behavioral deviation regarding all features and the distribution of behavioral deviation after two weeks, respectively. Encouragingly, it is found that after two weeks, the behavioral deviation under *S2* is much smaller than those under *S1*, i.e., some features' deviations are smaller than 2. In other words, users' touch gestures were much more stable under *S2* than those under *S1*.

For authentication accuracy, we applied the same five classifiers on the new sessions without re-training. That is, we used the already built behavioral model (before two weeks) for each classifier. It is found that SVM still could achieve a smaller AER under two situations, but the rate is much different, i.e., it reached a rate of 3.68% and 9.82% under *S2* and *S1*, respectively. The results validated that users' touch behavior could become relatively stable when they play with social networking applications, making it easier to build a robust authentication scheme for a long-term period.

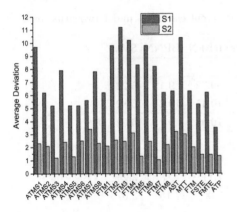

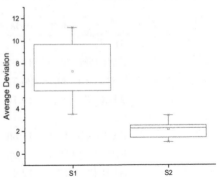

Fig. 9. The average behavioral deviation regarding all features after two weeks.

Fig. 10. The distribution of average deviation after two weeks.

5   Conclusion

How to design a robust scheme for a long-term period remains a challenge due to the inconsistent behavior. In this work, we advocate that users' touch behavior would become relatively stable when they interact with particular applications, and design a touch behavioral authentication scheme called *SocialAuth*. In the evaluation, we conducted a user study with 50 common Android phone users and considered two major situations for data analysis. We consider all recorded touch behavioral events under the first situation, whereas only consider the touch gestures when they use social networking applications under the second situation. It is found that an SVM classifier could reach an average error rate of 3.07% and 3.68% (before and after two weeks) under the second situation, versus a rate of 6.02% and 9.82% (before and after two weeks) under the first situation. The results demonstrated that with our scheme, users could achieve a much smaller behavioral deviation (more stable behavior) even after two weeks.

Acknowledgments. We would like to thank all anonymous reviewers for their helpful comments, and Jianying Zhou was supported by SUTD start-up research grant SRG-ISTD-2017-124.

References

1. Andrews, K.: Smartphone survey: the fascinating differences in the way we use our phones, 13 October 2017. http://www.abc.net.au/news/science/2017-10-13/smartphone-survey-results-show-fascinating-differences-in-usage/9042184
2. Aviv, A.J., Gibson, K., Mossop, E., Blaze, M., Smith, J.M.: Smudge attacks on smartphone touch screens. In: Proceedings of the 4th USENIX Conference on Offensive Technologies (WOOT), pp. 1–10 (2010)

3. Global Mobile Consumer Survey 2017 - Luxembourg, Accessed 12 Dec 2017. www. deloitte.com/lu/mobilesurvey
4. IDC: Smartphone vendor market share (2017). https://www.idc.com/promo/smartphone-market-share/vendor
5. Feng, T., et al.: Continuous mobile authentication using touchscreen gestures. In: Proceedings of the 2012 IEEE Conference on Technologies for Homeland Security (HST), pp. 451–456 (2012)
6. Frank, M., Biedert, R., Ma, E., Martinovic, I., Song, D.: Touchalytics: on the applicability of touchscreen input as a behavioral biometric for continuous authentication. IEEE Trans. Inf. Forensics Secur. 8(1), 136–148 (2013)
7. Karlson, A.K., Brush, A.B., Schechter, S.: Can i borrow your phone? Understanding concerns when sharing mobile phones. In: Proceedings of the 27th CHI, pp. 1647–1650 (2009)
8. Li, L., Zhao, X., Xue, G.: Unobservable re-authentication for smartphones. In: Proceedings of NDSS (2013)
9. Meng, Y., Wong, D.S., Schlegel, R., Kwok, L.: Touch gestures based biometric authentication scheme for touchscreen mobile phones. In: Kutyłowski, M., Yung, M. (eds.) Inscrypt 2012. LNCS, vol. 7763, pp. 331–350. Springer, Heidelberg (2013). https://doi.org/10.1007/978-3-642-38519-3_21
10. Meng, Y., Wong, D.S., Kwok, L.-F.: Design of touch dynamics based user authentication with an adaptive mechanism on mobile phones. In: Proceedings of the ACM Symposium on Applied Computing (SAC), pp. 1680–1687 (2014)
11. Meng, W., Li, W., Wong, D.S., Zhou, J.: TMGuard: a touch movement-based security mechanism for screen unlock patterns on smartphones. In: Proceedings of the 14th International Conference on Applied Cryptography and Network Security (ACNS), pp. 629–647 (2016)
12. Meng, W., Fei, F., Li, W., Au, M.H.: Harvesting smartphone privacy through enhanced juice filming charging attacks. In: Nguyen, P., Zhou, J. (eds.) ISC 2017. LNCS, vol. 10599, pp. 291–308. Springer, Cham (2017). https://doi.org/10.1007/978-3-319-69659-1_16
13. Nguyen, T.V., Sae-Bae, N., Memon, N.: DRAW-A-PIN: authentication using finger-drawn PIN on touch devices. Comput. Secur. 66, 115–128 (2017)
14. Sae-Bae, N., Memon, N., Isbister, K., Ahmed, K.: Multitouch gesture-based authentication. IEEE Trans. Inf. Forensics Secur. 9(4), 568–582 (2014)
15. Smith-Creasey, M., Rajarajan, M.: A continuous user authentication scheme for mobile devices. In: Proceedings of the 14th Annual Conference on Privacy, Security and Trust (PST), pp. 104–113 (2016)
16. Shahzad, M., Liu, A.X., Samuel, A.: Behavior based human authentication on touch screen devices using gestures and signatures. IEEE Trans. Mob. Comput. 1(10), 2726–2741 (2017)
17. Tari, F., Ozok, A.A., Holden, S.H.: A comparison of perceived and real shoulder-surfing risks between alphanumeric and graphical passwords. In: Proceedings of the 2nd Symposium on Usable Privacy and Security (SOUPS), pp. 56–66 (2006). ACM, New York
18. Data Mining Software in Java: WEKA-waikato environment for knowledge analysis. http://www.cs.waikato.ac.nz/ml/weka/

The Influence of Organizational, Social and Personal Factors on Cybersecurity Awareness and Behavior of Home Computer Users

Joëlle Simonet and Stephanie Teufel[✉]

International Institute of Management in Technology (IIMT),
University of Fribourg, 1700 Fribourg, Switzerland
{joelle.simonet,stephanie.teufel}@unifr.ch

Abstract. With the increased use of computers and network systems in a time of digitalization, the digital connectedness frames our daily life at work and at home. To ensure secure systems, all computer users should safely interact with these systems. Prior research indicates insufficient cybersecurity awareness of home computer users who are also difficult to reach as they are not necessarily part of organizational structures. This study therefore investigates organizational, social and personal determinants of an individual's cybersecurity awareness and its influence on cybersecurity behavior in the home environment, using partial least squares structural equation modeling based on survey data. The results show a low influence of the workplace and weak social influences, while the study confirms a significant effect of personal initiative and a strong effect of information systems knowledge on an individual's cybersecurity awareness. The results suggest that security strategies aimed at the general public should focus on improving the knowledge and understanding instead of making fear. The study provides valuable insights about cybersecurity awareness and its determinants contributing to the field of research. The findings can be used for reviewing cybersecurity strategies.

Keywords: Cybersecurity awareness · Cybersecurity behavior ·
Home computer user

1 Introduction

The trends of digitalization and increased interconnectedness have reached most areas of the daily life. These trends enhance the risk of cyber threats such as cybercrime or system failure. Computer users' interactions with such systems are critical to preserving a safe cyber environment as many of them do not possess a deep understanding of computers or cyber threats [2,27,31,39]. While the

© IFIP International Federation for Information Processing 2019
Published by Springer Nature Switzerland AG 2019
G. Dhillon et al. (Eds.): SEC 2019, IFIP AICT 562, pp. 194–208, 2019.
https://doi.org/10.1007/978-3-030-22312-0_14

management of cybersecurity has started in businesses, home users are often not aware of their responsibility [14,41]. Additionally, it demands the user's initiative to act secure [2,29]. The loosening of structures and decentralization, for example in smart grids [21] or with mobile working [7], call for safe user behavior in the home environment. Many studies have been conducted in the work context [7,12,18,25,33,38], the home user though has not received the same attention in research. Therefore, the present study aims to understand what factors influence a user's decision making for a safe (or unsafe) cyber behavior and to grasp what sources of information impact a computer user's cybersecurity awareness in his home environment.

The next section presents the theoretical background and the research model developed. In Sect. 3, the methodology used is presented, while the results of the analysis are listed in Sect. 4. A discussion of the findings (Sect. 5) and a conclusion (Sect. 6) round off this paper.

2   Theoretical Background and Research Model

A home computer user can learn about cybersecurity from various sources. The workplace of a user can function as a source by providing knowledge that the user might transfer to his home environment. Many organizations distribute security policies [18,25] or provide security training and awareness programs [12,18,41] explaining the correct use and interaction with computers and systems connected to the Web covering topics such as password management or phishing. Two streams differentiate how such security measures are implemented. While some authors suggest following the deterrence approach by creating fear-based campaigns [12,22], other researchers call for skills-based measures [18,24,39]. An all-encompassing approach towards cybersecurity in organizations is the promotion of an information security culture. According to [37], an information security culture should change employees' values in order to promote an intrinsic motivation for safe cyber behavior. As an intangible concept, information security culture has an impact on security awareness [34] but is in return nurtured through awareness [11,38]. In line with this, it is assumed that policy provision, security training and an information security culture at the workplace influence the cybersecurity awareness of a home computer user. The corresponding hypotheses are:

H1a: *Information Security Policy Provision (ISPP)* in the individual's workplace is positively related to the individual's *Cybersecurity Awareness (CSA)*.

H1b: *Security Training and Awareness Programs (SETA)* in the individual's workplace is positively related to the individual's *Cybersecurity Awareness*.

H1c: *Information Security Culture (ISC)* in the individual's workplace is positively related to the individual's *Cybersecurity Awareness*.

More informal determinants of a user's cybersecurity awareness can be found in the social environment of a home computer user. Especially since consequences of cybersecurity incidents are not always visible directly or at anytime, stories told by friends and family can act as vicarious examples and enhance a social learning process for cybersecurity issues [5, 23, 31].

International guidelines such as from the OECD [30] or national cybersecurity strategies target the society and thus include home computer users. It remains difficult though to reach out to those who are in loosely coupled structures [43]. Information provided by the public administration or reports distributed by the mass media can highlight the importance of cybersecurity and deliver security advice [18, 29, 31].

H2a: *Family and Friends Influence (FFI)* is positively related to the individual's *Cybersecurity Awareness*.

H2b: *Mass Media Influence (MMI)* is positively related to the individual's *Cybersecurity Awareness*.

H2c: *Public Administration Information (PAI)* is positively related to the individual's *Cybersecurity Awareness*.

Compared to the work environment, a home user is required to be self-initiative to learn about cybersecurity topics and take security-enhancing actions [2, 29, 39]. Being someone generally showing personal initiative is thus assumed to have a positive effect on awareness. In this context, having previous information systems knowledge is expected to be a strong determinant of cybersecurity awareness [14, 18, 31].

H3a: *Personal Initiative (PI)* is positively related to the individual's *Cybersecurity Awareness*.

H3b: *Information Systems Knowledge (ISK)* is positively related to the individual's *Cybersecurity Awareness*.

Understanding an individual's behavior or the factors influencing a decision to act are hard to grasp. Behavioral models such as the *Theory of Planned Behavior* [1] or the *Protection Motivation Theory* [35, 36] have been developed to investigate the cognitive processes involved. In the *Protection Motivation Theory*, the threat appraisal and the coping appraisal represent the two sides of the mediating process of an individual's intention to protect something (or someone) from a threat [36, 44]. Originally used for investigating health-related fears, the components of the model have been used to study cybersecurity fears many times [2, 17, 28, 39, 45]. Perceived vulnerability and perceived severity are elements of the threat appraisal, while perceived self-efficacy, perceived response efficacy and perceived costs constitute the coping appraisal. In the context of cybersecurity, the elements represent the understanding of a threat and the mental process an individual goes through before deciding to behave securely or not and therefore represent the construct of cybersecurity awareness (H5a-e).

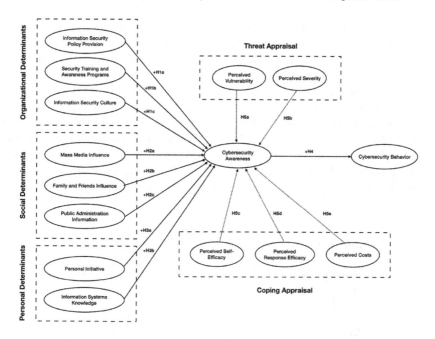

Fig. 1. Research model

H4: *Cybersecurity Awareness* is positively related to the individual's *Cybersecurity Behavior (CSB)*.

Figure 1 shows the research model summarizing the organizational, social and personal determinants, the multi-dimensional construct of cybersecurity awareness and cybersecurity behavior (for details, see [40]).

3   Methodology

This study draws on common methods in the domain of cybersecurity awareness and behavior research [2,18,29,39]. For the data collection, a survey was conducted, while the data analysis was performed with partial least squares structural equation modeling. Details about the analysis are given in Sect. 4.

The data collection process encompassed a self-report online questionnaire that was implemented via SoSci Survey, a Germany-based web tool for conducting online questionnaires [26]. Although using self-reported data can provoke a social desirability bias [8,9], it allows to capture the respondents' cognitive process [4] which was essential for this study. By allowing an anonymous completion of the questionnaire for which the respondent was not required to leave his familiar environment, the risk for social desirability bias was reduced [8].

The survey was sent out to employees of various organizations in Switzerland. This mode of distribution was chosen to ensure that participants work, which was necessary for being able to investigate the influence of the workplace. Organizations contacted are located in the French-, German- and Italian-speaking parts of Switzerland and are active in areas such as educational, health and social, IT-related businesses or public transport. The questionnaire was made available in German, English and French.

After a data collection of about five weeks, a total number of 562 participants started the questionnaire. Removing the unfinished cases and the records with more than 15% of missing data, suspicious response patterns and outliers as suggested by [16] results in 456 cases used for further analysis. Mean-value replacement is applied for the remaining missing data. Table 1 shows some demographic characteristics of the participants. A more detailed discussion of the sample can be found in [40].

Table 1. Demographic characteristics of the participants

Demographics	n	%		n	%
Sex			*Total*	*456*	*100*
Female	224	49.1			
Male	227	49.8	**Linguistic Region**		
Age			German-speaking	116	25.4
under 18	5	1.1	French-speaking	327	71.7
18-24	34	7.5	Italian-speaking	6	1.3
25-34	131	28.7	other / not CH	7	1.5
35-44	101	22.1	**Sector**		
45-54	109	23.9	Public Sector	410	89.9
55-64	72	15.8	Private Sector	25	5.5
over 64	3	0.7	Voluntary Sector	15	3.3

The survey is organized in eight sections covering the personal, social and organizational determinants, the variables constituting cybersecurity awareness, the construct of cybersecurity behavior as well as additional demographic questions such as age, gender or language region. This results in a total number of 53 items, all constructed as closed questions, corresponding to statements to which respondents indicate their level of agreement on a 5-point Likert scale. For three items, different minimum and maximum values are used.

The measures for the fourteen constructs are all adapted from previously validated constructs. A pretest was conducted to ensure the comprehensibility of the questionnaire. Some items were reworded and others exchanged before being subject of a second pretest. A general approach on cybersecurity actions was chosen to get an all-encompassing point of view and to avoid technology dependency and thus facilitate the repeatability of the study.

Information Security Culture, Friends and Family Influence, Information Systems Knowledge are considered reflective. The remaining constructs, *Information Security Policy Provision, Security Training and Awareness Programs,*

Mass Media Influence, Public Administration Information, Personal Initiative as well as *Cybersecurity Behavior* are considered formative. *Cybersecurity Awareness* is constructed as a reflective-formative second-order construct composed of the first-order constructs *Perceived Vulnerability, Perceived Severity, Perceived Self-Efficacy, Perceived Response Efficacy* and *Perceived Costs*. All constructs and the corresponding items in their final version can be found in Table 5 in the Appendix.

4   Analysis and Results

The model was analyzed with partial least squares structural equation modeling using the software SmartPLS 3.2.5 [32]. The analysis encompasses a first step of assessing the measurement models and a second step of evaluating the structural model. The analysis was conducted by following the guidelines proposed by [15] and [16]. For significance testing, 5000 bootstrap samples were used. Additionally, a mediation analysis was performed to investigate the awareness' mediating role.

4.1   Measurement Model Assessment

The proposed research model includes formative and reflective constructs, which require a different assessment. For reflective constructs, internal consistency and indicator reliability and the average variance extracted (AVE) are used to verify convergent validity. A composite reliability (CR) value between 0.7 and 0.9 indicates internal consistency reliability, higher values suggest high item similarity [16]. For the AVE, values above 0.5 are desired. Indicator reliability is assessed with the outer loading (OL) of the items, which indicate the strength of the path and should exhibit values above 0.7. Items with an outer loading between 0.4 and 0.7 can be kept in the model, while indicators with a lower loading should be removed [16]. All values are in the accepted ranges, except for *Information Systems Knowledge* that exhibits values above the desired values (see Table 2). Discriminant validity is assessed with the HTMT criterion as suggested by [15,19]. As all values are below 0.85, discriminant validity is established between all latent variables (see Table 3)

Formative constructs are assessed by looking at the variance inflation factor (VIF) for collinearity issues and the relative importance of each indicator. VIF values should be below five for all indicators. The items should exhibit significant outer weight or, if not, manifest outer loadings above 0.5. Indicators should be removed if neither the outer loading nor the outer weight is significant. The values are shown in Table 2. The items MMI3 and PAI3 were removed for further analyses. The reflective-formative second-order construct *Cybersecurity Awareness* is evaluated in the same manner.

Table 2. Reflective and formative constructs

Reflective				Formative		Outer weight			Outer loading		
Item	OL	CR	AVE	Item	VIF						
	l					w	t	p	l	t	p
ISC1	0.728	0.799	0.507	ISPP1	2.187	0.258	1.288	0.198	0.843*	8.413	0.000
ISC2	0.476			ISPP2	2.187	0.794*	3.586	0.000	0.985*	26.356	0.000
ISC3	0.77			SETA1	1.871	0.722*	2.081	0.037	0.965*	7.834	0.000
ISC4	0.825			SETA2	1.871	0.356	1.154	0.249	0.849*	4.508	0.000
FFI1	0.758	0.895	0.682	MMI1	1.503	0.344	1.278	0.201	0.732*	3.337	0.001
FFI2	0.853			MMI2	1.791	0.885*	2.786	0.005	0.863*	4.706	0.000
FFI3	0.821			MMI3^r	1.280	−0.471	1.889	0.059	0.033	0.201	0.841
FFI4	0.867			PAI1	1.752	0.914*	4.790	0.000	0.976*	18.604	0.000
ISK1	0.954	0.955	0.914	PAI2	2.230	0.202	0.992	0.321	0.665*	4.350	0.000
ISK2	0.957			PAI3^r	1.390	−0.251	1.555	0.120	0.105	0.883	0.377
PV1	0.83	0.874	0.698	PI1	1.826	0.353*	2.040	0.041	0.782*	8.640	0.000
PV2	0.833			PI2	1.891	0.538*	2.866	0.004	0.833*	10.476	0.000
PV3	0.844			PI3	1.629	0.460*	2.798	0.005	0.775*	8.175	0.000
PS1	0.708	0.864	0.682	PI4	1.366	−0.376*	2.487	0.013	0.214	1.696	0.090
PS2	0.929			PV	1.069	−0.007	0.243	0.808	−0.136*	2.034	0.042
PS3	0.825			PS	1.060	0.047	1.155	0.248	0.161*	2.217	0.027
PSE1	0.817	0.899	0.691	PSE	1.645	0.747*	12.019	0.000	0.952*	49.290	0.000
PSE2	0.833			PRE	1.305	0.310*	5.053	0.000	0.674	12.690	0.000
PSE3	0.852			PC	1.355	−0.125*	2.063	0.039	−0.570	10.116	0.000
PSE4	0.822			CSB1	1.312	0.225*	3.003	0.003	0.564*	7.955	0.000
PRE1	0.566	0.825	0.546	CSB2	1.396	0.207*	2.780	0.005	0.594*	9.237	0.000
PRE2	0.742			CSB3	1.310	0.507*	7.453	0.000	0.811*	19.250	0.000
PRE3	0.819			CSB4	1.373	0.160*	2.091	0.037	0.595*	9.044	0.000
PRE4	0.801			CSB5	1.137	0.070	1.246	0.213	0.257*	3.041	0.002
PC1	0.808	0.898	0.689	CSB6	1.108	0.019	0.458	0.647	0.217*	2.987	0.003
PC2	0.846			CSB7	1.063	0.104	1.645	0.100	0.214*	2.618	0.009
PC3	0.894			CSB8	1.275	0.249*	3.254	0.001	0.62*	9.484	0.000
PC4	0.768			CSB9	1.124	0.136*	2.079	0.038	0.329*	4.392	0.000

*Notes: w = weight, l = loading, t = t-value, p = p-value, *$p < 0.5$, ^{r} excluded item*

4.2 Structural Model Assessment

The structural model should exhibit no collinearity issues, indicated with VIF values below five, which is the case for all latent variables in the model. Estimated path coefficients that take on values between −1 and +1 indicate positive and negative effects one latent construct has on another. In the proposed model, except for the paths from *Information Security Culture* and *Security Training and Awareness Programs* to *Cybersecurity Awareness*, all coefficients are significant but exhibit great differences in strength. *Information Systems Knowledge*

Table 3. Discriminant Validity - HTMT criterion

	FFI	ISC	ISK	PC	PRE	PS	PV	PSE
FFI								
ISC	0.387							
ISK	0.118	0.079						
PC	0.179	0.104	0.313					
PRE	0.342	0.326	0.338	0.246				
PS	0.191	0.214	0.05	0.067	0.271			
PV	0.054	0.064	0.123	0.163	0.099	0.175		
PSE	0.265	0.228	0.671	0.593	0.538	0.104	0.198	

has the strongest effect on *Cybersecurity Awareness*, while the other exogenous variables show low to moderate effects (see Fig. 2). The path coefficient from *Cybersecurity Awareness* to *Cybersecurity Behavior* exhibits a moderate effect. The R^2 values for *Cybersecurity Awareness* and *Cybersecurity Behavior* indicate moderate explanation of the endogenous variables through the exogenous constructs. By performing multi-group analyses (PLS-MGA [20]), differences for users of different gender or language groups can be found. While women are influenced by *Mass Media* but not by *Public Administration Information*, it is the other way around for men (men: $P_{PAI->CSA} = 0.159$, p = 0.008, $P_{MMI->CSA} = 0.003$, p = 0.467; women: $P_{PAI->CSA} = -0.005$, p = 0.44, $P_{MMI->CSA} = 0.171$, p = 0.004). When comparing the German- and French-speaking people's influences, the German-speaking are influenced by *Security Training and Awareness Programs* and *Public Administration Information*, while the French-speaking are not influenced (DE: $P_{PAI->CSA} = 0.180$, p = 0.008; FR: $P_{PAI->CSA} = 0.030$, p = 0.228; DE: $P_{SETA->CSA} = 0.142$, p = 0.041; FR: $P_{SETA->CSA} = -0.072$, p = 0.057). Moreover, the awareness of people who have experienced a cybersecurity incident in the past year (NEX) is significantly influenced by *Mass Media Influence*, whereas people with no bad experiences are not influenced (NEX: $P_{MMI->CSA} = 0.200$, p = 0.010; no NEX: $P_{MMI->CSA} = 0.016$, p = 0.445).

4.3 Mediation Analysis

In order to evaluate the role of cybersecurity awareness as a mediator between the determinants and cybersecurity behavior, a mediation analysis was performed following the guidelines proposed by [46]. The evaluation includes looking at direct and indirect effects from the exogenous variables to the endogenous variable. Table 4 shows the results of the analysis. The results suggest *Cybersecurity Awareness* is only a full or partial mediator for *Information Security Policy Provision, Friends and Family Influence, Personal Initiative* and *Information Systems Knowledge*, while other variables only have a direct (*SETA, MMI*) or no effect (*ISC, PAI*) on *Cybersecurity Behavior*.

Table 4. Mediation analysis

Path	Indirect Path Coeff.			Direct Path Coeff.			Mediation
	P	t	p	P	t	p	
ISSP -> CSB	0.063*	2.979	0.001	0.002	0.053	0.479	Full Mediation
SETA -> CSB	−0.005	0.394	0.347	0.117*	2.346	0.009	No Mediation, Direct-Only
ISC -> CSB	0.021	1.310	0.095	−0.033	0.892	0.186	No Mediation, No Effect
FFI -> CSB	0.056*	2.796	0.003	0.059	1.432	0.076	Full Mediation
MMI -> CSB	0.031	1.595	0.055	−0.074*	1.720	0.043	No Mediation, Direct-Only
PAI -> CSB	0.026	1.591	0.056	0.027	0.815	0.208	No Mediation, No Effect
PI -> CSB	0.042*	2.232	0.013	0.142*	2.892	0.002	Partial Mediation
ISK -> CSB	0.205*	6.778	0.000	0.208*	3.962	0.000	Partial Mediation

*Notes: P = Path coefficient, t = t-value, p = p-value, *$p < 0.5$*

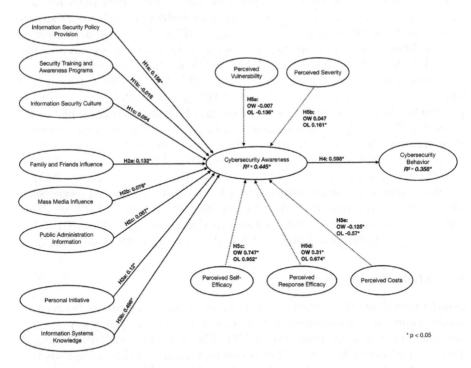

Fig. 2. Results - structural model evaluation

5   Discussion and Implications

The results of this study show diverse levels of impact of organizational, social and personal determinants on a user's cybersecurity awareness in his home environment. The main findings are:

- Weak influence of the workplace
- Weak to moderate social influences
- Personal initiative has a significant effect
- Strongest effect of information systems knowledge
- No significant contribution of threat appraisal to cybersecurity awareness

The limited workplace effects are in line with other studies [31,39]. It is not evident if security training and the information security culture are prevalent but not transferrable to the home environment or if they cannot be found. Mass media and public administration information exhibit disparate effects for men and women and for people who have experienced cybersecurity incidents in the past year. People from the different language groups react differently to various sources of information, emphasizing a potential cultural gap in how cybersecurity topics are handled and perceived in different cultural regions. The strong influence of information systems skills as well as the fact that the threat appraisal does not significantly contribute to cybersecurity awareness highlight the need for campaigns focusing in improving skills and understanding, confirming results of similar studies [17,39,45].

As with other studies, there are some limitations. The study relies on self-report data, which might contain a social desirability bias [10]. Additionally, the PLS-SEM method allows no goodness-of-fit measure for evaluating the fit of the model and the path estimation contains a measurement error resulting in a bias [16]. Although the sample exhibits a good balance of gender and age, works in diverse job areas, most participants work in the public sector. The influence of the workplace could be different in the private or voluntary sector.

While this study kept a generalized approach on most variables to ensure a holistic view, different types of mass media or the form of security information provided at the workplace should also be researched individually as they might lead to distinct user reactions as shown in [29,39]. Moreover, in order to create individualized and adapted campaigns, cultural differences should be investigated more closely. Considering the high potential in reaching broad masses of people, future research should investigate the reasons that inhibit a transfer of work-provided cybersecurity information to the private environment.

6   Conclusion

The human interaction with computer systems becomes increasingly important considering current trends in digitalization. This study investigates organizational, social and personal determinants of a home computer user's cybersecurity awareness and the factors impacting behavior. By providing valuable insights about cybersecurity awareness and behavior creation, the study contributes to research in the field of cybersecurity behavior and can act as a support for security practitioners while reviewing security strategies.

Appendix

Table 5. Overview survey measures

Information Security Policy Provision (ISPP)	[18]
ISPP1: Information security policies are written in a manner that is clear and understandable	
ISPP2: Information security policies are readily available for my reference	
Security Training and Awareness Programs (SETA)	[12,18]
SETA1: My organization provides training to help employees improve their assessment and knowledge of computer and information security issues	
SETA2: My organization educates employees on their computer security responsibilities	
Information Security Culture (ISC)	[34]
ISC1: My colleagues and I would warn each other if we saw one of us taking risks (e.g. insecure use of email, downloading malicious software, or risky password practices)	
ISC2: I have a good relationship with my colleagues and other members of my organization	
ISC3: My organization takes the view that information security is a collective responsibility	
ISC4: My colleagues and I have the same ambitions and visions in terms of protecting our information assets from cyber threats (e.g. unauthorized access to information assets, becoming infected with malicious software)	
Family and Friends Influence (FFI)	[29,42]
FFI1: My family members would approve of me practicing a safe cyber behavior	
FFI2: My family members expect me to practice a safe cyber behavior	
FFI3: My friends would approve of me practicing a safe cyber behavior	
FFI4: My friends expect me to practice a safe cyber behavior	
Mass Media Influence (MMI)	[29]
MMI1: The mass media suggest that I should practice a safe cyber behavior	
MMI2: Mass media reports influence me to practice a safe cyber behavior	
MMI3: I feel under pressure from the mass media to practice a safe cyber behavior	
Public Administration Information (PAI)	[6,29]
PAI1: The public administration suggests that I should practice a safe cyber behavior	
PAI2: The public administration influences me to practice a safe cyber behavior	
PAI3: I feel under pressure from the public administration to practice a safe cyber behavior	
Personal Initiative (PI)	[13]
PI1: In general, I actively attack problems (of any kind)	
PI2: Whenever something goes wrong, I search for a solution immediately	
PI3: I take initiative immediately when others don't	
PI4: Usually I do more than I am asked to	
Information Systems Knowledge (ISK)	[18]
ISK1: What is your general knowledge of computers?	
ISK2: What is your general knowledge of the Internet (e.g. Web, email systems)?	
Perceived Vulnerability (PV)	[9,17]
PV1: I believe that I am at risk of becoming a victim of a cyber security incident (e.g. phishing, malware)	
PV2: I believe that it is likely that I will become a victim of a cyber security incident (e.g. phishing, malware)	

(*continued*)

Table 5. (*continued*)

PV3: I believe that it is possible that I will become a victim of a cyber security incident (e.g. phishing, malware)	
Perceived Severity (PS)	[3, 28]
PS1: Having my computer infected by a virus as a result of opening a suspicious email attachment is a serious problem for me	
PS2: Having my confidential information accessed by someone without my consent or knowledge is a serious problem for me	
PS3: Loss of data resulting from hacking is a serious problem for me	
Perceived Response Efficacy (PRE)	[17, 39]
PRE1: I believe that protective software would be useful for detecting and removing malware	
PRE2: I believe that having passwords that are hard to guess and different for each of my accounts will help improve my security protection	
PRE3: I believe that keeping my operating systems and software updated will help improve my security protections	
PRE4: I believe that following online safety practices will help protecting me from online safety threats	
Perceived Costs (PC)	[28, 45]
PC1: Practicing safe cyber behavior is inconvenient	
PC2: Practicing safe cyber behavior is time-consuming	
PC3: Practicing safe cyber behavior would require considerable investment of effort other than time	
PC4: Practicing safe cyber behavior would require starting a new habit, which is difficult	
Perceived Self-Efficacy (PSE)	[39, 42]
PSE1: I feel comfortable practicing safe cyber security behavior	
PSE2: Practicing safe cyber security behavior is entirely under my control	
PSE3: I have the resources and knowledge to practice safe cyber security behavior	
PSE4: Practicing safe cyber security behavior is easy	
Cybersecurity Behavior (CSB)	[3, 28]
CSB1: I use different passwords for my different online accounts (e.g., social media, online banking)	
CSB2: I usually review privacy/security settings on my online accounts (e.g., social media, online banking)	
CSB3: I keep the software and operating system on my computer up-to-date	
CSB4: I watch for unusual computer behaviors/responses (e.g., computer slowing down or freezing, pop-up windows, etc.)	
CSB5: I do not open email attachments from people whom I do not know	
CSB6: I have never sent sensitive information (such as account numbers, passwords, social security number, etc.) via email or using social media	
CSB7: I make backups of important files on my computer	
CSB8: I always respond to any malware alerts that I receive	
CSB9: I do not click on short URLs unless I know where the links will really take me	

References

1. Ajzen, I.: From intentions to actions: a theory of planned behavior. In: Kuhl, J., Beckmann, J. (eds.) Action Control. SSSSP, pp. 11–39. Springer, Heidelberg (1985). https://doi.org/10.1007/978-3-642-69746-3_2
2. Anderson, C.L., Agarwal, R.: Practicing safe computing: a multimedia empirical examination of home computer user security behavioral intentions. MIS Q. **34**(3), 613–643 (2010). https://doi.org/10.2307/25750694
3. Anwar, M., He, W., Ash, I., Yuan, X., Li, L., Xu, L.: Gender difference and employees' cybersecurity behaviors. Comput. Hum. Behav. **69**, 437–443 (2017). https://doi.org/10.1016/j.chb.2016.12.040

4. Baldwin, W.: Information no one else knows: the value of self-report. In: Stone, A., Bachrach, C., Jobe, J., Kurtzman, H., Cain, V. (eds.) The Science of Self-report, 1st edn, pp. 15–20. Psychology Press, Mahwah (1999)

5. Bandura, A.: Social Learning Theory. General Learning Press, New York, NY (1971)

6. Belanche Gracia, D., Casaló Ariño, L., Flavián Blanco, C.: Understanding the influence of social information sources on e-government adoption. Inf. Res. **17**(3) (2012)

7. Blythe, J.: Cyber security in the workplace: Understanding and promoting behaviour change. In: Proceedings of CHItaly 2013 Doctoral Consortium, vol. 1065, pp. 92–101 (2013)

8. Bortz, J., Döring, N.: Forschungsmethoden und Evaluation für Human-und Sozial-wissenschaftler, 4th edn. Springer, Heidelberg (2006). https://doi.org/10.1007/978-3-540-33306-7

9. Crossler, R.E.: Protection motivation theory: understanding determinants to backing up personal data. In: 2010 43rd Hawaii International Conference on System Sciences (HICSS), pp. 1–10. IEEE (2010). https://doi.org/10.1109/HICSS.2010.311

10. Crossler, R.E., Johnston, A.C., Lowry, P.B., Hu, Q., Warkentin, M., Baskerville, R.: Future directions for behavioral information security research. Comput. Secur. **32**, 90–101 (2013). https://doi.org/10.1016/j.cose.2012.09.010, http://www.sciencedirect.com/science/article/pii/S0167404812001460

11. Da Veiga, A., Eloff, J.H.: A framework and assessment instrument for information security culture. Comput. Secur. **29**(2), 196–207 (2010). https://doi.org/10.1016/j.cose.2009.09.002

12. D'Arcy, J., Hovav, A., Galletta, D.: User awareness of security countermeasures and its impact on information systems misuse: a deterrence approach. Inf. Syst. Res. **20**(1), 79–98 (2009). https://doi.org/10.1287/isre.1070.0160

13. Frese, M., Fay, D., Hilburger, T., Leng, K., Tag, A.: The concept of personal initiative: operationalization, reliability and validity in two German samples. J. Occup. Organ. Psychol. **70**(2), 139–161 (1997). https://doi.org/10.1111/j.2044-8325.1997.tb00639.x

14. Furnell, S., Bryant, P., Phippen, A.: Assessing the security perceptions of personal internet users. Comput. Secur. **26**(5), 410–417 (2007). https://doi.org/10.1016/j.cose.2007.03.001

15. Hair, J., Hollingsworth, C.L., Randolph, A.B., Chong, A.Y.L.: An updated and expanded assessment of PLS-SEM in information systems research. Ind. Manag. Data Syst. **117**(3), 442–458 (2017). https://doi.org/10.1108/IMDS-04-2016-0130

16. Hair, J.F., Hult, T., Ringle, C., Sarstedt, M.: A Primer on Partial Least Squares Structural Equation Modeling, 2nd edn. Sage, Thousand Oaks (2017)

17. Hanus, B., Wu, Y.A.: Impact of users' security awareness on desktop security behavior: a protection motivation theory perspective. Inf. Syst. Manag. **33**(1), 2–16 (2016). https://doi.org/10.1080/10580530.2015.1117842

18. Häussinger, F.J., Kranz, J.J.: Information security awareness: its antecedents and mediating effects on security compliant behavior. In: International Conference on Information Systems (ICIS) (2013)

19. Henseler, J., Ringle, C.M., Sarstedt, M.: A new criterion for assessing discriminant validity in variance-based structural equation modeling. J. Acad. Mark. Sci. **43**(1), 115–135 (2015). https://doi.org/10.1007/s11747-014-0403-8

20. Henseler, J., Ringle, C.M., Sinkovics, R.R.: The use of partial least squares path modeling in international marketing. In: Sinkovics, R.R., Ghauri, P.N. (eds.) New Challenges to International Marketing, vol. 20, pp. 277–319. Emerald Group Publishing Limited (2009). https://doi.org/10.1108/S1474-7979(2009)0000020014

21. Hertig, Y., Teufel, S.: Prosumer communities: electricity as an interpersonal construct. In: 2016 International Conference on Smart Grid and Clean Energy Technologies (ICSGCE), pp. 89–94. IEEE (2016). https://doi.org/10.1109/ICSGCE.2016.7876032

22. Hickmann Klein, R., Mezzomo Luciano, E.: What influences information security behavior? A study with Brazilian users. JISTEM - J. Inf. Syst. Technol. Manag. **13**(3), 479–496 (2016). https://doi.org/10.4301/s1807-17752016000300007

23. Howe, A.E., Ray, I., Roberts, M., Urbanska, M., Byrne, Z.: The psychology of security for the home computer user. In: 2012 IEEE Symposium on Security and Privacy (SP), pp. 209–223. IEEE (2012). https://doi.org/10.1109/SP.2012.23

24. Kajtazi, M., Bulgurcu, B., Cavusoglu, H., Benbasat, I.: Assessing sunk cost effect on employees' intentions to violate information security policies in organizations. In: 2014 47th Hawaii International Conference on System Sciences (HICSS), pp. 3169–3177. IEEE (2014). https://doi.org/10.1109/HICSS.2014.393

25. Ki-Aries, D., Faily, S.: Persona-centred information security awareness. Comput. Secur. **70**, 663–674 (2017). https://doi.org/10.1016/j.cose.2017.08.001

26. Leiner, D.J.: Sosci survey (version 3.1.01-i) [computer software] (2018). http://www.soscisurvey.com

27. Muhirwe, J., White, N.: Cybersecurity awareness and practice of next generation corporate technology users. Issues Inf. Syst. **17**(2), 183–192 (2016)

28. Ng, B.Y., Kankanhalli, A., Xu, Y.C.: Studying users' computer security behavior: a health belief perspective. Decis. Support. Syst. **46**(4), 815–825 (2009). https://doi.org/10.1016/j.dss.2008.11.010

29. Ng, B.Y., Rahim, M.: A socio-behavioral study of home computer users' intention to practice security. In: PACIS 2005 Proceedings, pp. 234–247 (2005)

30. Organisation for Economic Co-operation and Development: OECD guidelines for the security of information systems and networks: Towards a culture of security (2002). http://www.oecd.org/sti/ieconomy/15582260.pdf

31. Rader, E., Wash, R.: Identifying patterns in informal sources of security information. J. Cybersecur. **1**(1), 121–144 (2015). https://doi.org/10.1093/cybsec/tyv008

32. Ringle, C.M., Wende, S., Becker, J.M.: Smartpls 3 (version 3.2.5) [computer software] (2015). http://www.smartpls.com

33. Rocha Flores, W., Antonsen, E., Ekstedt, M.: Information security knowledge sharing in organizations: investigating the effect of behavioral information security governance and national culture. Comput. Secur. **43**, 90–110 (2014). https://doi.org/10.1016/j.cose.2014.03.004

34. Rocha Flores, W., Ekstedt, M.: Shaping intention to resist social engineering through transformational leadership, information security culture and awareness. Comput. Secur. **59**, 26–44 (2016). https://doi.org/10.1016/j.cose.2016.01.004

35. Rogers, R.W.: A protection motivation theory of fear appeals and attitude change1. J. Psychol.: Interdiscip. Appl. **91**(1), 93–114 (1975). https://doi.org/10.1080/00223980.1975.9915803

36. Rogers, R.W.: Cognitive and physiological processes in fear appeals and attitude change: a revised theory of protection motivation. In: Cacioppo, J.T., Petty, R. (eds.) Social Psychophysiology: A Sourcebook, chap. 6, pp. 153–177. Guilford, New York (1983)

37. Schlienger, T., Teufel, S.: Information security culture. In: Ghonaimy, M.A., El-Hadidi, M.T., Aslan, H.K. (eds.) Security in the Information Society. IAICT, vol. 86, pp. 191–201. Springer, Boston, MA (2002). https://doi.org/10.1007/978-0-387-35586-3_15

38. Sherif, E., Furnell, S., Clarke, N.: Awareness, behaviour and culture: the ABC in cultivating security compliance. In: The 10th International Conference for Internet Technology and Secured Transactions (ICITST-2015), pp. 90–94. IEEE (2015). https://doi.org/10.1109/ICITST.2015.7412064

39. Shillair, R., Dutton, W.H.: Supporting a cybersecurity mindset: getting internet users into the cat and mouse game. SSRN Electron. J. (2016). https://doi.org/10.2139/ssrn.2756736

40. Simonet, J.: The Influence of Organizational, Social and Personal Factors on Cybersecurity Awareness and Behavior of Home Computer Users. Master's thesis, iimt, University of Fribourg (2018)

41. Talib, S., Clarke, N.L., Furnell, S.M.: An analysis of information security awareness within home and work environments. In: ARES 2010 International Conference on Availability, Reliability, and Security, pp. 196–203. IEEE (2010). https://doi.org/10.1109/ARES.2010.27

42. Taylor, S., Todd, P.A.: Understanding information technology usage: a test of competing models. Inf. Syst. Res. 6(2), 144–176 (1995). https://doi.org/10.1287/isre.6.2.144

43. Teufel, S., Teufel, B.: Crowd energy information security culture - security guidelines for smart environments. In: 2015 IEEE International Conference on Smart City/SocialCom/SustainCom (SmartCity), pp. 123–128 (2015). https://doi.org/10.1109/SmartCity.2015.58

44. Weinstein, N.D.: Testing four competing theories of health-protective behavior. Health Psychol. 12(4), 324–333 (1993). https://doi.org/10.1037//0278-6133.12.4.324

45. Woon, I., Tan, G., Low, R.: A protection motivation theory approach to home wireless security. In: Proceedings of the Twenty-Sixth International Conference on Information Systems (ICIS), pp. 367–380 (2005)

46. Zhao, X., Lynch, J., Chen, Q.: Reconsidering Baron and Kenny: myths and truths about mediation analysis. J. Consum. Res. 37(2), 197–206 (2010). https://doi.org/10.1086/651257

To Be, or Not to Be Notified

Eliciting Privacy Notification Preferences for Online mHealth Services

Patrick Murmann[1(✉)], Delphine Reinhardt[2], and Simone Fischer-Hübner[1]

[1] Karlstad University, Karlstad, Sweden
patrick.murmann@kau.se
[2] University of Göttingen, Göttingen, Germany

Abstract. Millions of people are tracking and quantifying their fitness and health, and entrust online mobile health (mhealth) services with storing and processing their sensitive personal data. Ex post transparency-enhancing tools (TETs) enable users to keep track of how their personal data are processed, and represent important building blocks to understand privacy implications and control one's online privacy. Particularly, privacy notifications provide users of TETs with the insight necessary to make informed decision about controlling their personal data that they have disclosed previously. To investigate the notification preferences of users of online mhealth services, we conducted an online study. We analysed how notification scenarios can be grouped contextually, and how user preferences with respect to being notified relate to intervenability. Moreover, we examined to what extent ex post notification preferences correlate with privacy personas established in the context of trust in and reliability of online data services. Based on our findings, we discuss the implications for the design of usable ex post TETs.

Keywords: Privacy · Transparency-enhancing tool · Usability · Personas · mHealth

1 Introduction

According to Cisco Systems, the worldwide number of wearable devices connected to the Internet will climb from 325M in 2016 to a projected 929M in 2021 [16]. This means that an increasing number of people track their health using personal activity trackers like fitness bracelets or smart watches. Such wearables allow their users to collect statistics about a plethora of physiological characteristics, and optionally enrich the data with location data and information about one's lifestyle. Hence, potentially seamless information are collected about a person's health, pinpointing it in time and space.

However, according to [7], the consequences of how mhealth data processed by data services are not fully transparent to the users of such services. Data subjects often lack the information necessary to make informed decisions about managing

G. Dhillon et al. (Eds.): SEC 2019, IFIP AICT 562, pp. 209–222, 2019.
https://doi.org/10.1007/978-3-030-22312-0_15

their personal data and to exercise their right of intervenability, especially in scenarios that involve third parties. Conversely, the EU General Data Protection Regulation (GDPR) [2] stipulates that users of data services must be able to control their personal data, and grant them a right of ex post transparency and intervenability rights to delete, rectify, block or export data or to object to data processing (GDPR Art. 15–20). However, exercising these intervenability rights implies that data subjects are aware of how their data are processed, and therefore depend on processes that are transparent and comprehensible as mandated by GDPR Art. 12–15. *Ex post transparency-enhancing tools* (TETs) facilitate transparency by informing about how a data subject's personal data have been processed by online services, e.g. by the means of privacy notifications [17]. As it has been shown in our previous work, however, existing privacy indicators of TETs often lack transparency themselves in that their settings are not always verifiable or customisable [11].

Seeking to infer viable predictors for the design of usable TETs, our contribution is to investigate the notification preferences of users of mhealth services in terms of (1) how data processing scenarios can be grouped contextually, (2) to what extent these preferences can be predicted by means of privacy personas, and (3) how intervenability relates to notification preferences. Ultimately, our goal is to help designers of TETs to provide users with default settings for receiving privacy notifications based on a user's predisposition. Receiving privacy notifications that are tailored to their individual needs will allow users of mhealth systems to make informed decisions about controlling their personal data they have disclosed previously. We postulate the following hypotheses:

H1. Users of online mhealth services have different notification preferences depending on the contextual cue underlying the notifications.
H2. There is a correlation between a user's privacy persona (Sect. 3.2) and her notification preferences.
H3. The ability to intervene with the processing of one's personal mhealth data has an impact on one's notification preferences.

Our paper is structured as follows: Sect. 2 discusses related work. Section 3 describes the methodology applied in our online study. Section 4 presents the results, while we discuss our findings in Sect. 5, before concluding this paper in Sect. 6.

2   Related Work

Our work shares similarities with [3,4,6,12,19]. Similar to [19], our study relies on the concept of personas and how they relate to behaviour and their consequences, but investigates the outcome of preferences for privacy notifications instead of behavioural intent. It is related to the work of Knijnenburg et al. [6] in that it accounts for multiple dimensions that lead to data subjects being grouped in terms of their privacy attitude. However, we do not seek to segment subjects according to their disclosing styles, but to establish a correlation between their disposition in terms of privacy and their notification preferences. Like Emami-Naeini et al. [12], we envision a privacy assistant that provides its users with customised notifications about personal data processing. However, the measure that reflects the independent variables in our study are not constituted solely by discomfort, but by the overall values captured by a particular privacy persona.

Related to ex post transparency, Harkous et al. [4] touched upon privacy indicators by providing insight about potential consequences based on history-based insight about data processing. Our work complements these indicators in that we aim to establish the circumstances under which users of TETs want to receive such notifications. Moreover, our previous work published in [3] indicates that participants have different notification preferences depending on whether they could intervene, i.e., do something about how their data were processed. Hence, we follow up this research by investigating whether and to what extent intervenability has an effect on users' notification preferences.

In summary, our study follows a new direction in terms of (1) how scenarios dealing with ex post personal data processing can be grouped conceptually, (2) how privacy personas established in the literature [9] relate to a user's preferences of being notified about such scenarios, (3) to what extent intervenability has an impact on these decisions, and (4) what implication these findings have for the design of usable TETs.

3   Methodology

Our study was implemented using an online questionnaire that consisted of three parts, which are addressed in Sects. 3.1, 3.2 and 3.3, respectively.

3.1   Demographics and Usage Behaviour

In addition to demographics, the first part collected information on the types of devices our participants owned, what they were using them for, and with whom they shared their data. These insights helped us better understand how our participants reflected the intended target audience of users of mhealth services.

3.2   Privacy Personas Based on Privacy Statements

The second part dealt with privacy statements, which reflected our participants' privacy personas according to Morton et al. [9]. We considered but ultimately disregarded alternative models established in the literature, such as Dupree et al.'s [1] segmentation based on qualitative research, as well as Westin's [18] tripartite, mostly linear classification of privacy personas whose conception pre-dates the advent of the Internet age.

We chose to segment users of online mhealth services according to their privacy attitude based on 15 statements described in Morton et al.'s study [8]. Test subjects assigned a total of 70 points as weights of 0–10 among 15 statements, which map to five triples that reflect the dominant factors of each of the five personas. The reason for choosing their segmentation over alternative approaches is twofold: Firstly, the methodology suggested by Morton et al. is based on quantitative research that does not require manual post-processing. Secondly, the narrative factors and themes identified by Morton et al. are generic in that they capture the notion of a user's trust in online services, which potentially map similarly to scenarios encountered in mhealth environments. We have therefore slightly adapted the original statements to reflect the particularities of the mhealth context. By doing so, we tried to capture the original meanings that reflect the five personas established by Morton et al.: Security Concerned (SC), Organisational Assurance Seekers (OAS), Crowd Followers (CF), Benefit Seekers (BS), and Information Controllers (IC). SC seek the use of technological means to ensure the security of their personal data. OAS look for formal indicators, such as privacy policies, that warrant their trust in a service. CF value the reputation of a service and heed the recommendations of trusted peers. BS are after useful benefits and are willing to give up personal data in exchange. IC seek to control the collection, access to and use of their personal data. The Flesch reading ease and Flesch-Kincaid grade level [5] were not affected by our changes and remained stable at 59.1 and 8.8, respectively. To mitigate the effects of cognitive fatigue and habituation, we randomised the order in which the statements were displayed. We have made available supplementary material about the study design and results on a dedicated website.[1]

3.3   Categories of Notification and Notification Scenarios

The third part of the study covered notification preferences, which describe what kind of scenarios related to personal data processing data subjects want to be notified about. We distinguished three categories of privacy notifications:

Breaches refer to "a breach of security leading to the accidental or unlawful destruction, loss, alteration, unauthorised disclosure of, or access to, personal data [...]" according to GDPR Art. 4 (12). Hence, breaches cover both accidental incidences and deliberate misappropriation of personal data by the data controller or by affiliated parties.

Consequences. This category seeks to clarify the consequences that arise for a user of an mhealth service due to the processing of her data. It covers consequences based on actual facts as well as hypothetical outcomes given the circumstances at hand. Consequences differ from breaches in that respective outcomes are the result of personal data processing that is compliant with the privacy law, or that pertain to a possible occurrence in the future.

Practical tips refer to customised guidance for a user intended to improve her online privacy. They are customised in that they pertain to personal situations, and therefore address matters to which users can relate. Practical tips may suggest a concrete change of behaviour or motivate action by notifying her to consider alternatives that would improve her data privacy.

---

[1] https://murmann.hotell.kau.se/notification-preferences/ [10].

Table 1. Notification scenarios

Breaches	1.	Your data are stored longer than is specified in the privacy policy that you have agreed to
	2.	Your data are processed differently from what is specified in the privacy policy that you have agreed to
	3.	Your data are shared with parties not covered in the privacy policy that you have agreed to
	4.	Your mhealth provider used software in which a critical software bug was detected, which made the service vulnerable to hacking and unauthorised access
	5.	Your data got lost on their way to your mhealth service provider
	6.	Your mhealth service provider was attacked by an unknown party on the Internet, which succeeded in copying some of the data
	7.	One of your mhealth provider's partners has access to data not intended for them
Consequences	8.	Recording both your location data (GPS) and your health data (like your pulse and blood pressure) allows someone with access to these data to know where you performed your activities, like the trails you hike or ride most frequently and how you performed along the way
	9.	Recording both your health data (like your pulse and blood pressure) and the time allows someone with access to these data to learn about your general life style, like how fit you are, your health risks and diseases
	10.	You receive customised advertisements about healthy food, sports products and insurances based on the data recorded using your device
	11.	Your mhealth provider shares your data with other companies for the purpose of profiling (analyse your data for patterns)
	12.	Your mhealth provider or their partners reside outside of Europe
	13.	Someone with access to your location data (GPS) and the location data of other users may learn when you have spent time together and what type of activities you have been performing
	14.	Your mhealth provider changed their privacy policy twice since you started using their services a couple of years ago. Each time, the policy stated that by continuously using their service users will agree to the terms and conditions
Tips	15.	Your current mhealth service provider shares your data with an online marketing company and an insurance company. There is a different service provider that offers you the same level of service quality and device compatibility but that does not share your data with third parties. You have the option to have your archived data transferred to the new provider and have them erased from the current one
	16.	Your mhealth device senses your pulse rate using the highest resolution possible. The device has an option to inform you that by using this setting you collect more data than is necessary to track your health reliably
	17.	You have stopped tracking your health and switched off your mhealth device. You have the option to be notified that you can download and/or erase your health data that are currently stored online by your mhealth service

We constructed 17 hypothetical scenarios with themes related to mhealth (Table 1). We relied on plain language and avoided technical terms that might be misinterpreted by lay persons. The overall Flesch reading ease of the scenarios was 54.4 and the Flesch-Kincaid grade level was 10.8.

Each scenario was displayed in random order on a dedicated screen that showed the narrative and two questions. In the first question, we asked the participant whether she wanted to be notified about it. Possible answers were 'Yes,' 'No' and 'I don't know'. If a participant selected the latter option, a secondary set of options appeared. This set captured the respondent's uncertainty and offered four follow-up options: "I don't understand the scenario," "I can't relate to the scenario," "I need further details to make an informed decision" and 'Other.' We deliberately excluded a free text field in lieu of 'Other' to prevent users from disclosing sensitive information.

The second question, "Does your ability to object to the processing of your data affect your choice above?" aimed at capturing the impact of intervenability on the choice to be notified. For the sake of comprehensibility, we substituted the verb 'to intervene' by 'to object' even though objection does not holistically reflect intervenability. Available options were 'Yes,' 'No' and 'I don't know,' the latter triggering the following secondary options: "I don't understand the question" indicated ambiguity of the task itself. "I don't know what it means to object in this context" meant that a respondent understood the concept of intervening, but felt unable to apply it, "I wouldn't know how to object" implied inability to exercise the legal right, and 'Other' covered everything else.

3.4 Online Survey

We implemented an online questionnaire and hosted it on a web server located in Germany. The answers of the participants were stored using pseudonyms, so that they could later be linked back to a participant ID assigned by the crowd sourcing platform for payment purposes. Once the participants had been paid for their work, these IDs were removed for the purpose of data minimisation. Before publishing the questionnaire, we conducted six independent user tests to evaluate its usability and to fix minor issues. The study was approved by the ethics committee of the University of Göttingen.

3.5 Recruitment

We recruited our participants through the crowd sourcing platform Prolific Academic Ltd[2] because all their data processing was conducted within the EU and its workers were reviewed as being reliable [13,14]. The population of workers were screened using three criteria: (1) 18+ years of age, (2) reside within the borders of the EU, the European Economic Area, or an European country in which data protections laws similar to the GDPR applies, and (3) own a mhealth device in the form of a fitness tracker. The high percentage of workers from the UK [15]

---

[2] https://prolific.ac/.

mirrored the population primarily reflected in the studies conducted by Morton et al. [8,9]. Our test subjects finished the questionnaire in roughly 20 min. Considering European standards for minimum wages, we paid the workers € 8.4/h, i.e., € 2.8 for 20 min. The study was published between August 17 and 18, 2018, during which time 300 submissions were gathered.

4   Results

4.1   Demographics and Usage Behaviour

82% of our participants were from the UK. The second largest groups were from Portugal and Spain with 3% each, and Italy with 2%. The rest hailed from all over Europe. 69% of our participants were female and 31% male, one participant identified as 'other.' Their age distribution was similar to the one published for the total population of workers available on Prolific [15].

The majority of our participants owned mainstream devices, such as fitness bracelets or smart watches. Only few owned breast belts or headbands. The predominant purpose reported was to track their fitness and motivate them to exercise. 47% used their device to track their geographic location. This implies that their devices are capable of processing GPS signals, a feature usually found only in premium price segments or in combination with mobile phones. Less than half our participants shared their data with relatives, and one third shared them with acquaintances denoted as friends. More than one third did not share their data at all. Roughly 5% of our participants were not using mhealth devices (anymore/yet). The majority were mid-term or long-term users who had been using their devices for four months or longer [10].

4.2   Privacy Persona Segmentation

Composition Analysis. Overall, the distributions of most statements were similar in terms of their means and quartiles. However, statement 10 was a noticeable exception in that its mean and quartiles differed significantly from the ones of statements 11 and 12. This peculiarity motivated a deeper in-between analysis of the triples that constituted each of the five personas established by Morton et al. [8].

In many cases, we observed a lack of coherence between the three statements of the triples, and in some cases we even detected negative correlations. The dataset contained a considerable number of extreme values, i.e. weights of zeros and tens assigned to individual statements. For statements 7, 9 and 10, zero was the weight assigned most frequently. Overall, 3.3% of our participants assigned their points exclusively to the two scores 0 and 10, and 9.7% relied on four or fewer different patterns to allot their 70 points among the privacy statements. In 17.3% of all 1500 (300 participants × 5 triples) sets of triples, two statements were very high (≥ 9), whereas the corresponding third one was very low (≤ 1). These patterns seemed random and contradicted the supposedly high coherence

Table 2. (a) Classification of the participants ($n = 300$) according to Morton et al. [8]. (b) Number of personas including duplicates

	(a)								(b)			
	SC	OAS	CF	BS	IC	Σ		#Personas	0	1	2	3
Incl. duplicates	79	60	37	44	45	265		Count	109	128	52	11
Excl. duplicates	37	24	19	25	23	128						

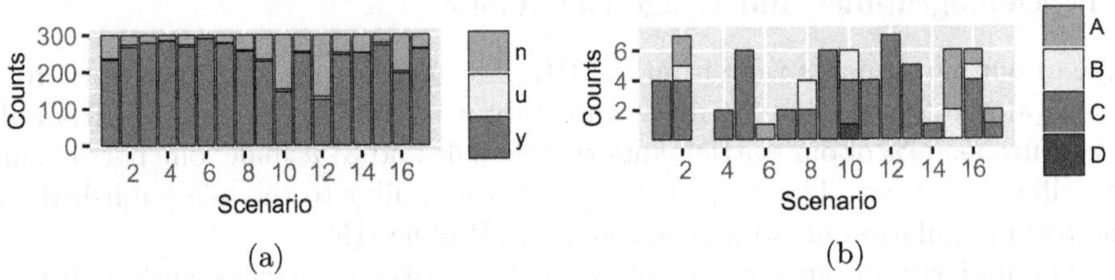

Fig. 1. (a) Counts of notification choices ($n = 300$) for scenarios 1–17: **yes**, **undecided**, **no**. (b) Counts of reasons for being undecided: **A.** Scenario unclear, **B.** Cannot relate to scenario, **C.** Need further details, **D.** Other.

of the statements constituting the triples. More importantly, it questioned the explanatory power they posed in terms of indicating privacy personas. The high amount of variance was also reflected in noticeably low values of Cronbach's α for all five triples ($\alpha = \{0.15, 0.25, 0.31, 0.30, -0.02\}$).

Classification. Irrespective of the incoherence, we carried out the classification according to Morton et al. [8] to ascertain the privacy personas of our participants (Table 2a). We noticed that 109 respondents could not be classified uniquely, and that an additional 63 were ambiguous in terms of being classified as more than one persona (Table 2b). Morton et al. designated such cases as 'unclear,' and specified clearly specified personas as cases with but a single dominant triple.

We therefore tried to establish alternative personas based on the privacy statements using both k-means analysis and principal component analysis. However, neither method yielded satisfactory results. In the latter case, even the Kaiser-Meyer-Olkin and Bartlett's tests failed due to the composition of the underlying data.

We repeated both analyses on an adjusted dataset, in which we removed cases with fewer than four different patterns used for weighing the 15 privacy statements. On average, it took our participants 218 s to allot their 70 points. We removed cases in which a respondent had spent less than 90 s on this sub task, which left us with a total of 240 cases. However, the outcomes of the analyses did not change and we therefore rejected the hypothesis that the noise inherent in the data was the result of superficiality on the part of some of the participants.

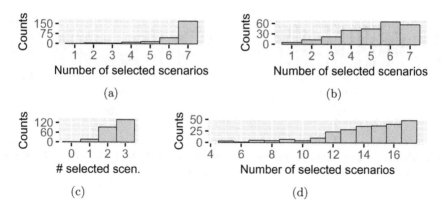

Fig. 2. Counts of sums of positive notification choices per category ($n = 250$): (a) Data breaches, (b) Consequences, (c) Practical tips, (d) Overall.

4.3 Notification Preferences

Per-Scenario Analysis. Most participants chose either 'yes' or 'no' as answers on the questions of whether they wanted to be notified (Fig. 1a). On average, less than five respondents per scenario chose 'I don't know' (Fig. 1b). The reason for doing so indicated most frequently was that the respondent needed further details to make a decision (option C). In most cases, the scenarios seemed to be comprehensible (few counts of option A), yet not always personally applicable (few counts of option B). Scenarios 10, 12 and 16 registered a noticeably low amount of positive choices. As for accidental events, scenarios dealing with software vulnerabilities (scenario 4) and cyber attacks (6) registered the highest counts overall, whereas data loss (5) registered slightly fewer counts and roughly at the mean of the breaches category. Scenarios about hypothetical data processing (8, 9, 13) registered average amounts. Scenarios related to location data (8, 13) registered counts that were slightly above the means of the categories of consequences and tips.

For the subsequent analysis, we removed from the dataset those participants who answered 'I don't know' on any scenario, which resulted in 250 out of the original 300 cases. As the cardinalities of the three categories of scenarios varied (7 breaches, 7 consequences, 3 tips), we introduced a metric of normalised sums for each group. They represented a uniform measure of the amount of positive choices irrespective of the cardinality of that group. Overall, 83% of our participants wanted to be notified about the circumstances described in the scenarios. 92% wanted to be notified about privacy breaches, 74% about consequences, and 83% about tips.

Figure 2 shows the sums of positive choices for (a) breaches, (b) consequences, (c) tips, and (d) throughout all scenarios. The counts reflect how many different scenarios of each respective category our respondents wanted to be notified about. As regards breaches, roughly two thirds (171) wanted to be notified about all seven scenarios, whereas few wanted to be notified about five or fewer scenar-

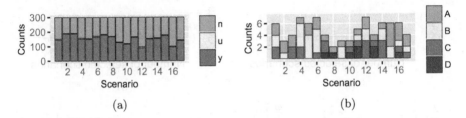

Fig. 3. (a) Counts of whether intervenability affects choice for scenarios 1–18: yes, undecided, no. (b) Counts of reasons for being undecided: **A.** Task unclear, **B.** Intervenability unclear, **C.** Unclear how to intervene, **D.** Other.

ios. For consequences, any combination of four to seven out of seven scenarios accounted for 84% of the cases. For practical tips, we observed that 93% wanted to be notified about at least two out of three scenarios. Overall, more than 86% of the respondents wanted to be notified about any 12 out of 17 scenarios.

Hence, we conclude that hypothesis **H1** holds in that our participants reported distinctively different patterns of notification preferences for each of the three categories of notifications.

Correlation with Morton et al.'s Privacy Personas. We conducted a frequency analysis of positive choices across the scenarios. Overall, all participants had high demands of being notified about privacy breaches. The only noticeable deviation was when the retention period of personal data storage was exceeded (scenario 1), which subjects classified as Crowd Followers and Benefit Seekers were less interested in. As for consequences, scenarios 10 and 12 registered noticeably low values for all personas, but especially among Benefit Seekers whose interest for being notified about consequences were generally low. Conversely, Organisational Assurance Seekers were among the ones who had the highest demands in this category. Getting the best service possible (scenario 15) seemed important for all personas.

We applied logistic regression to establish a model that helped us predict the notification settings of a subject based on her privacy persona, but failed due to high collinearity between the regression coefficients, which either resulted in high residuals and unusable models. Moreover, we investigated the statistical relationship between privacy personas and notification preferences by relying on general linear models. For multiple combinations of categories and personas both the Levene's test of homogeneity of variances and the post hoc tests failed. In most combinations, the models thus established were not significant for either all personas or just the ones clearly classified.

We therefore conclude that hypothesis **H2** does not hold in that we were unable to establish a holistic model that describes a relationship between notification preferences and privacy personas established by Morton et al. [8].

Intervenability. Our participants' opinion on whether their ability to intervene with the processing of their mhealth data impacted their choice for being notified is depicted in Fig. 3a. On average, barely 50% answered positive on the impact of their intervenability, which was broken down into 55% breaches, 46% consequences, and 45% tips. Scenarios dealing with deviating processes (scenario 2) or nameable affiliates (3, 7, 11, 15) registered high positive counts. Similar to notification, scenarios 10, 12 and 16 registered the fewest positive counts. On average, roughly five respondents per scenario chose 'I don't know' (Fig. 3b). The reason indicated most frequently was (A) "I don't understand the question" (35%) and (B) "I don't know *what it means* to object" (30%). Not knowing *how* to object (C) was registered in 22% of the cases. Only few respondents selected 'Other' for either notification or intervenability as their reason for being undecided. This suggests that the three other options captured the reasons for their hesitation satisfactorily.

To analyse the effect of intervenability on our participants' notification preferences, we cleared the dataset of any ambiguous cases related to either notification or intervenability, which resulted in 225 unambiguous cases with values of 'yes' or 'no' for both variables. We analysed the choices in each of the 17 scenarios using a cross tabulation of notification preferences and impact of intervenability [10]. The residuals of all scenarios were positive for identical choices, meaning the observed counts exceeded the expected counts. It indicated that the counts of identical choices (yes/yes or no/no) were high compared to the counts of deviant choices (yes/no or no/yes), which suggested that the two variables correlated.

We therefore conclude that hypothesis **H3** holds in that intervenability had an effect on our participants' notification preferences, even though the absolute positive counts were low.

5   Discussion

5.1   Segmentation of Ex Post Transparency Preferences

Since we were unable to establish a correlation between notification preferences and the privacy personas established by [8], we further seek to investigate the segmentation of notification preferences in the context of ex post transparency, which would allow us to derive alternative personas for this context.

For this purpose, we conducted a principal component analysis of the unambiguous data set ($n = 250$) obtained in Sect. 4.3. Using three factors [10], we found one cluster capturing scenarios for privacy breaches and for consequences in the form of privacy-related events (not classifying as a breach) that had actually taken place (scenarios $1, 2, 3, 5, 6, 7, 11, 14, 15, 17$, Cronbach's $\alpha = 0.75$), one cluster capturing scenarios that covered 'hypothetical' consequences that may occur in future (scenarios $8, 9, 12, 13, 16, \alpha = 0.62$), and one cluster capturing solely the scenario of targeted marketing-related privacy risks (scenario 10). Despite the overall relatively low alpha values, these results show promise in that privacy personas for ex post transparency based on privacy notifications may be established into personas, namely personas for those primarily interested to be notified about privacy-critical events that took place, those primarily interested in hypothetical privacy risks, and those primarily interested to be notified about consequences related to direct marketing and unsolicited messaging.

We seek to further investigate this research question in future work, in which we intend to refine the segmentation process based on notification preferences coded as ordinal variables rather than dichotomous variables.

5.2   Design Implications for TETs

It follows from Sect. 4.3 that the majority of our participants preferred to be notified over not being notified. This means that in cases when a user's preference cannot be determined equivocally, acting upon an event and sending a notification will be a reasonable default setting for a TET, which is in line with the data protection by default principle (Art. 25 GDPR). Nonetheless, we suggest that additional settings for notification clusters be available, which correspond to privacy personas for ex post transparency (Sect. 5.1). Depending on the outcome of a refined user segmentation in future work, these could, e. g., be presets related to targeted marketing, privacy breaches, incidents that have already taken place, and to hypothetical privacy risks that may occur in future.

Most respondents chose unambiguously in that they selected either 'yes' or 'no' for notification and intervenability. However, the few ambiguous choices we registered indicate that there is room for improvement in how notifications are framed, and in the amount and type of information they should contain to satisfy the demands of the target audience.

With respect to notifications, the ambiguity that was registered most frequently was a lack of details regarding the circumstances described in the scenarios. Consequently, respective respondents needed additional or more specific information to make informed decisions. To mitigate cognitive load, TETs will have to rely on multiple levels of detail to convey the full picture of sophisticated data processing scenarios. One way to accomplish this might be to start with a coarse-grained overview and provide details upon request [3], a gap that has been detected in the literature for many TETs [11]. The variety of preferences expressed by our participants supports our previous findings in that notification settings should be transparent and customisable [11].

As for intervenability, both the concept itself and how to leverage respective rights has been unclear for some respondents. Hence, TETs will have to provide such knowledge upon request. Ideally, TETs will guide users in exercising their legal right to manage their personal data and information privacy. For this, privacy notifications should be coupled with context-specific guidance on how users can react by easily exercising their intervenability rights, preferably electronically.

5.3   Limitations

Changing Morton et al.'s original statements [8] was carried out with great care, but we did not validate whether the original meanings have been preserved and carried over to the context of mhealth, nor whether the respondents interpreted them the way they were intended. Hence, the deviation from the original statements and the altered usage context may have had an impact on individual statements, and thus the privacy personas.

All data provided by our participants were self-reported information, both in terms of the filter criteria used to recruit them via Academic Prolific, and the statements made in the online survey. All data were examined in terms of plausibility by the first author before the participants were paid.

Technical terms, such as 'privacy policy' and the concept of what it means to intervene with the processing of one's personal data, were briefly described in the study. However, we did not verify whether the respondents had actually understood the legal concepts underlying these terms.

6   Conclusion

We have conducted an online study that aimed at assessing the notification preferences of users of online mhealth services in terms of being notified about privacy breaches, consequences and tips. Our objective was to investigate to what extent ex post TETs should be individualised and what are suitable defaults that provide their users with meaningful notification settings.

We have found that notification preferences can be grouped as distinctive categories related to the contextual cues underlying the privacy notifications. Moreover, the legal right of intervenability had an impact on our participants' choices to be notified, which implies that guidance of users in exercising this right in response to notifications should be investigated further in the future.

We have, however, been unable to ascertain a correlation between notification preferences and privacy personas as described by Morton et al. [8]. Nonetheless, our first statistical analyses showed that it is possible to elicit new privacy personas for ex post transparency scenarios based on notification preferences. This is a direction of research that we will further pursue in the future.

Acknowledgments. This research has received funding from the European Union's Horizon 2020 research and innovation programme under the Marie Skłodowska-Curie grant agreement No. 67573 and the SSF project SURPRISE.

The authors thank Dan Larsson and Erik Wästlund for advice on the study design, and advice on conducting and interpreting various statistical analyses.

References

1. Dupree, J.L., Devries, R., Berry, D.M., Lank, E.: Privacy personas: clustering users via attitudes and behaviors toward security practices. In: Proceedings of the ACM Conference on Human Factors in Computing Systems (CHI) (2016)
2. The European Parliament and the Council of the European Union. Regulation (EU) 2016/679 of the European Parliament and of the Council (2016)

3. Fischer-Hübner, S., Pettersson, J.S., Angulo, J., Edbom, J., Toresson, M., Andersson, H.: D:C-7.3 Report on end-user perceptions of privacy-enhancing transparency and accountability. Technical report D37.3, A4Cloud Project (2014)
4. Harkous, H., Rahman, R., Aberer, K.: Data-driven privacy indicators. In: Proceedings of the Symposium on Usable Privacy and Security (SOUPS) (2016)
5. Peter Kincaid, J., Fishburne Jr., R.P., Rogers, R.L., Chissom, B.S.: Derivation of new readability formulas (automated readability index, fog count and flesch reading ease formula) for navy enlisted personnel. Technical report, Institute for Simulation and Training, University of Central Florida (1975)
6. Knijnenburg, B.P., Kobsa, A., Jin, H.: Dimensionality of information disclosure behavior. Int. J. Hum.-Comput. Stud. **71**(12), 1144–1162 (2013)
7. Lowens, B., Motti, V.G., Caine, K.: Wearable privacy: skeletons in the data closet. In: Proceedings of the IEEE International Conference on Healthcare Informatics (ICHI) (2017)
8. Morton, A.: Individual privacy concern and organisational privacy practice - bridging the gap. Ph.D. thesis, University College London (2015)
9. Morton, A., Angela Sasse, M.: Desperately seeking assurances: segmenting users by their information-seeking preferences. In: Proceedings of the IEEE Annual International Conference on Privacy, Security and Trust (PST) (2014)
10. Murmann, P.: Supplementary material. https://murmann.hotell.kau.se/notification-preferences/. Accessed 13 Nov 2018
11. Murmann, P., Fischer-Hübner, S.: Tools for achieving usable ex post transparency: a survey. IEEE Access **5**, 22965–22991 (2017)
12. Naeini, P.E., et al.: Privacy expectations and preferences in an IoT world. In: Proceedings of the Symposium on Usable Privacy and Security (SOUPS) (2017)
13. Palan, S., Schitter, C.: Prolific.ac—a subject pool for online experiments. J. Behav. Exp. Finan. **17**, 22–27 (2018)
14. Peer, E., Brandimarte, L., Samat, S., Acquisti, A.: Beyond the Turk: alternative platforms for crowdsourcing behavioral research. J. Exp. Soc. Psychol. **70**, 153–163 (2017)
15. Prolific Academic Ltd., Prolific. https://www.prolific.ac/demographics. Accessed 27 Aug 2018
16. Statista. Number of connected wearable devices worldwide from 2016 to 2021. https://www.statista.com/statistics/487291/. Accessed 28 June 2018
17. Wagner, I., He, Y., Rosenberg, D., Janicke, H.: User interface design for privacy awareness in ehealth technologies. In: Proceedings of the IEEE Annual Consumer Communications & Networking Conference (CCNC) (2016)
18. Westin, A.F.: Social and political dimensions of privacy. J. Soc. Issues **59**(2), 431–453 (2003)
19. Woodruff, A., Pihur, V., Consolvo, S., Schmidt, L., Brandimarte, L., Acquisti, A.: Would a privacy fundamentalist sell their DNA for $1000... if nothing bad happened as a result? The Westin categories, behavioral intentions, and consequences. In: Proceedings of the Symposium on Usable Privacy and Security (SOUPS) (2014)

A Structured Comparison of the Corporate Information Security Maturity Level

Michael Schmid[1,2]([envelope]) [iD] and Sebastian Pape[1,3]([envelope]) [iD]

[1] Chair of Mobile Business & Multilateral Security, Goethe University Frankfurt,
Frankfurt, Germany
{michael.schmid,sebastian.pape}@m-chair.de
[2] Hubert Burda Media Holding KG, Munich, Germany
[3] Chair of Information Systems, University of Regensburg, Regensburg, Germany

Abstract. Generally, measuring the information security maturity is the first step to build a knowledge information security management system in an organization. Unfortunately, it is not possible to measure information security directly. Thus, in order to get an estimate, one has to find reliable measurements. One way to assess information security is by applying a maturity model and assess the level of controls. This does not need to be equivalent to the level of security. Nevertheless, evaluating the level of information security maturity in companies has been a major challenge for years. Although many studies have been conducted to address these challenges, there is still a lack of research to properly analyze these assessments. The primary objective of this study is to show how to use the analytic hierarchy process (AHP) to compare the information security controls' level of maturity within an industry in order to rank different companies. To validate the approach of this study, we used real information security data from a large international media and technology company.

Keywords: Information security · Information security management ·
ISO 27001 · Analytic hierarchy process · Information security controls ·
Capability maturity model · Security maturity model ·
Security metrics framework

1 Introduction

Information security can only be measured indirectly [6]; unfortunately there is still no gold standard. One way to indirectly measure it is to use metrics and KPIs [1] which aim to approximate the real status of information security. This approach is not always reliable [22]. Some information to build those metrics are obtained from technical systems (e.g. firewalls, intrusion detection/prevention systems, security appliances). However, most of these metrics and KPIs have to be quantified by humans and are therefore prone to errors.

© IFIP International Federation for Information Processing 2019
Published by Springer Nature Switzerland AG 2019
G. Dhillon et al. (Eds.): SEC 2019, IFIP AICT 562, pp. 223–237, 2019.
https://doi.org/10.1007/978-3-030-22312-0_16

This can lead to possible inaccuracies, measurement errors, misinterpretations, etc. [4]. If these metrics are then compared across the board, the information security managers face a major challenge. As a consequence, this could lead to bad decisions based on wrong conclusions. Moreover, by just comparing the metrics, without any weighting the specifics of the respective industry are not considered. Thus, a prioritisation within the comparison is not possible [3]. This problem is reinforced when the comparison of information security metrics between different companies or departments would take place [9], which is exactly on of the current challenges enterprises face today: How to compare their (sub-)companies of a specific industry (e.g. eCommerce) in terms of information security.

The main goal of this paper is to compare the effect of multiple factors in the information security assessment process. Aiming at achieving this goal, the analytic hierarchy process (AHP) is applied. The Analytical Hierarchy Process (AHP) is one of the most commonly used Multiple Criteria Decision Methods (MCDM), combining subjective and personal preferences in the information security assessment process [20]. It allows a structured comparison of the information security maturity level of companies with respect to an industry [25] and to obtain a ranking [13]. This allows us to define a separate weighting of information security metrics for each industry with to respect their specifics while using a standardized approach based on the maturity levels of the ISO 27001:2013 controls [12]. ISO 27001 was in particular selected, because this standard is shown to be mature, widespread and globally recognized. This minimizes the additional effort for collecting the required metrics. In this study, the maturity level is based on a hierarchical, multi-level model to analyze the information security gap for the ISO 27001:2013 security standard [20]. As a prerequisite for the comparison, we assume companies have implemented an information security management system (ISMS) in accordance with ISO 27001 [26].

To validate the approach of this study, we used real information security data (i.e. security controls' maturity level) from Hubert Burda Media (HBM) a large international media and technology company consisting of over 200 individual companies. This provides sufficient data with a high degree of detail in the area of information security. The result from our AHP-based approach is then compared with the perceived status of information security by experts.

The remainder of this work is structured as follows: In Sect. 2 we give a brief overview of related work. Section 3 describes our methodology when we developed our approach shown in Sect. 4. Our results are shown in Sect. 5 followed by a discussion and our conclusion in Sect. 6, respectively Sect. 7.

2   Background and Related Work

In addition to the differences in the assessment of information security, all assessment procedures have in common that the ratings of the maturity level and the weighting of weights remain separate judgements and are not allocated to a common overall value in the sense of an 'information security score'. It is therefore up to the evaluator to carry out the respective evaluation, as he or she is forced to choose between these two quantitative aspects of the evaluation, i.e. the ratings on the one hand and the weighting on the other [15]. In contrast to this, the works of Boehme [6] and Anderson [3] deal more with the economic impact of investments in information security. The focus of this work is to compare the degree of maturity within an industry. This could later lead to a monetary assessment of information security or maturity.

A solution which involves to merging ratings and weights and thus integrates different assessment measures at the same time offers multi-attribute decision-making procedures [8]. These are methods that offer support in complex decision-making situations, i.e. when a decision has to be made in favour of one of several options against the background of several decision criteria (so-called attributes).

The prerequisite for using the multi attribute decision procedure is, as described above, the determination of weights. A popular method of doing this is the Analytic Hierarchy Process (AHP) method developed by Saaty [23]. Nasser [2] describes how to measure the degree of maturity using AHP. In contrast to our paper which deals with the comparison of the maturity level within an industry, Nasser [20] focuses on the determination of inaccurate expert comparison judgement in the application of AHP.

Some recent works deals with this problem setting using the AHP but there exist further restrictions. Watkins [27] uses for his approach not the control maturity level and is only valid in the cyber security environment. Bodins' [5] approach is based on the comparison of the CIA-Triangle and not on ISO 27001-controls. Peters [21] has already shown the application of AHP in the domain of project management but did not use real data to validate the approach.

2.1 Multiple Criteria Decision Methods

Multi criteria decision problems which could be solved with a multiple-criteria decision analysis method (MCDM) are a class of procedures for the analysis of decision or action possibilities characterized by the fact that they do not use a single superordinate criterion, but a multitude of different criteria. Problems in evaluating multiple criteria consist of a limited number of alternatives that are explicitly known at the beginning of the solution process. For multiple criteria, design problems (multiple objective mathematical programming problems), the alternatives are not explicitly known. An alternative (solution) can be found by solving a mathematical model. However, both types of problems are considered as a kind of subclass of multi-criteria decision problems [17]. MCDM helps to determine the best solution from multiple alternatives, which may be in conflict with each other. There are several methodologies for MCDM such as: Analytical hierarchical process (AHP), Grey relational analysis (GRA), Technique for order preference by similarity to ideal solution (TOPSIS), Superiority and inferiority ranking (SIR), Simple additive weighting (SAW), and Operational competitiveness rating (OCRA) [7].

2.2　The Analytical Hierarchy Process

The AHP, is a method developed by the mathematician Saaty [24] to support decision-making processes. Because of its ability to comprehensively analyse a problem constellation in all its dependencies, the AHP is called 'analytical'. It is called a 'process' because it specifies how decisions are structured and analysed. In principle, this procedure is always the same, which makes the AHP an easy-to-use decision tool that can be used more than once and is similar to a routine treatment [16]. The goal of the Analytic Hierarchy Process method is to structure and simplify complex decision problems by means of a hierarchical analysis process in order to make a rational decision. The AHP breaks down a complex evaluation problem into manageable sub-problems.

3　Research Methodology

Many companies use the maturity level measurement of the controls from ISO standard 27001 to obtain a valid and reliable metric. The ISO standard is well established and the maturity assessment of the standard's controls is an adequate possibility to create a picture of the information security processes of a company. While this might be sufficient for a continuous improvement within the same company, a problem arises if one wants to compare the information security processes of different companies or departments. Depending on the field of industry, some of the processes might be more important than others.

The general aim of this approach is to determine which company within an industry is better or worse in a (sub)area of information security, in order to create transparency among the companies within an industry concerning information security. Positive effects of this approach would be the improvement or deterioration of the information security in a sector within an industry recognizable up to the question where the management should invest money economically for information security in order to improve a sector.

We define the requirements in the next subsection, then determine the proper algorithm and finally describe the data collection for our approach.

3.1　Requirements

The most important requirement is that the metrics we rely on should be easy to gather. Assuming that the investigated company is running an information security management system (ISMS), a natural approach is to rely on the controls of the ISO/IEC 27001 standard and their maturity level. Existing data (e.g. information security maturity level) should be used wherever possible. Furthermore, the approach should consider the environment of the industry in which the company is located. Additionally, the information gathering should be repeatable and stable. Comparing and evaluating over a long period should be possible as well as an overall as an comparison of security levels of business units or companies in a similar area. Finally, the approach should allow it to visualize and explain the results of the comparison and allow to derive the areas where companies could improve.

3.2 Algorithm Selection

Taking all requirements into account, our problem is a multi-dimensional decision problem, and thus can be addressed by a multiple-criteria decision analysis method (MCDM). Our comparison criteria (dimensions) are the ISO/IEC 27001 controls and we compare the different companies based on their corresponding maturity levels for each control. Thus, the MCDM needs discrete, quantitative input and a criteria weighting method. Since the underlying controls are hierarchically and therefore very structured, the chosen method/model should reflect that also.

This leads us to the analytical hierarchy process (AHP) as a best fit method in the above described context. The AHP is a mature structured technique for organizing and analyzing complex decisions, combining subjective and personal preferences. The AHP has been the most widely used technique of multi-criteria decision making during the last twenty five years [19]. The advantage of this method over the utility value analysis, for example, is that it goes beyond the evaluation of ideas and generates a clear selection recommendation. Its hierarchical structuring of decision making fits well to the ISO/IEC 27001 controls' hierarchy and the qualitative evaluation part of the AHP is very much in line with the maturity level for information security. Since the AHP compares the maturity level for each control company-wise, it naturally allows to understand where each company's security level is ranking related to each control. Additionally, the weight of each criteria (control) can be easily derived. In the concrete application case it is possible to compare the importance of individual controls of ISO 27001 very granularly with each other (pairwise). This is in particular necessary in order to be able to establish an industry reference. Furthermore, the AHP enables precise calculations of weights, in this case the information security maturity ratings of companies in a specific sector.

Thus, we used a paired comparison questionnaire based on the AHP to compare controls and their maturity level for an industry.

3.3 Data Collection

To test the above approach it is necessary to set up the model and verify it with real data. We need a maturity assessment of the ISO/IEC controls and to weight them according to the considered industry. We focused on the eCommerce industry for the following reasons:

- Available data from a large range of companies
- Excellent data quality and validity
- High actuality of the existing data
- Very good know-how available in the expert assessment of the industry.

Maturity Assessment of ISO/IEC 27001 Controls. We collected data from Hubert Burda Media (HBM), an international media and technology company (over 10,000 employees, more than 2 billion annual sales, represented in over 20 countries). This group is divided into several business units that serve various business areas (including print magazines, online portals, e-commerce, etc.). The business units consists of over 200 individual companies with about 30 of them being in the eCommerce industry. Each subsidiary operates independently of the parent corporation. There is a profit center structure, so the group acts as a company for entrepreneurs and the managing directors have the freedom to invest money into information security or choose the appropriated level of security.

We will briefly describe how this data is collected before going into more detail on the data used for the comparison. Each individual company in the group operates its own Information Security Management System (ISMS) in accordance with ISO/IEC 27001:2013, which is managed by an Information Security Officer (ISO) on site and managed by a central unit in the holding company. As part of the evaluation of the ISMS, the maturity level for the respective ISO 27001 controls is ascertained - very granularly at the asset level. The maturity level is collected/updated regular once a year as part of a follow-up.

First, the information values of the respective company (e. g. source code, customer data, payment data, etc.) are determined according to the protection goals of confidentiality, integrity and availability and assigned to a technical system (e. g. application, client, server, etc.).

Second, these technical systems undergo a threat analysis[1] of the assets in relation to the respective asset type as part of information security risk management. The threat analysis is classically evaluated with regard impact[2] and the probability of occurrence. This results in an aggregated risk value (1–5) for each asset after a pre-defined settlement. This risk value is later transferred to the control valuation as the *target maturity level*. In this way, a comparison is made between the protection requirements of the information values and the protection level of the respective (IT) system.

Third, the control evaluation is then carried out using the Cobit maturity level. The controls are dynamically selected[3] according to the previously evaluated threats. The Cobit maturity level is a 6-step evaluation scale (0–5) with which a continuous improvement can be measured and a potential improvement can be identified. This allows it to evaluate the actual maturity level per control and asset. The assessment of the current status of the controls is carried out by the information security officer of the respective company. The collected data is therefore not technical data but subjectively quantified data with a possible bias. Although, the evaluated data is reviewed by further experts, a complete review cannot be carried out due to resource limits. The target maturity level is already determined by the risk value/protection level of the system. This provides a clear picture of the ISMS status at a very granular asset level.

---

[1] Threat catalogue according to ISO/IEC 27005:2011.

[2] Referring to the protection goals of confidentiality, integrity and availability.

[3] By a predefined threat/control matrix.

Fourth, the picture is completed by the Cobit maturity analysis of the IT-/ISM processes[4]. For each of these processes, the controls (e. g. A.16 for incident management) are evaluated with an actual maturity level [10]. In the later evaluation (typically by means of a spider graphic) the complete ISO 27001 standard is evaluated with the aid of the Cobit degree of maturity [14].

The available data is very granular on asset level (application, client, server, etc.). However, although the companies are from the same industry, they do not necessarily have the same kind of assets. Thus, we decided to abstract from the assets and to aggregate the data at company level. To do this automatically, we used the mean value of all evaluated assets per control. For the following proof of concept, we only show data from 5 companies.

4   The Approach - the AHP-Implementation

In this section, we discuss how the AHP is applied to our comparison. The first step of the AHP, to model the problem as a decision hierarchy, we have already done by deciding that our decision-criteria will be the ISO/IEC 27001 controls. The goal is clearly defined: to find the subsidiary within the company with the best information security/level of maturity within an industry. Appendix A of ISO 27001 helps us to select criteria and sub criteria, which is divided into 14 Control Categories, 35 Control Objectives and 114 Controls (see Fig. 1).

The next step is the prioritization of all criteria and sub criteria (Sect. 4.1). This represents the domain specific part of the AHP calculations and it only needs to be done once per domain. It is followed by the evaluation of the alternatives (Sect. 4.2). The alternatives represent the agile part of the calculation. We describe in the corresponding section, how the evaluation can be directly derived from the maturity level of a company's control. Based on the individual evaluations and prioritizations of controls, the AHP uses a mathematical model to determine a precise weighting of all alternatives in relation to the respective criteria and assembles them in a percentage order (Sect. 4.3).

In the next subsections we describe in detail how the AHP was used and show how the applied AHP model was implemented in a statistical software (in this case in R).

4.1   Pairwise Comparison of the Control Categories and Controls

The characteristics of an industry have a significant influence on the pairwise comparison when comparing the individual controls. If the information security of companies is to be compared with each other, e.g. in the e-commerce sector, it will differ significantly from that of companies in other sectors, e.g. publishing or the manufacturing industry. On the one hand this is due to the different business models within the industries, because the IT strategy and the information

---

[4] Business Continuity Management, Compliance, Incident Management, Information Security Management, Organizational Information Security, Protection Requirement Assessment.

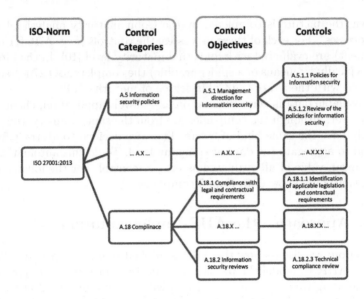

Fig. 1. Exemplary ISO 27001 Appendix A structure

security strategy are derived from the business strategy. On the other hand this is due to the different focus in information security. For example, the eCommerce industry is very focused on application development and (confidentiality) protection of customer data, whereas the highest commodity to be protected in the manufacturing industry is the availability of systems.

The decision-maker must compare each criterion with its pair and denotes which of the two criteria appears more important to him/her. This method of pairwise comparisons allows the decision-maker to elicit a very precise evaluation from the multitude of competing criteria. The comparisons must be carried out specifically for one industry (e. g. eCommerce). In the case of our hierarchy based on the ISO/IEC 27001 controls, 91 pairwise comparisons have to be made for the control categories and 208 for the controls, respectively. This leads to a ranking order in which the criteria are ranked according to their importance.

The comparison is done as follows: Each result of a pairwise comparison of two criteria entered in the evaluation matrix shows how much more significant a criteria is in relation to the criteria of the level above. To do this, refer to the scale in Table 1a. In order to make a comparison for one criteria, i.e. the control categories, we compare the individual control categories with each other. The authors made this comparison in a straight forward Excel spreadsheet. The assessment of the relative importance of the criteria at the criterion level can be found in Table 1b. These pairwise comparisons are always carried out by an expert with the background knowledge and with reference to the industry (here eCommerce). The comparison for the sub criteria, the controls, follows the same guidelines.

Table 1. AHP scores and their application

AHP Score	Verbal description
9	Extreme
8	preference
7	Very strong
6	preference
5	Strong
4	preference
3	Moderate
2	preference
1	Equal preference

Sub criteria A	Sub criteria B	A/B	Score
Control A.12.1.1[1]	Control A.12.1.2	B	$\frac{1}{7}$
Control A.12.1.1	Control A.12.1.3	B	$\frac{1}{7}$
Control A.12.1.1	Control A.12.1.4[4]	B	$\frac{1}{7}$
Control A.12.1.2[2]	Control A.12.1.3	B	$\frac{1}{3}$
Control A.12.1.2	Control A.12.1.4	A	3
Control A.12.1.3[3]	Control A.12.1.4	A	3

[1] Documented operating procedures [2] Change management
[3] Capacity management [4] Separation of development

(a) *Fundamental AHP Score*   (b) *AHP Comparison with sub criteria (Controls) from control group A.12.1*

4.2 Pairwise Evaluation of the Controls' Maturity Levels

The alternatives in our example are the information security maturity of 5 eCommerce companies of HBM. For each control and each company there is a corresponding maturity level based on the Cobit Maturity Model. 0 represents the worst and 5 the best result, always in relation to the evaluation of a control. As already discussed in Sect. 3.3, the maturity levels for each company were based on assets and we aggregated the maturity levels by calculating the average maturity level for each control over all evaluated assets of the respective company.

For the pairwise comparison, the gap between the comparative maturity levels of two companies' controls is considered to decide which company is doing better at a specific control. For that purpose, we need to map the 6-stage scale of the Cobit maturity grade gaps (see Table 2) to the 9-stage AHP score. The result is a table where each GAP Cobit interval represents an AHP score, which is verbally described. An exemplary calculation can be found in Table 2c). Alternative A (Company 1) is compared with the alternatives B (Company 2 to 5). A Cobit GAP -2 (i.g. 1–3) means hat Company 2 is 2 control maturity better than Company 1, the AHP score is, corresponding to the Cobit GAP interval, 4, respectively 1/4. This can be used to calculate which of the 5 companies performs best in Control A.5.1.1.

The step of comparing the companies' maturity levels for each control represents the business unit specific part of the analysis. Note that, due to our mapping of the GAP Cobit interval and the AHP score, this can be done fully automatic if the corresponding maturity levels are provided. The pairwise comparison, the calculation of the difference and the 'translation' to the GAP intervals is done in the statistics software R.

Table 2. Combined GAP of Cobit Maturity Model and AHP score

Cobit Maturity Model	Cobit level
Optimized	5
Managed and Measurable	4
Defined Process	3
Repeatable but Intuitive	2
Initial/Ad Hoc	1
Non-existent	0

AHP Score	Cobit GAP Interval	Verbal description
9	4.45 - 5.00	Extreme
8	3.89 - 4.44	preference
7	3.34 - 3.88	Very strong
6	2.78 - 3.33	preference
5	2.23 - 2.77	Strong
4	1.66 - 2.22	preference
3	1.12 - 1.65	Moderate
2	0.56 - 1.11	preference
1	0.00 - 0.55	Equal preference

Alt. A	Alt. B	Cobit GAP	Score
Co. 1	Co. 2	-2	$\frac{1}{4}$
Co. 1	Co. 3	1	2
Co. 1	Co. 4	-3	$\frac{1}{6}$
Co. 1	Co. 5	1	2

(a) *Maturity Model vs. level*   (b) *AHP Score vs. GAP Cobit level*   (c) *Comparison for Control A.5.1.1*

4.3   Calculation of the Comparison

As mentioned above, the actual calculation of the AHP is done with R. The implementation in R worked with the help of a YAML (Ain't Markup Language) script executed in R. The YAML script is a simplified markup language for data serialization. The YAML script contains all results of the pairwise comparison of criteria and sub criteria, as well as the maturity levels of the 114 controls of the 5 eCommerce companies. The decision hierarchy built up in the YAML script corresponds to the ISO standard. The decision hierarchy is then enriched with alternatives. The paired comparison of the alternatives is executed by a function of the R-package 'ahp' (version 0.2.12 from Christoph Glur) at script runtime for a simple data processing flow. The runtime of the script (with data from 5 companies) on an iMac (3.2 GHz Intel Core i5) was less than 10 s, indicating that it is efficient enough to handle large amounts of data easily.

5   Results of the Comparison

The AHP was used to compare the maturity level in order to find the company with the best information security within an industry (here eCommerce).

Prioritization of Controls. Here we show which priority the control categories (criteria) and controls (sub criteria) have in relation to the complete Appendix A of ISO 27001 over all. The pairwise comparison for the eCommerce industry shows that the controls of the control category 'A.14' have the highest priority (17.6%), followed by 'A.17' (14.7%) and 'A.12' (10.1%). Within control category 'A.14', controls 'A.14.2.8' (22.6%), 'A.14.2.7' (15.2%) and 'A.14.2.6' (11.8%) are the most important as shown in Fig. 2.

	Priority	Company3	Company5	Company1	Company4	Company2
Comparison eCommerce	100.0%					
A.14 System acquisition	17.6%					
A.14.2.8 System security testing	22.6%	16.7%	4.3%	32.7%	32.7%	13.7%
A.14.2.7 Outsourced development	15.2%	23.3%	27.6%	27.6%	17.0%	4.5%
A.14.2.6 Secure development environment	11.8%	19.8%	22.4%	22.4%	13.1%	22.4%
A.14.2.1 Secure development policy	8.0%	4.4%	39.6%	19.4%	19.4%	17.2%
A.14.1.2 Securing application services on public networks	7.4%	5.9%	23.5%	23.5%	23.5%	23.5%
A.14.1.3 Protecting application services transactions	7.0%	6.1%	5.4%	30.3%	30.3%	27.5%
A.14.1.1 Information security requirements analysis and specification	6.2%	21.6%	23.3%	23.3%	23.3%	8.5%
A.14.2.9 System acceptance testing	5.1%	17.5%	19.8%	19.8%	19.8%	23.2%
A.14.2.2 System change control procedures	4.9%	14.3%	28.6%	14.3%	28.6%	14.3%
A.14.2.4 Restrictions on changes to software packages	4.4%	28.6%	14.3%	14.3%	14.3%	28.6%
A.14.2.5 Secure system engineering principles	3.9%	20.0%	20.0%	20.0%	20.0%	20.0%
A.14.2.3 Technical review of applications after operating platform changes	3.5%	28.6%	14.3%	14.3%	14.3%	28.6%
A.17 Information security aspects of business continuity management	14.7%					
A.17.1.2 Implementing information security continuity	48.1%	40.0%	10.0%	20.0%	10.0%	20.0%
A.17.1.1 Planning information security continuity	40.5%	28.6%	14.3%	28.6%	14.3%	14.3%
A.17.1.3 Verify review and evaluate information security continuity	11.4%	40.0%	10.0%	20.0%	10.0%	20.0%
A.12 Operations security	10.1%					
A.12.1.3 Capacity management	13.9%	40.0%	21.9%	21.9%	11.4%	4.8%
A.12.4.1 Event logging	13.2%	35.9%	5.4%	19.6%	19.6%	19.6%
A.12.6.1 Management of technical vulnerabilities	12.4%	13.4%	36.7%	19.4%	19.4%	5.2%
A.12.4.2 Protection of log information	11.5%	24.4%	3.7%	23.7%	23.7%	24.4%
A.12.1.2 Change management	11.3%	20.0%	20.0%	20.0%	20.0%	20.0%
A.12.4.3 Administrator and operator logs	10.2%	5.3%	5.3%	30.5%	30.5%	26.4%
A.12.1.4 Separation of development	9.3%	9.2%	4.5%	29.7%	29.7%	27.0%
A.12.6.2 Restrictions on software installation	7.1%	6.6%	28.3%	12.2%	26.3%	26.7%
A.12.4.4 Clock synchronisation	5.6%	24.3%	12.2%	12.2%	6.7%	44.6%
A.12.1.1 Documented operating procedures	5.6%	5.3%	38.8%	21.7%	21.7%	12.5%

Fig. 2. Top3 control categories prioritized and companies ranked

Comparision of the Companies. The Control Category 'A.14' was used to exemplarily show the evaluation. Figure 2 also shows how the individual eCommerce companies weighting compare with each other in the control category 'A.14' in detail. Overall (cf. Fig. 3), Company3 (21.0%), Company5 (20.9%) and Company1 (20.5%) came out best in a direct comparison. The differences are marginal and only on closer inspection are there more pronounced differences observed at the control level. In relation to a control category e.g. of 'A.14', the maturity of Company1 (4.4%) and Company4 (4.0%) is better in detail, but considering the control category 'A.17', Company3 (5.2%) is clearly ahead of Company4 (1.7%).

	Weight	Company3	Company5	Company1	Company4	Company2
Comparison eCommerce	95.1%	21.0%	20.9%	20.5%	16.3%	16.2%
A.14 System acquisition	17.6%	3.0%	3.3%	4.4%	4.0%	3.0%

Fig. 3. Control category A.14 weight contribution and ranked companies

6   Discussion

Based on these results, we discuss the main findings as follows. The results show that with the pairwise comparison it is possible to obtain a priority for each individual control, and thus very granular, in the overall context of ISO/IEC 27001 for the eCommerce industry. The priorities of the larger control categories are also very helpful, as a quick comparison of priorities is possible here. The approach with the pairwise comparison by AHP meets all requirements of the methodology part. Similarly, it is shown that the weighting of the pairwise comparisons of the maturity level of eCommerce companies can be mapped very granularly to the controls of the ISO/IEC 27001 standard. It was also possible to derive the AHP score from the maturity levels automatically. This makes it easy to compare the rankings of the companies. The only effort which needs to be invested (for each industry) is the prioritization of the controls.

The results suggest that the approach works in conjunction with real data (the maturity levels of HBM's eCommerce companies) at least for the chosen area. The results of the comparison also withstand the reality that one of the authors observes in his daily professional life. The results also showed that the ranking results reflect the reality of at least the HBM eCommerce companies. However, it can be strongly assumed that the method is directly applicable to other companies with the same or similar results.

6.1   Limitations

For reasons of simplification and clarity, we have demonstrated the approach only with a small number of companies. But is easily possible to run the approach with the full set of HBM's companies and to extend it to other business units by readjusting the ISO/IEC 27001 controls' priorities.

The application of the AHP methodology is not undisputed in technical literature. At this point the authors consider some points of this criticism. On the one hand, these are points concerning the mathematical part of the AHP and on the other hand, the criticism is based on the procedure. In the model calculated above, the pairwise comparison of the criteria and sub criteria has been carried out by one person (with expert knowledge), which can be regarded as a very subjective survey of all pair comparisons. This assumes that there are high demands on the respondent due to the many pair comparisons, which is why there are often problems with validity [18]. This could lead to a limitation of the size of the decision model and is seen as a critical and possible optimization point of the AHP methodology in literature and practice [11].

If you take a closer look at the origin of the maturity level, you immediately notice that it is determined by the information security officer's self-disclosure. As with all quantification, the human factor, a lack ob objectivity or bias, cannot be excluded here. However, it can be largely validated by a team of experts. Another point concerns the type of data collection, the resulting prevailing data quality and possible imponderables in data evaluation. These issues could only be reduced but not completely eliminated by several iterations of quality assurance.

In the next chapter, some of the limitations will be discussed and further improvements of the methodology/model will be proposed.

7 Conclusion and Future Work

The results of the pairwise comparison suggest that AHP is very well suited to compare the information security maturity of different companies and to find the company with the best information security within an industry.

It has been proven that a comparison within the eCommerce industry is possible using this model and thus ranking the prioritization of control categories and, above all, the individual controls can follow. The AHP provides in this case a robust and comprehensive treatment for decision makers in both qualitative and quantitative ways as found in this study and it can be assumed that this will also work for other companies in the same environment. The real insight is to adapt the AHP or the data so that it works together. The AHP-model has shown how AHP might be used to assist decision maker evaluate information security in one branch. Very interesting, and also for validation, would be the pairwise comparison for other industries such as publishing houses, manufacturing industry. Companies with very different degrees of maturity could also be interesting here.

Some of the limitations mentioned above regarding the AHP methodology deal with the comparison of pairs. A possible improvement of the model would be to compare it with the help of a team of experts from the eCommerce industry. This would have the advantage that the pair comparison is subject to validation.

In future work, the focus will be on the details of implementing this model across a variety of different examples, as well as working on more expanded decision hierarchy with an additional level of sub criteria (control objectives). In addition, it would be interesting to calculate the approach with different aggregated data (min, max, median) in addition to the mean value and to observe the effects. Furthermore, it would be interesting to apply the AHP methodology to other industries (e. g. publishing, manufacturing industry etc.). Ultimately, this would provide the prerequisites for comparing information security across industries, comparing apples and pears, so to speak.

References

1. Abbas Ahmed, R.K.: Security metrics and the risks: an overview. Int. J. Comput. Trends Technol. **41**(2), 106–112 (2016)
2. Al-Shameri, A.A.N.: Hierarchical multilevel information security gap analysis models based on ISO 27001: 2013. Int. J. Sci. Res. Multidisc. Stud. **3**(11), 14–23 (2017)
3. Anderson, R., et al.: Measuring the cost of cybercrime. In: Böhme, R. (ed.) The Economics of Information Security and Privacy, pp. 265–300. Springer, Heidelberg (2013). https://doi.org/10.1007/978-3-642-39498-0_12
4. Axelrod, C.W.: Accounting for value and uncertainty in security metrics. Inf. Syst. Control J. **6**, 1–6 (2008)

5. Bodin, L.D., Gordon, L.A., Loeb, M.P.: Evaluating information security investments using the analytic hierarchy process. Commun. ACM **48**(2), 78–83 (2005)

6. Böhme, R.: Security metrics and security investment models. In: Echizen, I., Kunihiro, N., Sasaki, R. (eds.) IWSEC 2010. LNCS, vol. 6434, pp. 10–24. Springer, Heidelberg (2010). https://doi.org/10.1007/978-3-642-16825-3_2

7. Choo, K.K., Mubarak, S., Mani, D., et al.: Selection of information security controls based on AHP and GRA. In: Pacific Asia Conference on Information Systems, vol. 1, no. Mcdm, pp. 1–12 (2014)

8. Eisenführ, F., Weber, M.: Rationales Entscheiden, p. 415. Springer, Heidelberg (2003). https://doi.org/10.1007/978-3-662-09668-0

9. Gordon, L.A., Loeb, M.P.: The economics of information security investment. ACM Trans. Inf. Syst. Secur. **5**(4), 438–457 (2002)

10. Haufe, K.: Maturity based approach for ISMS. Ph.D. thesis, University Madrid (2017)

11. Ishizaka, A., Labib, A.: Review of the main developments in the analytic hierarchy process. Expert Syst. Appl. **38**(11), 14336–14345 (2011)

12. ISO/IEC 27001: Information Technology—Security Techniques—Information Security Management Systems—Requirements. International Organization for Standardization (2013)

13. Khajouei, H., Kazemi, M., Moosavirad, S.H.: Ranking information security controls by using fuzzy analytic hierarchy process. Inf. Syst. e-Bus. Manag. **15**(1), 1–19 (2017)

14. Le, N.T., Hoang, D.B.: Capability maturity model and metrics framework for cyber cloud security. Scalable Comput.: Pract. Exp. **18**(4), 277–290 (2017)

15. Lee, M.C.: Information security risk analysis methods and research trends: AHP and fuzzy comprehensive method. Int. J. Comput. Sci. Inf. Technol. (IJCSIT) **6**(February), 29–45 (2014)

16. Liu, D.L., Yang, S.S.: An information system security risk assessment model based on fuzzy analytic hierarchy process. In: 2009 International Conference on E-Business and Information System Security, pp. 1–4 (2009)

17. Majumder, M.: Impact of Urbanization on Water Shortage in Face of Climatic Aberrations. Springer, Singapore (2015). https://doi.org/10.1007/978-981-4560-73-3

18. Millet, I.: Ethical decision making using the analytic hierarchy process. J. Bus. Ethics **17**(11), 1197–1204 (1998)

19. Mu, E., Pereyra-Rojas, M.: Pratical Decision Making: An Introduction to the Analytic Hierarchy Process (AHP) Using Super Decisions (v2). Springer, Heidelberg (2017). https://doi.org/10.1007/978-3-319-33861-3

20. Nasser, A.A.: Measuring the information security maturity of enterprises under uncertainty using fuzzy AHP. Int. J. Inf. Technol. Comput. Sci. **4**(April), 10–25 (2018)

21. Peters, M.L., Zelewski, S.: Analytical Hierarchy Process (AHP) – dargestellt am Beispiel der Auswahl von Projektmanagement-Software zum Multiprojektmanagement. Institut für Produktion und Industrielles Informationsmanagement (2002)

22. Rudolph, M., Schwarz, R.: Security indicators – a state of the art survey public report. FhG IESE VII(043) (2012)

23. Saaty, T.L., Vargas, L.G.: Decision Making with the Analytic Network Process: Economic, Political, Social and Technological Applications with Benefits, Opportunities, Costs and Risks. Springer, Heidelberg (2006). https://doi.org/10.1007/0-387-33987-6

24. Saaty, T.L., Vargas, L.G.: Models, Methods, Concepts & Applications of the Analytic Hierarchy Process, vol. 175. Springer, Heidelberg (2012). https://doi.org/10.1007/978-1-4614-3597-6
25. Syamsuddin, I., Hwang, J.: The application of AHP to evaluate information security policy decision making. Int. J. Simul.: Syst. Sci. Technol. **10**(4), 46–50 (2009)
26. Vaughn, R.B., Henning, R., Siraj, A.: Information assurance measures and metrics - state of practice and proposed taxonomy. In: Proceedings of the 36th Annual Hawaii International Conference on System Sciences, HICSS 2003 (2003)
27. Watkins, L.: Cyber maturity as measured by scientific-based risk metrics. J. Inf. Walfare **14.3**(November), 60–69 (2015)

Predicting Students' Security Behavior Using Information-Motivation-Behavioral Skills Model

Ali Farooq[1]([⊠]), Debora Jeske[2], and Jouni Isoaho[1]

[1] University of Turku, Turku, Finland
{alifar,jouni.isoaho}@utu.fi
[2] University College Cork, Cork, Ireland
adminapsych@ucc.ie

Abstract. The Information-Motivation-Behavioral Skills (IMB) Model has shown reliability in predicting behaviors related to health and voting. In this study, we examine whether the IMB Model could predict security behavior among university students. Using a cross-sectional design and proxy IMB variables, data was collected from 159 Finnish students on their security threats' awareness (representing IMB's information variable), attitude toward information security and social motivation (replacing IMB's motivation variable), self-efficacy and familiarity with security measures (variables related to IMB's behavioral skills), and self-reported security behavior (IMB outcome variable). An analysis conducted with PLS-SEM v3.2 confirmed that the IMB Model was an appropriate model to explain and predict security behavior of the university students. Path analysis showed that behavioral skills measures predict security behavior directly, while students' information and motivation variables predicted security behavior through behavioral skills (self-efficacy and familiarity with security measures). The findings suggest that the security behavior of students can be improved by improving threat knowledge, their motivation and behavioral skills – supporting the use of the IMB Model in this context and combination with existing predictors.

Keywords: Information security · Threat knowledge · Security behavior · IMB Model

1 Introduction

While the Internet has brought a variety of benefits to us, we are also exposed to the dark side of the Internet due to the different information security threats [1]. To mitigate these security threats, organizations implement not only technical measures [2] but also, non-technical or educational measures, such as information security policies and security education, training and awareness programs (also known as SETA programs; [3–6]). In this paper, the term security is used synonymously with information security.

Educational institutions have also been concerned about information security since the arrival of the Internet [7–9]. Educational institutions, especially higher education institutions (HEIs), serve large populations of students, but also maintain the

© IFIP International Federation for Information Processing 2019
Published by Springer Nature Switzerland AG 2019
G. Dhillon et al. (Eds.): SEC 2019, IFIP AICT 562, pp. 238–252, 2019.
https://doi.org/10.1007/978-3-030-22312-0_17

technological infrastructures to support learning and research activities. HEIs often manage large computer centres which collect work-related and private information of students and staff as well as crucial research information [10]. If compromised, these resources can be misused by the malicious entities. For example, leveraging denial of service attacks, phishing attacks and identity theft of staff and students (e.g. Cobalt Dickens attacks in 2018), and selling products information for financial gains.

Unlike other organisations, HEIs have two distinct groups of personnel to support, employees and students, both of which are subject to security policies [11]. Users are regarded as the weakest link in the security [12] and many young adults transitioning from school to HEIs lack awareness of how their behavior impacts network security. It is therefore important that measures are taken to improve the security behavior of both staff and students in HEIs. In this regard, it is imperative to understand users' (both staff and student) security knowledge and behaviors in the HEI context to devise appropriate strategies. Fortunately, a number of theory-driven approaches have been used in the security research to explore which factors influence behavior to identify ways in which the security behaviors of the users may be improved [13, 14]. Among these approaches, the Protection Motivation Theory (PMT) [15–17] and Theory of Planned Behavior (TPB) [18, 19] has been used predominantly.

Information-Motivation-Behavioral Skills (IMB) Model was proposed in 1992 to predict health behavior [20]. The IMB Model posits that information and motivation are the key prerequisites towards a given behavior. These prerequisites connect to behavior through the behavioral skills of the person. Since then, the model has been effectively used for understanding users' behaviors as well as for designing interventions to improve users' behaviors in different domains (for example, health [21–23], voting [24] and recycling behaviors [25]). Considering Model's potential to effectively predict and change users' behavior, a few of security researchers have proposed the use of IMB model in the context of security and privacy as well (see [26–29]). However, it has not yet been tested empirically in this context as yet.

The purpose of this paper is to empirically test the applicability of the IMB model to predict users' security behaviors. In doing so, we seek to contribute to the existing research in two ways. First, we wish to add to the theory-driven research in security by considering the IMB Model in the context of managing users' security behavior in HEIs. Second, we would like to improve the information security efforts of HEIs by examining the applicability of the IMB Model [20] as a suitable model to predict the security behavior of university students. The IMB Model has been used effectively as a tool for developing behavioral change programs in contexts other than education [24].

This paper summarises our effort to test the predicting powers of IMB Model in the context of security behaviors with a set of Finnish University students in 2017. The article is structured as follows. Section 2 provides a description of IMB Model and the research model constructed from this model and other theory components. Section 3 outlines the methodology and data analysis. Section 4 describes the results, followed by the discussion in Sect. 5.

2 Theoretical Background: The IMB Model

The IMB Model consists of two predictors (information and motivation), one proposed mediator variable (behavioral skills) and one outcome variable (behavior) (Fig. 1).

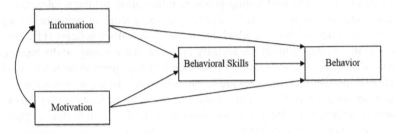

Fig. 1. The information-motivation-behavioral skills model and its constituent variables

The first predictor in the IMB Model is *information*. Information is a prerequisite to a correct and consistent enactment of given behavior [20, 23, 30]. An individual can hold accurate information (that will help in the performance of desired behavior) and inaccurate information (that may impede the desired behavior). Information in the context of information security and privacy may, for example, refer to awareness of the risks related to their use of various devices such as mobile phones [26]. Threat perception plays an important role in the selection of appropriate security measures [11, 31, 32]. The assumption, therefore, is that if a person is aware of information security (risks and threats), he or she will have a more positive attitude and intention to comply to security policies [3] and behave securely [33]. Crossler and Belanger [26] suggest that the IMB Model provides an important link between information (awareness of threats/risk) and the development of skills to behave securely. We propose that threat awareness of users about security threats constitutes a measure of information users have to inform their behavior.

The second predictor of behavior in the IMB Model is *motivation* which is considered a critical component for engaging in and maintaining required behaviors [20, 23, 30]. Fisher and Fisher [20, 23] posited that motivation includes both personal and social motivation. They operationalized personal motivation in terms of a user's personal attitude and socially derived motivation (which arises as a function of the perceived social support for performing a behavior). In terms of the attitude component, users are expected to engage in the desired behavior if they are highly motivated and have a positive attitude towards the desired behavior. In terms of the social motivation component, the assumption is that this may increase or decrease based on an individual's perceptions of the support one is getting from its surroundings to engage in a specific behavior. Such perceptions of support may be subject to both subjective norms [18] (e.g., a user's perception of what kind of security behaviors other peers advocate) and descriptive norms (rules which significant leaders or managers follow, advocate and endorse publicly). Therefore, in the context of security, personal motivation to engage in secure behavior is captured by the security attitude of individuals (as these would be strongly correlated), while social motivation may be a function of what users perceive to be the social support (which arise as a function of both subjective and descriptive norms) regarding security behaviors.

Behavioral skills is proposed as a mediator variable between the two predictors and security behavior in the IMB Model. An individual needs to possess the necessary skills to engage in certain behaviors, in both the health and security domain (see also [20, 23, 30]). However, there are mixed views on how behavioral skills can be measured, which led some authors to use self-efficacy measures to assess perceived capabilities to deal with challenges [24, 34–36]. An example here is the use of self-efficacy as a proxy measure for behavioral skills in terms of patients' health-related behavior (see work by Fisher and Fisher [20, 23]). An alternative approach to using self-efficacy alone is to include some form of knowledge assessment in addition to self-efficacy [25, 26]. Familiarity and actual knowledge have already been studied in the context of information security (for example [37] and [38]). Considering security measures familiarity here in addition to self-efficacy is important as users in a security scenario may find it difficult to select the appropriate behaviors when they are not familiar with the counter measures to combat a threat. In the context of HEIs, many users may need to not just feel capable to act in order to support security, but they also have to recognize specific threats as well as security measures that they should employ. As a result of these concerns and the existing research as well as the exploratory nature of the study, behavioral skills among HEI users may be dependent on a combination of self-efficacy (perceived capability) and subjective familiarity with security measures to counteract a threat - as both are needed for individuals to build and then engage the appropriate behavioral skills in response to potential threats.

Security behavior is proposed to be the dependent variable of the model. This includes specific behavior to counteract threats by employing specific measures such as using software against viruses, ransomware and identity theft.

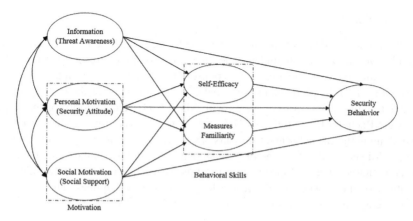

Fig. 2. Modified IMB model in the context of security behavior

In line with the discussion above, the following modified research model is proposed (Fig. 2) to examine the applicability of IMB Model in the context of information security.

3   Method

3.1   Survey Design and Procedure

Quantitative methodology was adopted for the study for which data was collected using a two-part online survey which was developed using a tool called Webropol. The constructs of information, motivation and behavioral skills and demographic were part of first part, whereas, security behavior construct was measured in the second part. Participants were recruited from a large, public university in the Southwest of Finland. Prospective participants were enrolled in a four week long blended learning course on cybersecurity at the time of the study (fall 2017).

First part was administered before the start of the course, whereas, second part was shared among the students two weeks after the completion of the course. To clarify the difference between the awareness (information) and familiarity (behavioral skills), both concepts were introduced to the participants in the survey before they answered items related to threat awareness and familiarity with security measures. Threat awareness was introduced as a fleeting degree of threat knowledge, where a person has heard of a threat but may not have experienced it personally, whereas, Familiarity with security measures was referred to as a degree of knowledge where person has known a security measure through personal experience or association implying a deeper understanding (for further clarification, refer to [37]). The questionnaire took 15–20 altogether. The participants were asked to create a unique ID to connect responses from the two-part survey. While no financial or academic benefits were provided, participants were entered into a prize draw for movie tickets.

3.2   Participants

All 376 enrolled students of the aforementioned course were invited to participate in the study. Out of which 169 students took the two-part survey (response rate = 45%). However, after removing incomplete responses, 159 responses were retained.

About 65% of the participants were male. The average age of the participants was 24 years (ranging from 18 to 63 years with SD = 6.94). The majority of the respondents were Bachelor level students (77%), while the rest were from a Master degree or above. Among the Bachelor level students, 45% were the 1st year, 15% were 2nd year, 6% were 3rd year, and 11% were 4th year students. About 69% of the participants were from computer science and information technology discipline, followed by 23% from the natural sciences, whereas, the rest belonged to other disciplines. The participants had an average internet experience of 14 years (SD = 4.10).

3.3 Measures

Several existing measures were utilized to assess the constructs chosen for information, motivation, behavioral skills and security behavior. Table 1 shows the operational definition of each construct. The detail of items and their sources is given in part-1 of supplement available here: https://goo.gl/AQs1XE.

Table 1. Constructs and operational definitions

Model constructs	Operationalized constructs	Operational definition
Information	Threat awareness	The extent to which a participant is aware of security threats
Motivation	Security attitude	The personal attitude a participant has towards security (personal motivation)
	Social support	The extent to which participant feels that others motivate for engaging in a secure behavior (social motivation)
Behavioral skills	Self-efficacy	The extent to which participant believe he/she is equipped to deal with security threats and exhibit secure behavior
	Measures familiarity	The extent to which participant thinks s/he is familiar with security measures to counteract familiar threats
Behavior	Security behavior	The extent to which participant follow prescribed security advice

Information. Threat awareness was chosen to provide a measure of threat information our participants were aware of. It was measured in terms of user awareness with 20 security threats (taken from [11, 37]). The list consisted of the following threats: Trojan, botnet, identity theft, cookies, virtual stalking, internet surveillance, theft/loss of devices, malware, shoulder-surfing, rogueware, theft/loss of cards and wallets, spyware, information leakage in social network, social engineering, data harvesting (applications), keylogger, virus, phishing, zero-day attack and email harvesting. In each case, participants were asked how aware they were with each threat. Answering options ranged from 1 = very poor to 5 = excellent.

Motivation. Motivation was measured in terms of two measures: attitude towards security (reflecting personal motivation) and social support (reflecting social motivation) [20]. Attitude was measured using four items, adapted from [39]. Social support was self-developed using work of [40, 41] and measured with the help of 3 items. Both constructs were measured on a 7-point scale (1 = strongly disagree to 7 = strongly agree).

Behavioral Skills. As suggested by [26], behavioral skills was measured in term of two proxies that influence skills use: the perceived self-efficacy to deal with a security challenge as well as familiarity with security measures to counteract specific threats. Self-efficacy was measured using six items (adapted from [42, 43]). The scale was similar to the one used for measuring personal motivation. The familiarity of security measures (measures familiarity) was assessed by familiarity of participants with top 20 security measures prescribed by the security experts [44]. The participants had 5-point scale (1 = not at all familiar to 5 = extremely familiar) to rate their familiarity with each measure.

Security Behavior. Self-reported security behavior of participants was measured with the help of self-developed scale consisting of 12 items. We measured the frequency of the engaging in behavior (1 = never, 2 = rarely, 3 = sometimes, 4 = often, 5 = always) related to updating operating system, anti-virus software, downloading software from trust sources, locking computer when stepping away, creating strong passwords, using unique passwords, use of password manager, use of two factor authentication, checking for https while web-surfing, be mindful for the popping up dialogue boxes, and avoid opening unexpected email attachments. Each item represents a piece of advice given by most of the security expert [45]. Measuring frequency of engagement has been considered better as compared to measuring degree of agreement. Egelman and Peer [46] used the same approach for developing security behavior intention scale (SeBIS).

Demographics. The questionnaire captured the following demographics: gender, age, education, discipline, work and Internet experience.

3.4 Data Analysis

For model testing, we used partial least squares structural equation modelling (PLS-SEM, Smart PLS 3.2) as it is particularly appropriate for estimating complex models using small sample size and non-normally distributed data [47–49]. The model is tested in two phases: (1) measurement model testing, and (2) structural model testing. For these phases, established guidelines were followed [48, 50, 51]. Reflective constructs consist of items that show a common cause where cause flows from constructs to items, whereas, formative constructs is a composite measure summarizing a common variation through a set of items. In case of the formative construct, the causal relationship flows from items to the construct (for further differences refer to [52]) According to Chin [52], in formative construct, removal of a single item can affect the construct negatively. Our model consists of both reflective and formative constructs. Reflective model variables were three measures related to the IMB proxies for motivation (attitude and social support) and behavioral skills (in this case, self-efficacy). Formative model variables included the measures related to IMB proxies for information (threat awareness), behavioral skills (familiarity with security measures) and security behavior. The main results are briefly summarized below. (The detailed results of these are available for additional review in part-2 of the supplement available here (https://goo.gl/AQs1XE).

The results of reliability and validity assessment for reflective variables showed that the Coefficient α and composite reliability (CR) for the three reflective constructs were higher than the recommended threshold of 0.70. However, the average variance explained (AVE) of self-efficacy was below the threshold (0.50) suggested by [48]. Five items, two items (SE2 and SE5) were associated with self-efficacy (behavioral skills), two (SA2 and SA3) with security attitude (personal motivation), and one (SS2) with social support (behavioral skills) had items loading less than 0.70. Removal of two low loading items, improved AVE of self-efficacy, however, removal of low loading items related to attitude and social support did not improve AVE of the respective variables. Therefore, as per guidelines [48, 50, 51], low loading items of attitude and social support were retained. In the final model, both security attitude (personal motivation) and self-efficacy (behavioral skills) were measured with the help of four, whereas, social support (social motivation) was measured using three items. The HTMT ratio was below 0.85 for all the constructs, giving evidence of discriminant validity.

The quality of formative constructs was measured by assessing collinearity diagnosis and significance of formative items. In this regards, guidelines of [48] were followed. VIF for measures familiarity (behavioral skills) and security behavior was between 0.20 and 3, which was within the required threshold (VIF should between 0.20 and 5.0). However, two items (TA3 and TA5) of threat awareness (information) had values higher than the 5 (6.41 and 7.20 respectively). Removing of INFO5 from the model brought VIF for TA3 to 3.38, which was acceptable as per guidelines. Therefore, TA5 was removed from further analysis. In addition, six items (TA 18, MF11, SB1, SB4, SB10 and SB12) from measures capturing threat awareness (information), measures familiarity (behavioral skills), and security behavior did not fulfill significance criteria. Therefore, following the formative construct quality assessment, seven items were dropped, two from the measure for threat awareness (information, leaving 18 items), and one from the measure for familiarity with security measures (behavioral skills, leaving 19 items) and four from security behavior (leaving eight items).

4 Results

4.1 Validation of IMB Model

The standardized path coefficients (β), the coefficient of determination (R2) and significance ($p < .05$) of the individual paths in the estimated path analysis are shown in Fig. 3. We also checked the collinearity of the structural model with the help of predictor construct's tolerance (VIF) and found to be between 1.02 and 1.68. As per [48], VIF coefficient between 0.2 and 5 shows lack of collinearity significant levels and effects in the structural model.

As shown in Fig. 3, there were significant direct paths to security behavior from security attitude (IMB's motivation measure; $\beta = 0.28$, $p = 0.024$), self-efficacy (IMB's behavioral skills proxy; $\beta = 0.21$, $p = 0.035$), and familiarity with measures (also IMB's behavioral skills; $\beta = 0.23$, $p < 0.055$). There were indirect effects of information and security attitude (IMB's motivation proxy) on security behavior through self-efficacy (behavioral skills; $\beta = 0.34$, $p < .001$; $\beta = 0.28$, $p < .001$) as well as measures familiarity (IMB's behavioral skills; $\beta = 0.39$, $p = .001$; $\beta = 0.38$, $p = .001$). Social support (IMB proxy for motivation) did not have any direct or indirect significant path to security behavior. Moreover, information significantly

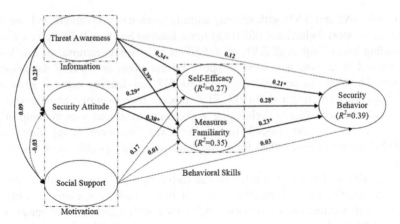

Fig. 3. IMB model constructs with path coefficients and determination coefficients. Dark arrows show a significant prediction, dotted arrows show insignificant prediction, and double edge arrows show correlations.

correlate with both the personal and social motivation constructs, that is, security attitude (r = 0.23, p < .05) and social support (r = 0.09, p < .05). For detail statistics on structural model and correlation consult Tables 2 and 3.

Table 2. Structural model statistics for IMB Model constructs. The path coefficient (β), adjusted coefficient of determination R^2, significance tested at p < 0.05 with effect size f^2. Non-significant relationships are highlighted in italic.

Path[a]	VIF	β	t	R^2	p	Sig.	f^2
TA → SE	1.07	0.34	3.06		0.002	Y	0.158
SA → SE	1.06	0.28	3.93		0.024	Y	*0.109*
SS → SE	*1.01*	*0.17*	*1.70*	0.27	*0.090*	*N*	*0.042*
TA → MF	1.07	0.39	3.46		0.001	Y	0.225
SA → MF	1.06	0.38	3.06		0.002	Y	0.216
SS → MF	*1.01*	*0.01*	*0.11*	0.35	*0.915*	*N*	*0.000*
TA → SB	*1.40*	*0.11*	*0.67*		*0.501*	*N*	*0.020*
SA → SB	1.35	0.28	2.26		0.024	Y	0.097
SS → SB	*1.05*	*0.03*	*0.29*		*0.769*	*N*	*0.001*
MF → SB	1.66	0.23	1.92		0.050	Y	0.057
SE → SB	1.47	0.21	2.11	0.39	0.035	Y	0.052

Note. [a]TA = Threat awareness (Information), SA = Security Attitude (personal motivation), SS = Social Support (social motivation), SE = Self-efficacy (behavioral Skills), MF = Measures familiarity (behavioral Skills), SB = Security behavior. Effect size(f^2): 0.02 = low, 0.15 = Medium, 0.35 = large]

Table 3. Correlation matrix of measures (in relation to IMB constructs)

Constructs	1	2	3	4	5	6
Threat awareness (information)	1					
Security attitude (personal motivation)	**0.23**	1				
Social support (social motivation)	**0.09**	−0.03	1			
Self-efficacy (behavioral skills)	0.42	0.36	0.19	1		
Measures familiarity (behavioral skills)	0.49	0.47	0.03	0.44	1	
Security behavior	0.39	0.50	0.08	0.48	0.51	1

5   Discussion

This study examined the applicability of IMB Model in a slightly modified format in a HEI information security context. Our study shows behavioral skills (both self-efficacy and familiarity with security measures) and personal motivation (security attitude) directly predicted the security behavior of the students. Also, our proxy variables (all selected on the basis of information, motivation, and behavioral skills in the IMB Model) explained 39% of the variance in security behavior in our student sample. Furthermore, information (threat awareness) and personal motivation (security attitude) were positively associated with behavioral skills (self-efficacy as well as measures familiarity). We also found that information (threat awareness) and personal motivation (represented by security attitude) indirectly affected security behavior - through the behavioral skills variables (which operated as a mediator). Social motivation (captured in the form of social support) does not have a direct or indirect relationship with security behavior. Information and motivation variables (personal and social) also correlated with one another.

To evaluate our contribution to the study of information security in the educational context, two points need to be clarified beforehand. First, our results need to be interpreted in the context of existing theory and previous findings. Mayer et al. [53] found that self-efficacy has a reliable but weak to medium positive effect on behavioral intention in three different studies. However, in the case of attitude measures, these have been shown to a reliable medium effect on behavioral intention in security studies [53]. The results in our study confirmed the previous findings: self-efficacy had a small effect size on the security behavior ($f2 = 0.05$). However, in contrast to the medium effect reported for attitude [19], we found that attitude in our sample has a small effect size ($f2 = 0.01$) as well. As our results pertain to behavior rather than intention, it is difficult to compare these effects directly. However, given the often noted disconnect between behavioral intention and behavior (as intention may not always lead to behavior), higher effect sizes may be expected for intention rather than behavior - which may not always align with one's intention.

5.1   Recommendations

The previous two points lead us to the following recommendations for those responsible for managing information security training in HEIs. According to our results based on the IMB Model, constructs related to information (threat awareness) and motivation (based on security attitude and social norms) are crucial factors for students to acquire skills to engage in information security behaviors. However, practical knowledge is important in addition to information and motivation to employ security measures [26]. HEIs should focus on all IMB related constructs simultaneously to achieve improved security behavior. This means tackling threat awareness, personal motivation (in form of security attitudes) and behavioral skills (such as self-efficacy and familiarity with measures to counteract security threats) together, rather than just picking one of the three to improve security behavior.

Training and other interventions based on the IMB Model may improve students' security-related knowledge by: (a) increasing their access to information (e.g., through training about threats and measures), (b) raising their motivation (by increasing the perceived relevance and providing resources (social support) to perform secure behaviors), and (c) providing them with opportunities to gain and test their behavioral skills as this will increase their confidence and capability to do so when an actual threat emerges which requires immediate counteraction. All three may then hopefully improve the security behaviors of the students in HEIs, reducing institutional vulnerability to threats while also giving the students the skills to act securely when they transition into the workplace and use employer systems.

5.2   Limitations and Future Research

The study is not without limitations. For example, our cross-sectional sample was recruited from a pool of students who enrolled on a security-related course. This suggests they may have been more interested in information security compared to those who selected other courses instead. Moreover, the majority of the students were Bachelor level students belonging to computer science, information technology and engineering (STEM) disciplines. Therefore, our findings may not translate to students' behavior outside these STEM areas that may lack familiarity with threat and also behavioral skills. Furthermore, some methodological issues arise also. First, in this paper, behavioral skills (familiarity with security countermeasures) and security behavior were assessed using a self-report measures rather than objective indicators. And second, considering the nature and types of security measures, the aforementioned variables were measured as formative variables.

This leads us to four areas worthy of more investigation. One, further research is needed to establish the generalizability of our findings to other non-STEM samples. Two, moving from subjective to objective measures would hopefully be possible when assessing security skills and behaviors. As proposed in the recommendations section, it would be helpful to run security tests and countermeasure exercises (similar to health and safety trial runs) to ensure students know how to effectively respond to a threat not just in theory but also in practice. Such trials would also, if captured, generate more objective data about the behavioral skills and actual security behavior of students. If the study is replicated with the help of an IT support center, for example, actual security behavior may be captured by the IT system through the interaction of users with the system, circumventing the need for self-reported behavioral measures.

And three, it would be interesting to see the longitudinal effect of information, motivation and skills on security behavior. We had a brief interval (two weeks) between our assessment of motivation, information and behavioral skill on the one hand, and security behavior on the other. The relationship between the constructs may change over time, particularly if training is provided following the first round.

Four, security behavior may be measured in numerous ways. A thoroughly designed construct may improve the predictability of the IMB Model. In this study, security behavior was measured by asking participants to indicate which of the twelve security recommendations they follow (see also [45]). Considering that there are more than twelve measures that a user requires for his/her information security, a concise list of important security behaviors may be identified, and security measure may be operationalized accordingly. We believe using formative structure for a holistic security behavior variable will be suitable.

6 Conclusion

The purpose of the study was to examine the predictability of a slightly modified IMB Model in the context of information security. We tested IMB model using SmartPLS SEM, and found that indeed proxies for information (threat awareness), motivation (security attitude and social support) and behavioral skills (self-efficacy and familiarity with threat measures) enabled us to predict the security behavior of 159 university students. The results showed that students with higher threat awareness, more positive security attitude, higher self-efficacy, and familiarity with security measures engaged in more secure behavior. This work proves empirically that IMB model can be used to study security behaviors of the students.

References

1. Kim, W., Jeong, O.-R., Kim, C., So, J.: The dark side of the Internet: attacks, costs and responses. Inf. Syst. **36**, 675–705 (2011)
2. Aurigemma, S., Panko, R.: A composite framework for behavioral compliance with information security policies. In: 45th Hawaii International Conference on System Sciences, pp. 3248–3257. IEEE (2012)
3. Bulgurcu, B., Cavusoglu, H., Benbasat, I.: Information security policy compliance: an empirical study of rationality-based beliefs and information security awareness. MIS Q. **34**, 523–548 (2010)
4. Pahnila, S., Siponen, M., Mahmood, A.: Employees' behavior towards IS security policy compliance. In: 2007 40th Annual Hawaii International Conference on System Sciences (HICSS 2007), p. 156b. IEEE (2007)
5. Abraham, S.: Information security behavior: factors and research directions. In: AMCIS 2011 (2011)
6. D'Arcy, J., Hovav, A., Galletta, D.: User awareness of security countermeasures and its impact on information systems misuse: a deterrence approach. Inf. Syst. Res. **20**, 79–98 (2009)

7. Kerievsky, B.: Security and confidentiality in a university computer network. ACM SIGUCCS Newsl. **6**, 9–11 (1976)
8. Ingerman, B.L., Yang, C.: Top-ten IT issues, 2011. Educ. Rev. **46**, 24 (2011)
9. Al-Janabi, S., Al-Shourbaji, I.: A study of cyber security awareness in educational environment in the Middle East. J. Inf. Knowl. Manag. **15**, 1650007 (2016)
10. Katz, F.H.: The effect of a university information security survey on instruction methods in information security. In: Proceedings of the 2nd Annual Conference on Information Security Curriculum Development - InfoSecCD 2005, p. 43. ACM Press, New York (2005)
11. Farooq, A., Kakakhel, S.R.U., Virtanen, S., Isoaho, J.: A taxonomy of perceived information security and privacy threats among IT security students. In: 10th International Conference for Internet Technology and Secured Transactions, ICITST 2015, pp. 280–286. IEEE (2016)
12. Savitz, E.: Humans: the weakest link in information security. https://www.forbes.com/sites/ciocentral/2011/11/03/humans-the-weakest-link-in-information-security/#77a4bb46de87
13. Lebek, B., Uffen, J., Neumann, M., Hohler, B., Breitner, M.H.: Information security awareness and behavior: a theory-based literature review. Manag. Res. Rev. Inf. Manag. Comput. Secur. **37**, 1049–1092 (2014)
14. Howe, A.E., Ray, I., Roberts, M., Urbanska, M., Byrne, Z.: The psychology of security for the home computer user. In: 2012 IEEE Symposium on Security and Privacy, pp. 209–223. IEEE (2012)
15. Rogers, R.W.: A protection motivation theory of fear appeals and attitude change. J. Psychol. **91**, 93–114 (1975)
16. Maddux, J.E., Rogers, R.W.: Protection motivation and self-efficacy: a revised theory of fear appeals and attitude change. J. Exp. Soc. Psychol. **19**, 469–479 (1983)
17. Sommestad, T., Karlzén, H., Hallberg, J.: A meta-analysis of studies on protection motivation theory and information security behaviour. Int. J. Inf. Secur. Priv. **9** (2015)
18. Ajzen, I.: The theory of planned behavior. Organ. Behav. Hum. Decis. Process. **50**, 179–211 (1991)
19. Sommerstad, T., Karlzen, H., Hallberg, J.: The theory of planned behavior and information security policy compliance. J. Comput. Inf. Syst. 1–10 (2017)
20. Fisher, J.D., Fisher, W.A.: Changing AIDS-risk behavior. Psychol. Bull. **111**, 455–474 (1992)
21. Robertson, A.A., Stein, J.A., Baird-Thomas, C.: Gender differences in the prediction of condom use among incarcerated juvenile offenders: testing the information-motivation-behavior skills (IMB) model. J. Adolesc. Health **38**, 18–25 (2006)
22. Fisher, W.A., Williams, S.S., Fisher, J.D., Malloy, T.E.: Understanding AIDS risk behavior among sexually active urban adolescents: an empirical test of the information–motivation–behavioral skills model. AIDS Behav. **3**, 13–23 (1999)
23. Fisher, J.D., Fisher, W.A., Harman, J.J.: An information-motivation-behavioral skills model of adherence to antiretroviral therapy. Health Psychol. **25**, 462–473 (2006)
24. Glasford, D.E.: Predicting voting behavior of young adults: the importance of information, motivation, and behavioral skills. J. Appl. Soc. Psychol. **38**, 2648–2672 (2008)
25. Seacat, J.D., Northrup, D.: An information–motivation–behavioral skills assessment of curbside recycling behavior. J. Environ. Psychol. **30**, 393–401 (2010)
26. Crossler, R.E., Bélanger, F.: The mobile privacy-security knowledge gap model: understanding behaviors. In: 50th Hawaii International Conference on System Sciences (2017)
27. Khan, B., Alghathbar, K.S., Khan, M.K.: Information security awareness campaign: an alternate approach. In: Kim, T.-h., Adeli, H., Robles, R.J., Balitanas, M. (eds.) ISA 2011. CCIS, vol. 200, pp. 1–10. Springer, Heidelberg (2011). https://doi.org/10.1007/978-3-642-23141-4_1

28. Mariani, M.G., Zappalà, S.: PC virus attacks in small firms: effects of risk perceptions and information technology competence on preventive behaviors. TPM Test. Psychom. Methodol. Appl. Psychol. **21**, 51–65 (2014)
29. Pattinson, M.R., Anderson, G., Analyses, A.: End-user risk-taking behaviour: an application of the IMB model. In: 6th Annual Security Conference (2007)
30. Fisher, J.D., Fisher, W.A., Misovich, S.J., Kimble, D.L., Malloy, T.E.: Changing AIDS risk behavior: effects of an intervention emphasizing AIDS risk reduction information, motivation, and behavioral skills in a college student population. Health Psychol. **15**, 114–123 (1996)
31. Huang, D.-L., Rau, P.-L.P., Salvendy, G.: Perception of information security. Behav. Inf. Technol. **29**, 221–232 (2010)
32. Yeh, Q.-J., Chang, A.J.-T.: Threats and countermeasures for information system security: a cross-industry study. Inf. Manag. **44**, 480–491 (2007)
33. Farooq, A., Isoaho, J.J., Virtanen, S., Isoaho, J.J.: Information security awareness in educational institution: an analysis of students' individual factors. In: Proceedings of the 14th IEEE International Conference on Trust, Security and Privacy in Computing and Communications, TrustCom 2015, pp. 352–359. IEEE (2015)
34. Chang, T., et al.: A study on the information-motivation-behavioral skills model among Chinese adults with peritoneal dialysis. J. Clin. Nurs. **27**, 1884–1890 (2018)
35. Compeau, D., Higgins, C.A., Huff, S.: Social cognitive theory and individual reactions to computing technology: a longitudinal study. MIS Q. **23**, 145 (1999)
36. Compeau, D.R., Higgins, C.A.: Application of social cognitive theory to training for computer skills. Inf. Syst. Res. **6**, 118–143 (1995)
37. Jeske, D., van Schaik, P.: Familiarity with Internet threats: beyond awareness. Comput. Secur. **66**, 129–141 (2017)
38. Kruger, H., Drevin, L., Steyn, T.: A vocabulary test to assess information security awareness. Inf. Manag. Comput. Secur. **18**, 316–327 (2010)
39. Taylor, S., Todd, P.A.: Understanding information technology usage: a test of competing models. Inf. Syst. Res. **6**, 144–176 (1995)
40. Zimet, G.D., Dahlem, N.W., Zimet, S.G., Farley, G.K.: The multidimensional scale of perceived social support. J. Pers. Assess. **52**, 30–41 (1988)
41. Hupcey, J.E.: Clarifying the social support theory-research linkage. J. Adv. Nurs. **27**, 1231–1241 (1998)
42. Anderson, C.L., Agarwal, R.: Practicing safe computing: a multimethod empirical examination of home computer user security behavioral intentions. MIS Q. **34**, 613–643 (2010)
43. Thompson, N., McGill, T.J., Wang, X.: "Security begins at home": determinants of home computer and mobile device security behavior. Comput. Secur. **70**, 376–391 (2017)
44. Reeder, R., Ion, I., Consolvo, S.: 152 simple steps to stay safe online: security advice for non-tech-savvy users. IEEE Secur. Priv. **15**, 55–64 (2017)
45. Ion, I., Reeder, R., Consolvo, S.: "…no one can hack my mind": comparing expert and non-expert security practices. In: 2015 Symposium on Usable Privacy and Security, pp. 327–340 (2015)
46. Egelman, S., Peer, E.: Scaling the security wall: developing a security behavior intentions scale (SeBIS). In: Proceedings of the 33rd Annual ACM Conference on Human Factors in Computing Systems - CHI 2015, pp. 2873–2882. ACM Press, New York (2015)
47. Ringle, C.M., Smith, D., Reams, R.: Partial least squares structural equation modeling (PLS-SEM): a useful tool for family business researchers. J. Fam. Bus. Strateg. **5**, 105–115 (2014)
48. Hair Jr., J.F., Hult, G.T., Ringle, C., Sarstedt, M.: A Primer on Partial Least Squares Structural Equation Modeling (PLS-SEM). Sage Publishers, Thousand Oaks (2016)

49. Lowry, P.B., Gaskin, J.: Partial least squares (PLS) structural equation modeling (SEM) for building and testing behavioral causal theory: when to choose it and how to use it. IEEE Trans. Prof. Commun. **57**, 123–146 (2014)
50. Hair, J.F., Black, W.C., Babin, B.J., Anderson, R.E., Tatham, R.L.: Multivariate Data Analysis. Prentice Hall, Upper Saddle River (2010)
51. Henseler, J., Ringle, C.M., Sarstedt, M.: A new criterion for assessing discriminant validity in variance-based structural equation modeling. J. Acad. Mark. Sci. **43**, 115–135 (2015)
52. Chin, W.W.: The partial least squares approach to structural equation modeling. In: Marcoulides, G.A. (ed.) Modern Methods for Business Research, pp. 295–336 (1998)
53. Mayer, P., Kunz, A., Volkamer, M.: Reliable behavioural factors in the information security context. In: Proceedings of the 12th International Conference on Availability, Reliability and Security, ARES 2017, pp. 1–10. ACM Press, New York (2017)

Why Do People Pay for Privacy-Enhancing Technologies? The Case of Tor and JonDonym

David Harborth$^{(\boxtimes)}$ ⓘ, Xinyuan Cai, and Sebastian Pape ⓘ

Chair of Mobile Business and Multilateral Security,
Goethe University Frankfurt, Frankfurt am Main, Germany
david.harborth@m-chair.de

Abstract. Today's environment of data-driven business models relies heavily on collecting as much personal data as possible. One way to prevent this extensive collection, is to use privacy-enhancing technologies (PETs). However, until now, PETs did not succeed in larger consumer markets. In addition, there is a lot of research determining the technical properties of PETs, i.e. for Tor, but the use behavior of the users and, especially, their attitude towards spending money for such services is rarely considered. Yet, determining factors which lead to an increased willingness to pay (WTP) for privacy is an important step to establish economically sustainable PETs. We argue that the lack of WTP for privacy is one of the most important reasons for the non-existence of large players engaging in the offering of a PET. The relative success of services like Tor corroborates this claim since this is a service without any monetary costs attached. Thus, we empirically investigate the drivers of active users' WTP of a commercial PET - JonDonym - and compare them with the respective results for a donation-based service - Tor. Furthermore, we provide recommendations for the design of tariff schemes for commercial PETs.

Keywords: Privacy · Privacy-enhancing technologies · Pricing ·
Willingness to pay · Tor · JonDonym

1 Introduction

Perry Barlow states: "The internet is the most liberating tool for humanity ever invented, and also the best for surveillance. It's not one or the other. It's both" [1]. One of the reasons for surveilling users is a rising economic interest in the internet [2]. However, users who have privacy concerns and feel a strong need to protect their privacy are not helpless, they can make use of privacy-enhancing technologies (PETs). PETs allow users to improve their privacy by eliminating or minimizing personal data disclosure to prevent unnecessary or unwanted processing of personal data [3]. Examples of PETs include services which allow anonymous communication, such as Tor [4] or JonDonym [5]. There has been lots of research on Tor and JonDonym [6, 7], but the large majority of it is of technical nature and does not consider the user. However, the number of users if crucial for this kind of services. Besides the economic point of view which suggests that more users allow a more cost-efficient way to run those services, the quality of the offered service is depending on the number of users

G. Dhillon et al. (Eds.): SEC 2019, IFIP AICT 562, pp. 253–267, 2019.
https://doi.org/10.1007/978-3-030-22312-0_18

since an increasing number of (active) users also increases the anonymity set. The anonymity set is the set of all possible subjects who might cause an action [8], thus a larger anonymity set may make it more difficult for an attacker to identify the sender or receiver of a message.

In the end, the sustainability of a service not only depends on the number of active users but also on a company or organization with the intention of running the service. One intention certainly is a well working business model. As a consequence, it is crucial to not only learn about the users' intention to use a PET, but also to understand the users' willingness to pay (WTP) for a service. Determining factors to understand the users' WTP along with a suitable tariff structure is the key step to establish economically sustainable services for privacy. The current market for PET providers is rather small, some say the market even fails [9]. We argue that the lack of WTP for privacy is one of the most important reasons for the non-existence of large players engaging in the offering of a PET. Earlier research on WTP often works with hypothetical scenarios (e.g. with conjoint-analyses) and concludes that users are not willing to pay for their privacy [10, 11]. We tackle the issue based on actual user experiences and behaviors and enhance the past research by analyzing two existing PETs with active users, with some of them already paying or donating for the service. Tor and JonDonym are comparable with respect to their functionality and partially with respect to the users' perceptions about them. However, they differ in their business model and organizational structure. Therefore, we investigate the two research questions:

RQ1: Which factors influence the willingness to pay for PETs?
RQ2: What are preferred tariff options of active users of a commercial PET?

The remainder of the paper is structured as follows: Sect. 2 briefly introduces the anonymization services Tor and JonDonym and lists related work on PETs and users' willingness to pay. In Sect. 3, we present the research hypotheses and describe the questionnaire and the data collection process. We present the results of our empirical research in Sect. 4 and discuss the results and conclude the paper in Sect. 5.

2   Theoretical Background and Related Work

Privacy-Enhancing Technologies (PETs) is an umbrella term for different privacy protecting technologies. Borking and Raab define PETs as "a coherent system of ICT measures that protects privacy [...] by eliminating or reducing personal data or by preventing unnecessary and/or undesired processing of personal data; all without losing the functionality of the data system" [12]. In the following sections, we describe Tor and JonDonym as well as related work with respect to WTP for privacy.

2.1   Tor and JonDonym

Tor and JonDonym are low latency anonymity services which redirect packets in a certain way in order to hide metadata (the sender's/receiver's internet protocol (ip) address) from passive network observers. Low latency anonymity services can be used for interactive services such as messengers. Due to network overheads this still leads to increased latency which was evaluated by Fabian et al. [13] who found associated usability issues when using Tor. Technically, Tor – the onion router – is an overlay network where the users' traffic is encrypted and directed over several different servers (relays). The chosen traffic routes should be difficult for an adversary to observe, which means that unpredictable routes through the Tor network are chosen. The relays where the traffic leaves the tor network are called "exit nodes" and for an external service the traffic seems to originate from those. JonDonym is based on user selectable mix cascades, with two or three mix servers in one cascade. For mix networks route unpredictability is not important so within one cascade always the same sequence of mix servers is used. Thus, for an external service the traffic seems to originate from the last mix server in the cascade. As a consequence, other usability issues may arise when websites face some abusive traffic from the anonymity services [14] and decide to restrict users from the same origin. Restrictions range from outright rejection to limiting the users' access to a subset of the services functionality or imposing hurdles such as CAPTCHA-solving [15]. For the user it appears that the website is not function properly. Tor offers an adapted browser including the Tor client for using the Tor network, the "Tor Browser". Similarly, the "JonDoBrowser" includes the JonDo client for using the JonDonym network. Although technically different, JonDonym and Tor are highly comparable with respect to the general technical structure and the use cases. However, the entities who operate the PETs are different. Tor is operated by a non-profit organization with thousands of voluntarily operated servers (relays) over which the encrypted traffic is directed. Tor is free to use with the option that users can donate to the Tor project. The actual number of users is estimated with approximately 2,000,000 active users [4]. JonDonym is run by a commercial company. The mix servers used to build different mix cascades are operated by independent and non-interrelated organizations or private individuals who all publish their identity. The service is available for free with several limitations, like the maximum download speed. In addition, there are different premium rates without these limitations that differ with regard to duration and included data volume. Thus, JonDonym offers several different tariffs and is not based on donations. The actual number of users is not predictable since the service does not keep track of this.

From a research perspective, there are some papers about JonDonym, e.g. a user study on user characteristics of privacy services [16]. Yet, the majority of work is about Tor. Most of the work is technical [6], e.g. on improvements such as relieved network congestion, improved router selection, enhanced scalability or reduced communication/computational cost of circuit construction [17]. There is also lots of work about the security respectively anonymity properties [18, 19] and traffic correlation [20].

2.2 Related Work

Previous non-technical work on PETs mainly considers usability studies and does not primarily focus on WTP. For example, Lee et al. [21] assess the usability of the Tor Launcher and propose recommendations to overcome the found usability issues. Further research suggests zero-effort privacy [22, 23] by improving the usability of the service. In quantitative studies, we already investigated privacy concerns and trust on

JonDonym [24] and Tor [25, 26] based on Internet users' information privacy concerns (IUIPC) [27] and could extent the causal model by "trust in the service" which plays a crucial role for the two PETs. Some experiments suggest that users are not willing to pay for their privacy [10, 11]. In contrast to these experiments, we surveyed actual users – some of them already paying or donating for the service. Grossklags find contradicting behavior of users when it comes to WTP to protect information and "willingness to accept" compensation for revealing information [28]. Further work covers selling personal data [29, 30] e.g. on data markets [31] or experiments on the value of privacy [32]. Some work tries to explain the privacy paradox with economic models [33] or discusses the right of the users to know the value of their data [34]. However, all of these are focused on the value of certain data or privacy and not on the users' WTP for privacy. Cranor et al. investigate how actual users use their privacy preferences tool [35]. Spiekermann investigate the traits and views of actual users of the predecessor of JonDonym, AN.ON/JAP, a free anonymity service [16]. However, since the tools were free, none of them investigated the users' WTP. Following a more high-level view, some research addresses the markets for PETs. Federrath claims that there is a market for PETs but they have to consider law enforcement functionality [36]. Rossnagel analyzes PET markets based on diffusion of innovations theory about anonymity services [9] and concludes a market failure. Schomakers et al. do a cluster analysis of users and find three groups with different attitudes towards privacy and argue that each of the groups need distinct tools [37]. In the same line, further research concludes that one should focus on specific subgroups for the adoption of Tor [38]. Following a market perspective, Boehme et al. analyze the condition under which it is profitable for sellers in e-commerce environments to support PETs, assuming that without PETs they could increase their profit with price discrimination [39].

3   Methodology

In this section we present the research hypotheses, the questionnaire and the data collection process. The demographic questions were not mandatory to fill out. This was done on purpose since we assumed that most of the participants are highly sensitive with respect to their personal data and could potentially react to mandatory demographic questions by terminating the survey. Consequently, the demographics are incomplete to a large extent. Therefore, we had to resign from a discussion of the demographics in our research context.

The statistical analysis of the research data is conducted with the open-source software R. First of all, we focus solely on JonDonym and compare the differences of average preferences for alternative tariff schemes. Thereby, we differentiate between participants stating to use JonDonym in the free of charge option those stating to use it with one of the available premium tariffs. Due to non-normality of the data, we use the non-parametric test Wilcoxon rank sum test to determine whether preferences for newly designed tariffs differ from each other among different types of users. We designed these new tariffs in collaboration with the chief executive of the JonDos

GmbH in order to provide realistic pricing schemes which are economically viable and sustainable for the company. We used the paired Wilcoxon test to determine whether users' preferences for one tariff are statistically significantly different from the other tariffs. The Wilcoxon rank sum test is also called Mann-Whitney-U-Test. It is a non-parametric test of the null hypothesis that the mean of one sample will be different from the mean from a second sample. The paired Wilcoxon test is also called the Wilcoxon signed-rank test which is a similar nonparametric test used for dependent samples [40, 41]. In order to illustrate the difference in preferences among two types of users, i.e. free users and premium users, we use boxplots to visualize the descriptive statistics of the two samples [42]. A boxplot is a method for graphically depicting groups of numerical data through their quartiles. Boxplots are non-parametric. They display variation in samples of a statistical population without making any assumptions of the underlying statistical distribution. The upper line of the box is the first quartile, the band inside the box is the second quartile (the median) and the bottom line of the box is the third quartile.

3.1   Research Model and Hypotheses for the Logistic Regression Model

As a last step, we conduct a logistic regression to find out which factors influence users' willingness to pay for privacy (in our case willingness to pay for JonDonym and willingness to donate to Tor). We used the logistics regression to build the model because our dependent variable is a binary variable. A linear regression is not an appropriate model here due to the violation of the assumption that the dependent variable (WTP) is continuous, with errors which are normally distributed [43]. The probit regression is also not suitable because it assumes that our dependent variable is not normally distributed. Willingness to pay for JonDonym is defined as the binary classification of JonDonym users' actual behavior.

$$willingness\,to\,pay = \begin{cases} 0,\ if\ the\ respondent\ uses\ a\ free\ tariff \\ 1,\ if\ the\ respondent\ uses\ a\ premium\ tariff \end{cases} \quad (1)$$

Accordingly, willingness to donate is defined as the binary classification of Tor users' actual behavior.

$$willingness\,to\,donate = \begin{cases} 0,\ if\ the\ respondent\ has\ never\ donated \\ 1,\ if\ the\ respondent\ has\ donated \end{cases} \quad (2)$$

The independent variables are risk propensity (RP), frequency of improper invasion of privacy (VIC), trusting beliefs in online companies (TRUST), trusting beliefs in JonDonym ($TRUST_{PET}$) and knowing of Tor/JonDonym (TOR/JD) or not. Thus, our research model is as follows:

$$WTP/WTD_i = \beta_0 + \beta_1 RP_i + \beta_2 VIC_i + \beta_3 TRUST_i + \beta_4 TRUST_{PET,i} + \beta_5 TOR/JD_i + \varepsilon_i$$
$$(3)$$

Risk propensity measures the risk aversion of the individual, i.e. the higher the measure, the more risk-averse the individual [44]. Literature finds that a risk aversion can act as a driver to protect an individual's privacy [45]. Thus, we hypothesize:

> *H1: Risk propensity (RP) has a positive effect on the likelihood of paying or donating for PETs.*

Privacy victim (VIC) measures how often individuals experienced a perceived improper invasion in their privacy [27]. Results of past research dealing with perceived bad experiences with privacy indicate that such experiences can cause individuals to protect their privacy to a larger extent [46]. Thus, we hypothesize:

> *H2: The more frequent users felt that they were a victim of an improper breach of their privacy, the more likely they are to pay or donate for PETs.*

The construct *trust in online companies* assesses individuals trust in online companies with respect to handling their personal data [27]. Results in the literature suggest that a higher trust in online companies has a positive effect on the willingness to disclose personal information. Following this finding, we argue that users who have a higher level of trust in online companies, are less likely to spend money for protecting their privacy. Therefore, we hypothesize:

> *H3: The more users trust online companies with handling their personal data, the less likely they are to pay or donate for PETs.*

Trust in JonDonym/Tor is adapted from Pavlou [47]. Trust can refer to the technology (in our case PETs (Tor and JonDonym)) itself as well as to the service provider. Since the non-profit organization of Tor evolved around the service itself [4], it is rather difficult for users to distinguish which label refers to the technology itself and which refers to the organization. The same holds for JonDonym since JonDonym is the only main service offered by the commercial company JonDos. Therefore, we argue that it is rather difficult for users to distinguish which label refers to the technology itself and which refers to the company. Thus, we decided to ask for trust in the PET (Tor and JonDonym, respectively), assuming that the difference to ask for trust in the organization/company is negligible. Literature shows that trust in services enables positive attitudes towards interacting with these services [24–26, 47]. In line with these results, we argue that a higher level of trust in the PET increases the likelihood to spend money for it. Thus, we hypothesize:

> *H4: The more users trust the PET, the more likely they are to pay or donate for it.*

Lastly, we included a question about whether users of Tor/JonDonym know JonDonym/Tor. We included this question due to previous findings about a substituting effect of Tor with regard to the WTP for JonDonym [48]. Users of JonDonym partially stated that they would only spend money for a premium tariff, if Tor was not existent. Thus, we wanted to include this factor as a control variable in our analysis and hypothesize:

> *H5: The likelihood of JonDonym users to pay for a premium tariff decreases, if they are aware of Tor (we do not expect a similar effect for Tor users).*

3.2 Data Collection

We conducted the studies with German and English-speaking users of Tor and JonDonym. For each service, we administered two questionnaires. Partially, items for the German questionnaire had to be translated since some constructs are adapted from the English literature. To ensure content validity of the translation, we followed a rigorous translation process. First, we translated the English questionnaire into German with the help of a certified translator (translators are standardized following the DIN EN 15038 norm). The German version was then given to a second independent certified translator who retranslated the questionnaire to English. This step was done to ensure the equivalence of the translation. Third, a group of five academic colleagues checked the two English versions with regard to this equivalence. All items were found to be equivalent [49]. The items for all analyses can be found in the appendix.

We installed the surveys on a university server and managed it with the survey software LimeSurvey (version 2.72.6) [50]. For Tor, we distributed the links to the English and German version over multiple channels on the internet. An overview of every distribution channel can be found in an earlier paper based on the same dataset [26]. In sum, 314 participants started the questionnaire (245 English version, 40 English version posted in hidden service forums, 29 German version). Of those 314 approached participants, 135 (105 English version, 13 English version posted in hidden service forums, 17 German version) filled out the questionnaires completely. After deleting all participants who answered a test question in the middle of the survey incorrectly, 124 usable data sets remained for the following analysis. For JonDonym, we distributed the links to the English and German version with the beta version of the JonDonym browser and published them on the official JonDonym homepage. In sum, 416 participants started the questionnaire (173 English version, 243 German version). Of those 416 approached participants, 141 (53 English version, 88 German version) remained after deleting unfinished sets and all participants who answered a test question incorrectly.

4 Results

We present the results of our empirical analyses in this section. In the first part, we discuss the analysis of the current tariff structures (JonDonym) and donation statistics (Tor). Furthermore, we assess preferences of JonDonym users regarding new alternative tariff schemes. In the second part, we show the results of the logistic regression model with the factors influencing the willingness to pay (JonDonym)/to donate (Tor).

4.1 Tariff Analysis for JonDonym

Among the 141 JonDonym users in of our survey, 85 users use a free tariff. 56 users are using JonDonym with a paid tariff. Among the 124 Tor users of our survey, 93 of them have never donated to Tor. Among donating users, the amounts of donation are arbitrary. The payment structure of JonDonym and descriptive statistics for the donations to Tor are shown in Table 1. It can be seen that roughly 1/3 of the participants spend money for JonDonym (25%) and Tor (39.72%). To analyze potential tariff optimizations for JonDonym, we asked about users' preferences for three general tariff structures, namely a high-data-volume tariff (TP1), a low-price tariff (TP2) and a low-anonymity tariff (TP3). In addition, we designed five new tariffs. TRN4 is the tariff with the lowest data volume per month and TRN5 is the tariff with highest data volume per month. The specific wording of the tariff options can be found in the appendix.

Table 1. Tariff and donation statistics of JonDonym and Tor users

JonDonym		Tor	
Tariff option	N = 141	Tariff option	N = 124
Free of charge option	85	No donation	93
Volume-M (1500 MB/12 months 10€)	28	Donation	31
Volume-L (5000 MB/24 months 30€)	19	Min. donation	0.00
Flat-M (monthly 2 GB/6 months/50€)	5	Median donation	100.00
Flat-L (monthly 5 GB/6 months/100€)	4	Mean donation	301.40
Volume-S (650 MB/6 months 5€)	0	Max. donation	4500.00

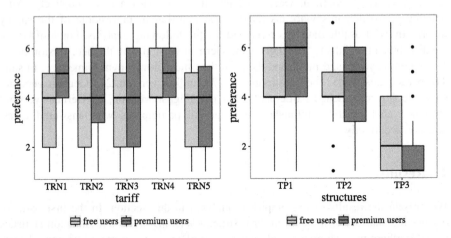

Fig. 1. Users' preference for alternative tariff structures (left side) and users' preferences for tariff structures (right side), free users = 85, premium users = 56

Figure 1 shows the boxplots for the preferences for the five new tariff options (TRN) differentiated between free and premium users as well as three alternative tariff structures (TP). The median preferences of free users for the five tariffs are neutral (preference = 4). However, the mean preference of free users for TRN4 is slightly higher compared to the other options. In comparison, premium users have a higher preference for TRN1 and TRN4. In a next step, we analyze whether the differences illustrated with the boxplots between options for the different groups (full sample, premium users, free users) are statistically significant (Table 2). Our results indicate that the whole sample of users shows the highest preference for TRN4 and the second highest preference for TRN1. The remaining tariffs, i.e. TRN2, TRN3 and TRN5 are favored least of all. However, this contradicts with the conclusion that the total users show the highest preference for TP1. Thus, it makes sense to split the sample and look at free and premium users. Premium users show the highest preference for TRN1 and TRN4, the second highest preference for TRN2 and TRN3, and the least preference for TRN5. Thus, they show a higher preference for 100 GB tariffs. This is in line with the conclusion that premium users have the highest preference for TP1. Free users show a neutral preference for all five tariffs except for TRN4 (slightly higher).

Table 2. Paired Wilcoxon tests for the five new tariffs and three tariff structures

New tariffs/structures		reject H_0: $X = Y$ N = 141 Total users	reject H_0: $X = Y$ N = 56 Premium users	reject H_0: $X = Y$ N = 85 Free users
X	Y			
TRN1	TRN2	Yes*	Yes**	No
TRN1	TRN3	Yes**	Yes*	No
TRN1	TRN4	Yes*	No	Yes***
TRN1	TRN5	Yes*	Yes**	No
TRN2	TRN3	No	No	No
TRN2	TRN4	Yes***	No	No
TRN2	TRN5	No	No	No
TRN3	TRN4	Yes***	Yes*	Yes**
TRN3	TRN5	No	No	No
TP1	TP2	Yes*	Yes***	No
TP1	TP3	Yes ***	Yes***	Yes***
TP2	TP3	Yes ***	Yes***	Yes***

*significance level of paired Wilcoxon test: *p < 0.05, **p < 0.01, ***p < 0.001*

Table 2 also presents the results for the differences in preferences for the tariff structures (TP). The results indicate that the 141 users have a higher preference for a high-data-volume tariff compared to a low-price tariff (TP1 vs. TP2). The results are similar for the sub-group of premium users. They have the same preference order as the whole sample of users. However, free users have the same preference for TP1 and TP2.

4.2 Factors Influencing Willingness to Pay for Privacy

Before analyzing the results in detail, we have to assess whether the independent variables correlate with each other (multicollinearity), since this would negatively impact the validity of our model. We test for multicollinearity by calculating the variance inflation factor (VIF) for all independent variables. None of the variables has a VIF larger than 1.7, indicating that multicollinearity is not an issue for our sample.

The results of the logistic regression model can be seen in Table 3. We highlighted statistically significant results in bold face. For JonDonym, RP and $TRUST_{PET}$ are the only statistically significant independent variables in the model. Surprisingly, RP has a negative coefficient, indicating that more risk-averse users are less likely to choose a premium tariff for JonDonym. This empirical result is in contrast to hypothesis 1, thus we cannot confirm this hypothesis derived from results of the literature and the associated rationale. Reasons for this contradictory result can be manifold. For example, there might be unobservable variables not included in the model which impact the relationship between RP and WTP. Hypotheses 2, 3 and 5 cannot be confirmed as well due to insignificant coefficients. In contrast to this, hypothesis 4 can be confirmed. Given the average marginal effect (avg. marg. effect), our result indicates that a one unit increase in trust in JonDonym increases the likelihood of choosing a premium tariff by 12.17%. This result is statistically significant at the 0.1% level. Hypothesis 4 can also be confirmed for the logistic regression model for Tor users with a slightly larger average marginal effect size of 12.45%. The variable VIC is statistically significant at the 1% level with a marginal effect of 5.33%. This indicates that bad experiences with privacy breaches lead to a higher probability of donating money to Tor, and thereby, supporting the Tor project financially. No other hypotheses can be confirmed for Tor.

Table 3. Results of the logistic regression model

	WTP for JonDonym		WTD for Tor		Difference
	Coef.	Avg. marg. effects	Coef.	Avg. marg. effects	Avg. marg. effects
(Intercept)	−0.0376	−0.0081	6.1455***	−0.9768	0.9687
RP	**−0.4967****	**−0.1067**	−0.1492	−0.0237	−0.083
VIC	−0.0397	−0.0085	**0.3352****	**0.0533**	−0.0618
TRUST	−0.0868	−0.0187	−0.1222	−0.0194	0.0007
$TRUST_{PET}$	**0.5661*****	**0.1217**	**0.7835*****	**0.1245**	−0.0028
TOR/JD	−0.5792	−0.1245	0.488	0.0776	−0.2021

$*p < 0.05$, $**p < 0.01$, $***p < 0.001$

5   Discussion and Conclusion

With respect to research question 1, our results show that PET providers should focus on building a strong reputation since trust in the PET is the strongest factor influencing the probability of spending money for privacy for both, JonDonym and Tor. In addition, we can observe that Tor users are more likely to donate for the service if they were a victim of a privacy breach or violation in their past.

Our second research question is about an optimized design of tariff options for users of commercial PETs based on the case of JonDonym. Here, we can see that the results differ when looking at different groups of users, which is in line with former research [37].

Users who use JonDonym with the free option, are indifferent with respect to the newly introduced tariffs as well as the general tariff structures (high volume vs. low price vs. low anonymity). However, some of them tend to prefer the tariff option with the lowest price with an included high-speed volume of 40 GB the most. Thus, free users would prefer the cheapest tariff, if they were to decide for paying at all. Practically, this implies that commercial PET providers should try to offer options with a relatively low monetary barrier to convert as many free users as possible into paying ones. The already paying users prefer high-volume tariffs over the other options.

Limitations of this study are the following. First, our sample only includes a relatively small number of active users of both PETs. This sample size is sufficient for the sake of our statistical analyses. However, the results about the current payment and donation numbers provide only a rough idea about the actual distribution. In addition, it is very difficult to gather data of actual users of PETs since it is a comparable small population that we could survey. It is also relevant to mention that we did not offer any financial rewards for the participation. A second limitation concerns possible self-report biases (e.g. social desirability). We addressed this issue by gathering the data fully anonymized. Third, mixing results of the German and English questionnaire could be a source of errors. On the one hand, this procedure was necessary to achieve the minimum sample size. On the other hand, we followed a very thorough translation procedure to ensure the highest level of equivalence as possible. Thus, we argue that this limitation did not affect the results to a large extent. However, we cannot rule out that there are unobserved effects on the results due to running the survey in more than one country at all. Lastly, demographic questions were not mandatory to fill out due to our assumption that these types of individuals who use Tor or JonDonym are highly cautious with respect to their privacy. Thus, we decided to go for a larger sample size considering that we might have lost participants otherwise (if demographics had to be filled out mandatorily). However, we must acknowledge that demographic variables might be relevant confounders in the regression model explaining the WTP of PET users.

Future work should aim to determine the relation between paying users and the groups Schomakers et al. [37] identified. In addition, researchers can build on our results by implementing such tariff options for commercial PET services in practice and investigate whether users are more prone to spend money for their privacy protection. Furthermore, it is relevant for commercial PET providers to differentiate themselves against free competitors as Tor in our example. This can be done by providing a higher level of usability in terms of ease of use, performance and compatibility with other applications [25, 48]. If commercial PET providers cannot create a unique selling point (USP) compared to free services, it is very unlikely that they establish a successful monetarization strategy in the market. Therefore, it is necessary to investigate how a USP for a commercial PET provider can look like and assess it in the field with active users of existing PETs as well as non-users.

Appendix - Questionnaire

A. Constructs and Questions for both PETs

Risk Propensity (RP)	**Trust in the PET (JonDonym / Tor) (TRUST$_{PET}$)**
1. I would rather be safe than sorry.	1. JonDonym / Tor is trustworthy.
2. I am cautious in trying new/ different products.	2. JonDonym / Tor keeps promises and commitments.
3. I avoid risky things.	3. I trust JonDonym / Tor because they keep my best interests in mind.

Trust in Online Companies (TRUST)

1. Online companies are trustworthy in handling information.
2. Online companies tell the truth and fulfill promises related to information provided by me.
3. I trust that online companies would keep my best interests in mind when dealing with information.
4. Online companies are in general predictable and consistent regarding the usage of information.
5. Online companies are always honest with customers when it comes to using the provided information.

Privacy Victim (VIC)

How frequently have you personally been the victim of what you felt was an improper invasion of privacy? (7-point frequency scale from "Never" to "Very frequently")

Knowledge about Tor (TOR)/JonDonym (JD)

Do you know the anonymization service Tor/JonDonym? (Yes/No)

B. Specific Questions for JonDonym

Current Tariff - Please choose your current tariff of JonDonym.

1. Free of charge option	4. Volume-S (650 MB / 6 months 5€)
2. Flat-M (monthly 2GB / 6 months / 50€)	5. Volume-M (1500 MB / 12 months 10€)
3. Flat-L (monthly 5GB / 6 months / 100€)	6. Volume-L (5000 MB / 24 months 30€)

Tariff Preference (TP)

1. I would use JD regularly with a data volume ten times higher than before (at the same price).
2. If the price decreased by half, I would use JonDonym regularly.
3. I would perceive a service with a lower anonymization level for half the price more attractive than JonDonym.

Tariff New (TRN)

1. Monthly 100 GB with a duration of 12 months for 100€ (total price)
2. Monthly 100 GB with a duration of 3 months for 30€ (total price)
3. Monthly 100 GB with a duration of 12 months for 10€ per month

4. Monthly 40 GB with a duration of 3 months for 5€ per month
5. Monthly 200 GB with a duration of 12 months for 15€ per month

C. Specific Questions for Tor
Donation to Tor
Did you ever donate money to the Tor project? (Yes/No)
Donation Amount
How much money did you donate to the Tor project? (open field with number only)
If not stated otherwise, constructs are measured based on a 7-point Likert scale ranging from strongly disagree to strongly agree.

References

1. Ball, J.: Hacktivists in the frontline battle for the internet. https://www.theguardian.com/technology/2012/apr/20/hacktivists-battle-internet
2. Bédard, M.: The Underestimated Economic Benefits of the Internet. In: Regulation Series. The Montreal Economic Institute (2016)
3. van Blarkom, G.W., Borking, J.J., Olk, J.G.E.: PET: Handbook of Privacy and Privacy-Enhancing Technologies (2003)
4. The Tor Project: Tor. https://www.torproject.org
5. JonDos Gmbh: Official Homepage of JonDonym. https://www.anonym-surfen.de
6. Saleh, S., Qadir, J., Ilyas, M.U.: Shedding light on the dark corners of the internet: A survey of tor research. J. Netw. Comput. Appl. **114**, 1–28 (2018)
7. Montieri, A., Ciuonzo, D., Aceto, G., Pescapé, A.: Anonymity services Tor, I2P, JonDonym: classifying in the dark. In: International Teletraffic Congress, pp. 81–89 (2017)
8. Pfitzmann, A., Hansen, M.: A terminology for talking about privacy by data minimization: anonymity, unlinkability, undetectability, unobservability, pseudonymity, and identity management, pp. 1–98. Tech. Univ. Dresden (2010)
9. Rossnagel, H.: The market failure of anonymity services. In: Samarati, P., Tunstall, M., Posegga, J., Markantonakis, K., Sauveron, D. (eds.) WISTP 2010. LNCS, vol. 6033, pp. 340–354. Springer, Heidelberg (2010). https://doi.org/10.1007/978-3-642-12368-9_28
10. Grossklags, J., Acquisti, A.: When 25 cents is too much: an experiment on willingness-to-sell and willingness-to-protect personal information. In: WEIS (2007)
11. Beresford, A.R., Kübler, D., Preibusch, S.: Unwillingness to pay for privacy: a field experiment. Econ. Lett. **117**, 25–27 (2012)
12. Borking, J.J., Raab, C.: Laws, PETs and other technologies for privacy protection. J. Inf. Law Technol. **1**, 1–14 (2001)
13. Fabian, B., Goertz, F., Kunz, S., Müller, S., Nitzsche, M.: Privately waiting – a usability analysis of the tor anonymity network. In: Nelson, M.L., Shaw, M.J., Strader, T.J. (eds.) AMCIS 2010. LNBIP, vol. 58, pp. 63–75. Springer, Heidelberg (2010). https://doi.org/10.1007/978-3-642-15141-5_6
14. Singh, R., et al.: Characterizing the nature and dynamics of Tor exit blocking. In: 26th USENIX Security Symposium (USENIX Security), Vancouver, BC, pp. 325–341 (2017)
15. Chirgwin, R.: CloudFlare shows Tor users the way out of CAPTCHA hell. https://www.theregister.co.uk/2016/10/05/cloudflare_tor/
16. Spiekermann, S.: The desire for privacy: insights into the views and nature of the early adopters of privacy services. Int. J. Technol. Hum. Interact. **1**, 74–83 (2005)

17. Alsabah, M., Goldberg, I.: Performance and security improvements for tor: a survey. ACM Comput. Surv. (CSUR) **49**(2), 1–36 (2016). Article no. 32
18. Koch, R., Golling, M., Rodosek, G.D.: How anonymous is the tor network? A long-term black-box investigation. Computer (Long. Beach. Calif.) **49**, 42–49 (2016)
19. Juarez, M., Elahi, T., Jansen, R., Diaz, C., Galvez, R., Wright, M.: Poster: fingerprinting hidden service circuits from a tor middle relay. In: Proceedings of IEEE S&P (2017)
20. Johnson, A., Wacek, C., Jansen, R., Sherr, M., Syverson, P.: Users get routed: traffic correlation on tor by realistic adversaries. In: ACM CCS, pp. 337–348 (2013)
21. Lee, L., Fifield, D., Malkin, N., Iyer, G., Egelman, S., Wagner, D.: A usability evaluation of tor launcher. In: Proceedings on Privacy Enhancing Technologies, pp. 90–109 (2017)
22. Herrmann, D., Lindemann, J., Zimmer, E., Federrath, H.: Anonymity online for everyone: what is missing for zero-effort privacy on the internet? In: Camenisch, J., Kesdoğan, D. (eds.) iNetSec 2015. LNCS, vol. 9591, pp. 82–94. Springer, Cham (2016). https://doi.org/10.1007/978-3-319-39028-4_7
23. Harborth, D., et al.: Integrating privacy-enhancing technologies into the internet infrastructure. arXiv Prepr. arXiv:1711.07220 (2017)
24. Harborth, D., Pape, S.: JonDonym users' information privacy concerns. In: Janczewski, L.J., Kutyłowski, M. (eds.) SEC 2018. IAICT, vol. 529, pp. 170–184. Springer, Cham (2018). https://doi.org/10.1007/978-3-319-99828-2_13
25. Harborth, D., Pape, S.: Examining technology use factors of privacy-enhancing technologies: the role of perceived anonymity and trust. In: Twenty-Fourth Americas Conference on Information Systems, New Orleans, USA (2018)
26. Harborth, D., Pape, S.: How privacy concerns and trust and risk beliefs influence users' intentions to use privacy-enhancing technologies - the case of tor. In: Hawaii International Conference on System Sciences Proceedings, Hawaii, US (2019)
27. Malhotra, N.K., Kim, S.S., Agarwal, J.: Internet users' information privacy concerns: the construct, the scale, and a causal model. Inf. Syst. Res. **15**, 336–355 (2004)
28. Grossklags, J.: Experimental economics and experimental computer science: a survey. In: Workshop on Experimental Computer Science, ExpCS 2007 (2007)
29. Acquisti, A.: The economics of personal data and the economics of privacy. In: Texte La Conférence Donnée En Décembre, pp. 1–24 (2010)
30. Benndorf, V., Normann, H.T.: The willingness to sell personal data. Scand. J. Econ. **120**, 1260–1278 (2018)
31. Li, C., Li, D.Y., Miklau, G., Suciu, D.: A theory of pricing private data. ACM Trans. Database Syst. (TODS) **39**(4), 1–28 (2014). Article no. 34
32. Preibusch, S.: The value of privacy in web search. In: WEIS (2013)
33. Cofone, I.N.: The value of privacy: keeping the money where the mouth is. In: 14th Annual Workshop on the Economics of Information Security, pp. 1–31 (2015)
34. Malgieri, G., Custers, B.: Pricing privacy - the right to know the value of your personal data. Comput. Law Secur. Rev. **34**(2), 289–303 (2018)
35. Cranor, L.F., Arjula, M., Guduru, P.: Use of a P3P user agent by early adopters. In: Proceedings of the ACM Workshop on Privacy in the Electronic Society, WPES 2002, pp. 1–10 (2002)
36. Federrath, H.: Privacy enhanced technologies: methods – markets – misuse. In: Katsikas, S., López, J., Pernul, G. (eds.) TrustBus 2005. LNCS, vol. 3592, pp. 1–9. Springer, Heidelberg (2005). https://doi.org/10.1007/11537878_1
37. Schomakers, E.M., Lidynia, C., Vervier, L., Ziefle, M.: Of guardians, cynics, and pragmatists - a typology of privacy concerns and behavior. In: IoTBDS, pp. 153–163 (2018)
38. Roßnagel, H., Zibuschka, J., Pimenides, L., Deselaers, T.: Facilitating the adoption of tor by focusing on a promising target group. In: Jøsang, A., Maseng, T., Knapskog, S.J. (eds.)

NordSec 2009. LNCS, vol. 5838, pp. 15–27. Springer, Heidelberg (2009). https://doi.org/10.1007/978-3-642-04766-4_2

39. Böhme, R., Koble, S.: On the viability of privacy-enhancing technologies in a self-regulated business-to-consumer market: will privacy remain a luxury good? Dresden (2007)

40. Wilcoxon, F.: Individual comparisons by ranking methods. Biom. Bull. **1**, 80–83 (1945)

41. Mann, H.B., Whitney, D.R.: On a test of whether one of two random variables is stochastically larger than the other. Ann. Math. Stat. **18**, 50 (1947)

42. Benjamini, Y.: Opening the box of a boxplot. Am. Stat. **42**, 257–262 (1988)

43. McKelvey, D., Zavorina, W.: A statistical model for the analysis of ordinal level dependent variables. J. Math. Sociol. **4**, 103–120 (1975)

44. Donthu, N., Gilliland, D.: Observations: the infomercial shopper. J. Advert. Res. **36**, 69–76 (1996)

45. Frik, A., Gaudeul, A.: The relation between privacy protection and risk attitudes, with a new experimental method to elicit the implicit monetary value of privacy. CEGE discussion papers, Number 296. SSRN (2016). http://papers.ssrn.com/abstract=2874202

46. Christofides, E., Muise, A., Desmarais, S.: Risky disclosures on Facebook: the effect of having a bad experience on online behavior. J. Adolesc. Res. **27**, 714–731 (2012)

47. Pavlou, P.A.: Consumer acceptance of electronic commerce: integrating trust and risk with the technology acceptance model. Int. J. Electron. Commer. **7**, 101–134 (2003)

48. Harborth, D., Pape, S.: Explaining technology use behaviors of privacy-enhancing technologies: the case of Tor and JonDonym. Submitted to IEEE European Symposium on Security and Privacy (EuroS&P 2019) (2019)

49. Harborth, D., Pape, S.: German translation of the concerns for information privacy (CFIP) construct. SSRN (2018). https://ssrn.com/abstract=3112207

50. Schmitz, C.: LimeSurvey Project Team. http://www.limesurvey.org. Accessed 12 Dec 2018

Crypto and Encryption

Arcana: Enabling Private Posts on Public Microblog Platforms

Anirudh Narasimman[1], Qiaozhi Wang[2], Fengjun Li[2], Dongwon Lee[3], and Bo Luo[2(✉)]

[1] IBM Corporation, San Francisco, CA, USA
anirudh.narasimman1@ibm.com
[2] University of Kansas, Lawrence, KS, USA
{qzwang,fli,bluo}@ku.edu
[3] Penn State University, University Park, PA, USA
dlee@ist.psu.edu

Abstract. Many popular online social networks, such as Twitter, Tumblr, and Sina Weibo, adopt too simple privacy models to satisfy users' diverse needs for privacy protection. In platforms with no (i.e., completely open) or binary (i.e., "public" and "friends-only") access control, users cannot control the dissemination boundary of the content they share. For instance, on Twitter, tweets in "public" accounts are accessible to everyone including search engines, while tweets in "protected" accounts are visible to *all* the followers. In this work, we present *Arcana* to enable fine-grained access control for social network content sharing. In particular, we target the Twitter platform and introduce the "private tweet" function, which allows users to disseminate particular tweets to designated group(s) of followers. Arcana employs Ciphertext-Policy Attribute-based Encryption (CP-ABE) to implement social circle detection and private tweet encryption so that access-controlled tweets are only readable by designated recipients. To be stealthy, Arcana further embeds the protected content as digital watermarks in image tweets. We have implemented the Arcana prototype as a Chrome browser plug-in, and demonstrated its flexibility and effectiveness. Different from existing approaches that require trusted third-parties or additional server/broker/mediator, Arcana is light-weight and completely transparent to Twitter – all the communications, including key distribution and private tweet dissemination, are exchanged as Twitter messages. Therefore, with small API modifications, Arcana could be easily ported to other online social networking platforms to support fine-grained access control.

1 Introduction

Over the past decade, our communication and information sharing behaviors have been gradually and inevitably changed by online social networks (OSNs). The convenience of OSNs drives users to share their social, cultural, professional, and even personal content online. Some OSNs such as Facebook and Google+

© IFIP International Federation for Information Processing 2019
Published by Springer Nature Switzerland AG 2019
G. Dhillon et al. (Eds.): SEC 2019, IFIP AICT 562, pp. 271–285, 2019.
https://doi.org/10.1007/978-3-030-22312-0_19

provide access control functions for posting, which allow users to specify a dissemination boundary (e.g., public, friends, customized lists of friends, etc.) for each post. Unfortunately, quite a few popular OSNs, such as Twitter, Sina Weibo and Tumblr, provide no or very limited access control. For example, in Twitter, all "public tweets" are accessible to all Twitter users and search engines. The only privacy protection mechanism is to "protect" the entire account so that all tweets are only accessible to followers. Without fine-grained access control for tweets, it is impossible to specify a customized audience for each tweet.

The popularity of Twitter especially among the younger generation and its openness have introduced serious privacy concerns. Public tweets with private content are not only a treasure for stalkers and adversaries [36], but also may cause unexpected social incidents. For example, a student who posts a tweet complaining about a professor may only want to share it with her close friends. It can be embarrassed if her classmates see the post. Also, when one holds a birthday party with close friends but not coworkers, she may feel her party pictures would upset the uninvited and thus hesitates to post. In fact, these situations are not uncommon. Survey shows that users have target followers in their minds before posting [32]. But due to the absence of fine-grained privacy protection functions, users may decide to post with different accounts or not to post at all. The lack of usable privacy protection causes complication and frustration, which may eventually affect the overall usage of OSNs. Meanwhile, for users who decide to ignore the privacy expectations and post publicly for convenience, the absence of privacy protection may lead to significantly negative repercussions, such as private information leakage and damage of public image. For example, it is reported that college-admissions officers looked for applicants' activities on OSNs to check signs of bad behavior that may lead to bad publicity or a legal investigation [2]. Therefore, there are strong needs for a fine-grained privacy protection mechanism for open OSNs such as Twitter to balance the sharing and exposure of online social activities. In the rest of the paper, we use *private tweets* to denote OSN content that is intended to be shared with only a subset of followers instead of the public.

In this paper, we present *Arcana*, a fine-grained privacy enforcement mechanism for open social networks. We present our design and implementation based on Twitter, while the Arcana system can be ported to other social network platforms with small modifications on the interface with the OSN. Arcana allows users to assign followers into groups, and designate one or more groups of followers as the audience for each individual tweet. Such access-controlled tweets are first encrypted by Ciphertext-Policy Attribute-Based Encryption (CP-ABE) [4,33], and then embedded into a (random) host image as an invisible watermark [21]. Only followers in the designated groups could decrypt the message and see the plaintext tweet. The mechanism could be implemented as a browser plug-in, or a wrapper app for Twitter. All storage and communication are enabled by the native capabilities of Twitter. First, the keys are distributed over one-time messages using Twitter's Direct Message function. Secondly, the private tweets are disseminated as a regular (image) tweets through the standard tweet API.

Unlike other existing approaches [3, 12, 28], we do not need a trusted third-party, a private server, nor any other platform with storage or access control capabilities. With the "private tweet" function of Arcana, a user can conveniently send a private tweet to a designated subset of followers and maintain full control of it (i.e., she can delete it later). Lastly, Arcana converts the private tweet as an invisible watermark so that it seems like a regular tweet to the unintended users.

Our Contributions are Three-fold: (1) We design the *Arcana* system to enable private posts on a public microblogging platform without the assistance of a trusted third-party. (2) Arcana provides flexible, fine-grained attribute-based privacy protection so that a customized dissemination boundary could be specified for each private post. A key technology and innovation in Arcana is platform-independent, therefore, it could be adopted in any online social network platform. (3) We have implemented a first prototype of Arcana as a Chrome plug-in for Twitter and demonstrated its effectiveness.

2   Problem and Preliminaries

2.1   Problem Statement

It is observed that users often post private messages on Twitter [32], which are (implicitly) intended for a small group of friends or followers [24]. Such arbitrary subset of followers is known as a social circle, in which the message sender has complete control over the circle boundary [17, 19, 30, 34]. We define the tweets designated to pre-selected social circles as *private tweets*, which can be considered as a person-to-group communication. However, designed to be an open network, Twitter lacks effective support to private posting. By default, all the tweets are open to public (with the exception of a very coarse-grained "protected tweets" function). The only private communication mechanism on Twitter is "private message", which is a costly person-to-person communication between two users. To post a private message to a subset of N followers, N copies need to be sent. So, it is very laborious and expensive (in terms of communication and storage) to post to a large social circle or to many social circles at the same time. Therefore, it is important to develop a fine-grained privacy enforcement mechanism to enable private posting in open microblogging platforms.

Design Goals. To tackle the problem, we present the Arcana system, which allows the users to disseminate private tweets to any arbitrary social circle. It provides fine-grained access control functions directly on top of the Twitter platform, without involving any other third party. In particular, Arcana is expected to provide the following key features: (1) *confidentiality:* only the intended followers are able to view the plaintext content of the private tweet; (2) *fine-grained:* any arbitrary dissemination boundary could be specified; (3) *self-sufficiency:* Arcana should not rely on any trusted third-party, external storage, mediation, or authentication platform other than Twitter.

2.2   Preliminary: CP-ABE

The Ciphertext-Policy Attribute-based Encryption (CP-ABE) scheme [4] allows a user to define dynamic access policies without knowing specific receivers in the system beforehand. In particular, each user is associated with a set of *attributes*, which are reflected in her secret key. The *access policies* are embedded in the ciphertext, which define the attributes that an authorized user should own. Therefore, one could specify an access structure over the set of attributes, and encrypt a message accordingly so that only the authorized users with attributes satisfying the access structure can decrypt it. In practice, access structures are described by monotonic access trees, whose leaves are the pre-defined attributes and non-leaf nodes are threshold gates. CP-ABE is resilient to collusion attacks. A message can be decrypted by a group of collusive users only when at least one of them could decrypt it on her own. In general, the CP-ABE scheme consists of four fundamental algorithms: Setup, KeyGen, Encrypt, and Decrypt [4]:

Setup: The setup algorithm generates a public key (a.k.a. the public parameters) PK and a master key MK. In particular, it chooses two randoms $\alpha, \beta \in Z_p$, a bilinear group G_0 of prime order p with a generator g, and the bilinear map $e : G_0 \times G_0 \to G_1$ to generate $PK = (G_0, g, h = g^\beta, e(g, g)^\alpha)$ and $MK = (\beta, g^\alpha)$.

KeyGen(MK, S): Assume a user has a set of attributes $S = S_1 \cup ... \cup S_m$, the key generation algorithm takes S and the master key MK as the input to generate the secret key SK, which is used to decrypt the messages designated for attributes S or a subset of S.

Encrypt(PK, M, $\mathbb{A}$): Message M is encrypted with public parameters PK and an access structure $\mathbb{A}$ defined over all possible attributes.

Decrypt(CT, SK): Ciphertext CT contains the access structure $\mathbb{A}$. The decrypt algorithm decrypts CT with SK if S satisfies $\mathbb{A}$.

Due to space limit, we do not include more details of the CP-ABE scheme but refer interested readers to [4,13,18,35].

2.3   Preliminary: Digital Watermarking

Digital watermarking embeds information into a carrier signal such as images, videos, and text documents. Such watermarks are visually imperceptible in order to hide certain information. In Arcana, we adopt the Least Significant Bit (LSB) watermarking approach [7,10] for its efficiency and simplicity. Besides LSB-based watermarking schemes that embed information into the spatial domain, other watermarking mechanisms insert hidden bits in frequency coefficients or the noisy blocks of the image. Please refer to [8] for more details of digital watermarks and other watermarking mechanisms.

In our work, an image is considered as an array of pixels, in which each pixel is represented by a numerical value, such as 8 bits for 256-gray-scale images, or 3×8 bits for 16M-color images. The LSB method substitutes the least significant bit of the carrier signal (the right-most bit of the pixel) with the bit value

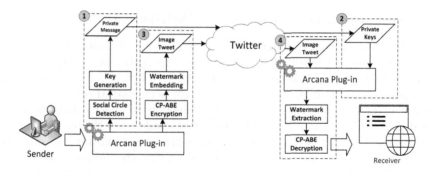

Fig. 1. Overview of the Arcana system.

from the embedded message. For example, if a greyscale pixel is represented as 0b00011100 and the information to be carried is 0b1, the new pixel value becomes 0b00011101. Ideally, we can insert 1 bit of watermark in each pixel of the image with no robustness consideration. Thus, with a common small-size image of 240 × 160 pixels, we can embed 4.8 MB information, which is sufficient for common tweets (280 characters) and encrypted private tweets.

3   Arcana: Private Posts on an Open Microblog Platform

3.1   Arcana Overview

Arcana uses CP-ABE to encrypt tweets and invisible digital watermarking to embed ciphertext in host images. As shown in Fig. 1, Arcana performs four main tasks: (1) The user (e.g., Alice) first invokes Arcana to categorize all her followers into social circles. In response, Arcana sets up the CP-ABE environment, generates cryptographic keys and distributes them to the followers through Twitter *direct messages*. (2) If a follower is assigned with at least one attribute by Alice, he will receive a private message which contains the cryptographic keys. (3) Alice sends the plaintext of the private tweet along with the designated social circle(s) to Arcana, which will encrypt the message, embed it into an image as an invisible watermark, and post to Twitter on behalf of Alice via an API call. (4) On the receiver side, Arcana detects if a received image in Bob's Twitter feed contains a private tweet watermark. If so, it extracts the binary watermark string (if exist), attempts to decrypt it with Bob's private key, and displays the recovered text together with the host image to Bob.

To be consistent with our prototype that implements Arcana as a Chrome extension, we illustrate Arcana as a browser plug-in in Fig. 1. In fact, Arcana can be implemented as a Twitter client app for mobile devices, similar to Tweetbot or Twitterrific. Finally, it is worth noting that Arcana does not affect normal Twitter functions. If the sender uses Arcana but her follower does not, the follower cannot tell if Arcana is in use or view any private tweet.

3.2   Social Circle Detection and Key Distribution

To start using Arcana, the first step is to assign roles (i.e., attributes) to each follower for access control. However, manually assigning attributes to a large number of followers is tedious and labor-intensive. Therefore, automatic social circle detection is desired. As social circles have been studied recently to enable privacy protection [30], several automatic detection approaches have been proposed [22,23]. In Arcana, we employ the SCSC multi-view clustering approach [22] to cluster the followers into non-overlapping groups and provide suggested social circles to the sender, who can manually refine the groups and assign an *attribute* to each social circle, such as *family, coworker, close friend.* Our scheme allows overlaps among social circles, e.g., a follower may be both *coworker* and *close friend.*

Arcana then invokes the Setup and Key Generation processes of CP-ABE to construct the cryptographic environment and generates the public, master and secret keys. In particular, a secret key SK_i is generated for follower f_i based on the attribute set S_i assigned by Alice. Arcana uses Twitter's *Direct Messages* function via an API call to distribute keys ($\{PK, SK_i\}$) encoded in Base91. It also embeds a note to indicate the key distribution message to Alice's followers. The follower stores the received keys locally, which are only used to decrypt private tweets. Since the keys are identified by the sender of the Direct Messages, they cannot be spoofed as long as Twitter is not compromised. Since Arcana does not store any personal information locally (in browser or on the client machine), Alice receives her cryptographic environment via Direct Message.

The key generation and distribution process is a one-time communication. Once the keys are distributed to the followers, they will be used to decrypt all future private tweets. If Alice wants to assign a new user to an existing circle, she needs to extract the master key MK from her own DM, and invoke **KeyGen**(MK, S) to generate and distribute a new secret key for this follower. It is worthy noting that this process does not affect any existing follower.

3.3   Posting a Private Tweet

To post a private tweet, the sender first submits the textual content of the tweet (plaintext M) to Arcana and specifies the access structure $\mathbb{A}$. Then, Arcana loads the cryptographic environment and invokes CP-ABE to encrypt this message as: $C = \textbf{Encrypt}(PK, M, \mathbb{A})$.

A naive approach would be directly converting the binary string C into a text string (e.g., using Base64 or Base91) and posting it to Twitter. However, this approach has two significant drawbacks: (1) the length of the CP-ABE ciphertext is significantly longer than the short plaintext C. For instance, the length of a Base91 encoded CP-ABE ciphertext of a regular tweet message ranges between 700 to 2000 characters, while each tweet is limited to 280 characters. Hence, directly posting this encoded ciphertext would take 3 to 8 tweets. (2) Although Base64 or Base91 encoded ciphertext are ASCII strings, they are not readable to users. Figure 2 (a) shows part of the Base91 encoded ciphertext for message

---

Algorithm 1. Posting a private tweet.

Data: Private message M; set of attributes **Attr**.
Result: Private tweet posted on Twitter.

1 **begin**
2   **PK** $\longleftarrow loadKey()$ /* Load public parameters. */
3     $\mathbb{A} \longleftarrow CreatAccessStructure(\textbf{Attr})$
4     $C \longleftarrow Encrypt(\textbf{PK}, M, \mathbb{A})$ /* CP-ABE encryption. */
5     $C' \longleftarrow strcat(C, MagicToken)$
6     $I \longleftarrow LoadImage()$ /* Load host image. */
7     $I' \longleftarrow Embed(I, C')$ /* Embed watermark. */
8   TwitterAPI.post(I')

"Dinner at my place!" Posting a large number of tweets like this will create an awkward socialization situation. To tackle the usability problem, Arcana embeds the ciphertext into a host image as an invisible watermark, and posts the image to Twitter. Besides the ciphertext, Arcana also adds a special *Magic Token* (predefined binary string) to indicate the existence of Arcana watermark and mark the end point of the watermark. To a regular user, the encoded image looks no different from any other regular image.

The process for posting a private message is shown in Algorithm 1. First, the CP-ABE public parameters are loaded from Alice's own Direct Messages [L2]. Then, the CP-ABE access structure is created from user specified attributes [L3]. In [L4-5], the plaintext tweet is encrypted with the public parameters and the access structure, with a "Magic Token" appended to the end of the ciphertext. Next, the host image is loaded from the Arcana image library or uploaded by the user [L6]. Finally, the ciphertext is embedded in the host image and posted to Twitter [L7-L8].

3.4 Viewing Private Tweets

The decryption process contains three major steps: (1) loading private keys from Twitter Direct Messages, (2) filtering image tweets to identify and extract watermarks, and (3) decrypting and displaying the private tweet(s).

The decryption process is shown in Algorithm 2. In [L4], when a user Bob loads Twitter timeline, Arcana will scan through the Direct Messages to identify key distribution messages, and extract credentials and keys from such messages: $\{U_j, PK_j, SK_j\}$, where user ID U_j is directly extracted from the sender field of the messages. Next, in [L6], each image tweet being loaded into the timeline is filtered by Arcana. We only scan candidate images posted by some users from whom we have received secret keys ($\textbf{U} = \bigcup U_j$) [L7], and scan through the image until the binary Magic Token [L8]. Then, the corresponding keys are loaded and the watermark is extracted from the host image [L9-L10]. If the decryption process is a success, the plaintext of the private tweet is displayed together with the host image in the users browser [L11-L13].

Algorithm 2. Displaying a private tweet.

Data: Twitter timeline **T**.
Result: Plaintexts of private tweets displayed on Twitter.

```
1  begin
2  │  foreach dm in Twitter Direct Message do
3  │  │  if dm is Arcana key distribution then
4  │  │  │  {U_j, PK_j, SK_j} ⟵ ExtractKeys(dm)
5  │  │  └  j++
6  │  foreach Image tweet I in Twitter timeline do
7  │  │  if I.owner ∉ U then break
8  │  │  if VerifyMagicToken(I) then
9  │  │  │  {PK_j, SK_j} ⟵ LoadKey(U, I.owner)
10 │  │  │  C ⟵ ExtractWatermark(I, MagicToken)
11 │  │  │  M ⟵ Decrypt(PK_j, C, SK_j)
12 │  │  │  if M is text then
13 │  │  │  └  Display(M)
```

3.5 Experiment and Performance Analysis

We have developed a proof-of-concept prototype of Arcana as a chrome extension. The interface for posting a private tweet is shown in Fig. 2(b).

In the example, the user has configured three attributes (social circles): *Family*, *Friends*, and *Colleagues*. To post a new private tweet, she first identifies the access policy by selecting one or more attributes, selects a host image from the library, and enters the text content of the private tweet. Figure 2(c) shows the view of the private tweets from an unauthorized follower's perspective. As expected, the user who does not have any designated attribute only sees a regular image, but cannot tell it is a private tweet.

We then created a test user account "twittertest", who had three social circles, and posted three different messages from this account. Each of these private tweets are only intended for a subset of his followers, as follows:

Tweet message	Attributes (circles)
Dinner at my place.	Family, Friends
New version of ubuntu is out.	Colleagues, Friends
We have a new ceo.	Colleagues, Family

The test user "twittertest" has three followers: *Faye* is in the user's social circle "Family", *Frank* is in circle "Friends", and *Cole* is in "Colleagues". When these three followers log in to Twitter on browsers with Arcana, they see different private tweets based on their attributes. Figure 3(a) shows part of Faye's

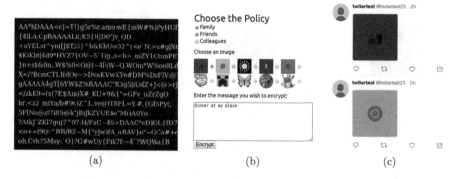

Fig. 2. Private tweets: (a) Part of Base91 encoded CP-ABE ciphertext; (b) Screenshot in Arcana: posting a private tweet "Dinner at my place", which is only visible to Family and Friends; (c) View of the private tweets from an unauthorized follower.

twitter feed processed by Arcana. In this example, she can only see the messages distributed to "Family" i.e. "**Dinner at my place**" and "**We have a new ceo**". The second tweet cannot be decrypted by Faye's secret key, hence, no plaintext is displayed with the image. Similarly Frank and Cole only see messages corresponding to their social circles, as shown in Fig. 3(b) and (c). This example has very few policies for the sake of convenience in demonstration, but it can be extended to add more host users, followers, attributes, and private tweets.

Efficiency and Usability. In our experiments with a commodity Dell desktop, it takes approximately 2 s to encrypt and embed a 100-character tweet. As we have explained, private tweets could be very long, since a small image could carry relatively large amount of information. Encrypting and embedding a 1000-character message will take approximately 4 s. At the recipient's end, Arcana runs in the background, after the user logs in to Twitter from a browser with Arcana. The plaintext of private tweets are added to the Twitter feed on-the-fly. The decryption time is almost linear to the number of images. In our experiments, it took approximately 11 s to recover 20 private tweets. As Twitter loads tweets incrementally, Arcana only needs to examine and decrypt the top tweets, while more tweets are added on demand.

Arcana currently supports an "OR" (or "UNION") logic among the set of attributes. CP-ABE is capable of handling complicate logic of the attributes. For example, Alice could define that a private tweet is for followers who are (**Family** OR (**Friends** AND **Colleagues**)). It is relatively straightforward to extend Arcana's user interface to support complicate logic with operations such as AND OR and NOT.

4 Security Analysis and Discussions

Next, we present the threat model and examine Arcana's security guarantees.

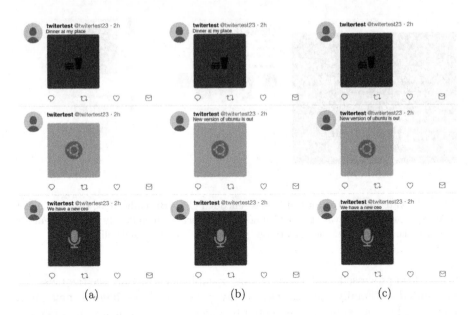

Fig. 3. Screenshots of Arcana: (a) Faye's view after decryption; (b) Frank's view after decryption; (c) Cole's view after decryption.

Threat Model. We assume the Twitter platform is trusted and the data carried by Twitter private messages is secure. The Twitter users, however, can be honest-but-curious or malicious. So, we consider two types of intruders: (1) Eve is a normal Twitter user (not necessarily Alice's friend) who attempts to learn the plaintext content of the private tweet messages that she is not authorized to read. (2) Chuck is a malicious user who attempts to impersonate Alice to send private tweets so that Alice's followers think it comes from Alice. We assume that neither Eve nor Chuck compromises other users' twitter accounts.

Key Security. Cryptographic keys are generated in Arcana (in Chrome extension in our prototype), transmitted through Twitter's Direct Messages (DM), and stored in DM. In key generation, the memory copy of keys and key materials are destroyed once they are distributed. Key confidentiality during key transmission and storage relies on the confidentiality of Twitter DM, which is considered trustworthy. Keys in DM are considered secure, since the key distribution messages are rarely read by users, and shoulder surfers can hardly remember the keys, which appear like a long random strings. Keys are only accessible to Arcana when a user logs in to a Twitter account, hence, access to the keys relies on the access to the Twitter account. When keys are utilized to decrypt private tweets, they are erased after the user leaves Twitter. In summary, neither Eve nor Chuck is able to learn secret/master keys that are not distributed to them.

Data Confidentiality. Data security in Arcana refers to the confidentiality of the private tweets, i.e., only authorized followers are allowed to view the plaintext of authorized tweets. In general, data security relies on five design features:

- Arcana binds credentials with Twitter accounts – anyone with access to Alice's Twitter account could access the keys generated by Alice or distributed to Alice, and then access the corresponding private tweets. Security of credential binding (i.e., authentication of followers) relies on the authentication and account security in Twitter, which is considered trustworthy.
- Chuck may obtain the public parameters of Alice's key, and use it to encrypt messages. However, in Arcana, sender identity is bound to the Twitter account that posts the image. Hence, as long as Chuck does not compromise Alice's account, he cannot impersonate Alice to post the private tweet.
- The confidential dissemination of private tweets against eavesdroppers relies on the semantic security of CP-ABE cryptosystem. When Eve does not have the attributes and the corresponding secret key, CP-ABE guarantees that she could not learn any information from the encrypted private tweet.
- Arcana does not save any sensitive information on the client computer. Thus, getting access to the client computer does not lead Eve to the access of any Arcana keys/messages. However, a privileged user (root) is able to access or dump Arcana's memory content while Arcana is running and the keys are in memory. This is unavoidable for any web-based application.
- The security of Arcana is self-contained. Arcana does not use any third party service or trusted server for storage, authentication, key management/distribution, or encryption/decryption. Moreover, Arcana does not communicate to anyone except Twitter. That says, unlike other approaches in the literature, no third party is capable of accessing any sensitive information or private tweets.

Key Refreshment/Revocation. To remove member(s) from a group, we choose to generate and redistribute new private keys instead of revoking old private keys. The old keys have already been used to decrypt old private tweets by the members to-be-removed, hence, user's private information has already leaked to the users in old groups. Dynamic attributes and efficient key refreshment mechanisms are actively studied in cryptography community, new methods could be adopted in Arcana to enable efficient key revocation/refreshment [13,35,37].

Forwards and Replies. In Arcana, directly forwarding a private tweet would not compromise data confidentiality, since the encryption is not broken during forwarding. Unauthorized users cannot decrypt private content from the original image tweet or from a forwarded tweet. Meanwhile, if the image is altered, the watermark may be broken, hence, the sensitive information is destroyed, but never leaked. Currently, Arcana does not encrypt the replies to private tweets.

Side-channel. The *Magic Token* (Algorithm 1) may become a side-channel to expert followers who have intensive knowledge of Arcana. That is, if a follower manually examines the image and finds the Magic Token, he knows that he has been excluded from the audience of the private tweet. This may cause an issue in socialization. However, we would like to claim that: (1) it requires intensive knowledge of Arcana detect the side-channel, which is beyond the capability of regular Twitter users. (2) It is difficult, if not impossible, to completely avoid this side-channel. Arcana will need a mechanism to extract the watermark, which could always be invoked by the attacker to detect the existence of the watermark. And (3) this side-channel does not leak any sensitive information.

Arcana for Other Online Social Networking Platforms. There exist other social networking platforms that do not have any access control functions, or only provide a coarse-grained privacy system, such as Tumblr or Sina Weibo. With limited modification of I/O to fit into the API of the specific OSN platform, Arcana could be deployed to work with almost any social networking platform that supports a private message function and an image posting function.

5   Related Works

Preserving user's privacy on OSNs has been intensively researched in recent years. The approach, frame and purpose of defending differ from case to case. In general, the literature of social network privacy research could be organized into two categories: *privacy models* and *privacy enforcement mechanisms*.

Privacy Models. Various privacy protection models have been proposed for social networks, such as [6,9,15]. There are also automatic tools to help users configure their privacy settings, such as *Privacy Wizards* [14], *A3P* [31]. Arcana is different from them that we introduce fine-grained privacy protection to open OSNs that employ very simple access models, such as Twitter or Tumblr.

Privacy Enforcement. Privacy enforcement can be classified into two categorizes: (1) *New designs of OSN architecture.* Privacy-aware OSNs protect users' privacy by storing private information on their own computers or distribute them across multiple administrative domains. Hence, they are decentralized, in stead of relying on one commercial OSN provider. There are also proposals of privacy-preserving SN architectures that add a layer of encryption outside the centralized server to prevent the service provider from peeking into users' information. This line of methods have been widely studied and there are several representative instances, such as Safebook [11], PeerSoN [5], Cachet [25], Hummingbird [12], and Twister [16]. However, their cost and availability disadvantages [27] are the major drawbacks. Detailed studies of decentralized privacy preserving OSNs are available at [1,26]. (2) *Reforming commercial OSN platforms.* [3] allows users to share encrypted text, images and files. It relies on trusted third parties for key management and storage. Meanwhile, sharing an item to a group requires sharing group keys or publishing multiple ciphertexts encrypted with different keys. Twitsper [29] implements a wrapper for Twitter to allow users to share private whispers. It uses Twitter direct messages to transmit whispers, which means N

copies of a whisper need to be sent in order to share it to N friends. Meanwhile, Twitsper employs an external supporting server to help manage whispers and replies. Twitterize [20] is an Android app that introduces anonymity and confidentiality for Twitter, through implementing an overlay network on top of Twitter, and using Android SQLite DB to store tweets and keys. Twitterize requires a secure channel for key distribution (QR code scanning or NFC), and significantly amount of local configuration and storage. It also posts Base64-encoded ciphertext on Twitter, which will be confusing to non-Twitterize users.

In summary, existing privacy-preserving social network approaches in the literature require (trusted) supporting servers (e.g., for authentication/authorization [12], supporting group tweet functions [29], content storage [3]), a proprietary social network infrastructure (e.g., [6,11]), or an additional secure channel (e.g., for key distribution in [3,20]). The main novelty of our proposed Arcana lies in the fact that it does not need any of these.

6 Conclusion and Future Work

Arcana provides a private tweet function to enable fine-grained access control on Twitter. It allows the user to categorize followers into groups and post private tweets to selected groups. In its implementation, Arcana encrypts private tweets with CP-ABE, and embeds the ciphertexts as invisible watermarks in host images. Arcana is self-contained that all storage, authentication, key management, and encryption/decryption are handled locally or via Twitter. It does not need a trusted third party or an additional broker/mediator. We have implemented Arcana as a Chrome extension and discussed its privacy guarantees.

Arcana may be extended to enable fine-grained access control for any online social network platform. The extension requires modifications in the I/O interface to work with underlying SN platform, which is our future plan.

Acknowledgements. This work is sponsored in part by the National Science Foundation under NSF CNS-1422206 and DGE-1565570, and the National Security Agency (NSA) Science of Security (SoS) Initiative.

References

1. Bahri, L., Carminati, B., Ferrari, E.: Decentralized privacy preserving services for online social networks. Online Soc. Netw. Media **6**, 18–25 (2018)
2. Bauerlein, M.: Your dream college can see twitter, too. https://www.bloomberg.com/view/articles/2013-11-08/your-dream-college-can-see-facebook-too
3. Beato, F., Ion, I., Čapkun, S., Preneel, B., Langheinrich, M.: For some eyes only: protecting online information sharing. In: ACM CODASPY (2013)
4. Bethencourt, J., Sahai, A., Waters, B.: Ciphertext-policy attribute-based encryption. In: IEEE Symposium on Security and Privacy. IEEE (2007)
5. Buchegger, S., Schiöberg, D., Vu, L.-H., Datta, A.: Peerson: P2p social networking: early experiences and insights. In: ACM SNS Workshop (2009)

6. Carminati, B., Ferrari, E., Perego, A.: Enforcing access control in web-based social networks. ACM TISSEC **13**(1), 6 (2009)
7. Chan, C.-K., Cheng, L.: Hiding data in images by simple LSB substitution. Pattern Recognit. **37**(3), 469–474 (2004)
8. Cheddad, A., Condell, J., Curran, K., Mc Kevitt, P.: Digital image steganography: survey and analysis of current methods. Signal Process. **90**(3), 727–752 (2010)
9. Cheng, Y., Park, J., Sandhu, R.: A user-to-user relationship-based access control model for online social networks. In: Cuppens-Boulahia, N., Cuppens, F., Garcia-Alfaro, J. (eds.) DBSec 2012. LNCS, vol. 7371, pp. 8–24. Springer, Heidelberg (2012). https://doi.org/10.1007/978-3-642-31540-4_2
10. Cox, I.J., Miller, M.L., Bloom, J.A., Honsinger, C.: Digital Watermarking, vol. 53. Springer, Heidelberg (2002)
11. Cutillo, L., Molva, R., Strufe, T.: Safebook: a privacy-preserving online social network leveraging on real-life trust. IEEE Commun. Mag. **47**(12), 94–101 (2009)
12. De Cristofaro, E., Soriente, C., Tsudik, G., Williams, A.: Hummingbird: privacy at the time of twitter. In: IEEE S&P (2012)
13. Doshi, N., Jinwala, D.: Updating attribute in CP-ABE: a new approach (2012). arXiv preprint arXiv:1208.5848
14. Fang, L., LeFevre, K.: Privacy wizards for social networking sites. In: International World Wide Web Conference (WWW) (2010)
15. Fong, P.W.L., Anwar, M., Zhao, Z.: A privacy preservation model for facebook-style social network systems. In: Backes, M., Ning, P. (eds.) ESORICS 2009. LNCS, vol. 5789, pp. 303–320. Springer, Heidelberg (2009). https://doi.org/10.1007/978-3-642-04444-1_19
16. Freitas, M.: Twister: the development of a peer-to-peer microblogging platform. Int. J. Parallel Emergent Distrib. Syst. **31**(1), 20–33 (2016)
17. Gao, B., Berendt, B.: Circles, posts and privacy in egocentric social networks: an exploratory visualization approach. In: IEEE/ACM ASONAM (2013)
18. Goyal, V., Pandey, O., Sahai, A., Waters, B.: Attribute-based encryption for fine-grained access control of encrypted data. In: ACM CCS (2006)
19. Kairam, S., Brzozowski, M., Huffaker, D., Chi, E.: Talking in circles: selective sharing in google+. In: ACM CHI (2012)
20. Karthick, A., Murali, D., Kumaraesan, A., Vijayakumar, K.: Twitterize: anonymous micro-blogging in computer systems and applications. Adv. Nat. Appl. Sci. **11**(6 SI), 96–103 (2017)
21. Katzenbeisser, S., Petitcolas, F.: Information Hiding Techniques for Steganography and Digital Watermarking. Artech house, Norwood (2000)
22. Lan, C., Yang, Y., Li, X., Luo, B., Huan, J.: Learning social circles in ego-networks based on multi-view network structure. IEEE TKDE **29**(8), 1681–1694 (2017)
23. Leskovec, J., Mcauley, J.J.: Learning to discover social circles in ego networks. In: Advances in Neural Information Processing Systems, pp. 539–547 (2012)
24. Madejski, M., Johnson, M., Bellovin, S.M.: The failure of online social network privacy settings. Columbia University, Technical report CUCS-010-11 (2011)
25. Nilizadeh, S., Jahid, S., Mittal, P., Borisov, N., Kapadia, A.: Cachet: a decentralized architecture for privacy preserving social networking with caching. In: ACM CoNEXT (2012)
26. Paul, T., Famulari, A., Strufe, T.: A survey on decentralized online social networks. Comput. Netw. **75**, 437–452 (2014)
27. Shakimov, A., Varshavsky, A., Cox, L.P., Cáceres, R.: Privacy, cost, and availability tradeoffs in decentralized osns. In: ACM WOSN (2009)

28. Singh, I., Butkiewicz, M., Madhyastha, H.V., Krishnamurthy, S.V., Addepalli, S.: Enabling private conversations on twitter. In: ACSAC, pp. 409–418 (2012)
29. Singh, I., Butkiewicz, M., Madhyastha, H.V., Krishnamurthy, S.V., Addepalli, S.: Twitsper: tweeting privately. IEEE Secur. Priv. **11**(3), 46–50 (2013)
30. Squicciarini, A., Karumanchi, S., Lin, D., Desisto, N.: Identifying hidden social circles for advanced privacy configuration. Comput. Secur. **41**, 40–51 (2014)
31. Squicciarini, A.C., Sundareswaran, S., Lin, D., Wede, J.: A3p: adaptive policy prediction for shared images over popular content sharing sites. In: Proceedings of the 22nd ACM Conference on Hypertext and Hypermedia (2011)
32. Vitak, J., Blasiola, S., Patil, S., Litt, E.: Balancing audience and privacy tensions on social network sites. Int. J. Commun. **9**, 20 (2015)
33. Waters, B.: Ciphertext-policy attribute-based encryption: an expressive, efficient, and provably secure realization. In: Catalano, D., Fazio, N., Gennaro, R., Nicolosi, A. (eds.) PKC 2011. LNCS, vol. 6571, pp. 53–70. Springer, Heidelberg (2011). https://doi.org/10.1007/978-3-642-19379-8_4
34. Watson, J., Besmer, A., Lipford, H.R.: +Your circles: sharing behavior on Google+. In: SOUPS. ACM (2012)
35. Xu, Z., Martin, K.M.: Dynamic user revocation and key refreshing for attribute-based encryption in cloud storage. In: IEEE TrustCom (2012)
36. Yang, Y., Lutes, J., Li, F., Luo, B., Liu, P.: Stalking online: on user privacy in social networks. In: ACM CODASPY, pp. 37–48. ACM (2012)
37. Zhao, Y., Ren, M., Jiang, S., Zhu, G., Xiong, H.: An efficient and revocable storage CP-ABE scheme in the cloud computing. Computing, 1–25 (2018). https://doi.org/10.1007/s00607-018-0637-2

Fast Keyed-Verification Anonymous Credentials on Standard Smart Cards

Jan Camenisch[1], Manu Drijvers[1], Petr Dzurenda[2], and Jan Hajny[2(✉)]

[1] Dfinity, Zurich, Switzerland
{jan,manu}@dfinity.org
[2] Brno University of Technology, Brno, Czech Republic
{dzurenda,hajny}@feec.vutbr.cz

Abstract. Cryptographic anonymous credential schemes allow users to prove their personal attributes, such as age, nationality, or the validity of a ticket or a pre-paid pass, while preserving their privacy, as such proofs are unlinkable and attributes can be selectively disclosed. Recently, Chase et al. (CCS 2014) observe that in such systems, a typical setup is that the credential issuer also serves as the verifier. They introduce keyed-verification credentials that are tailored to this setting. In this paper, we present a novel keyed-verification credential system designed for lightweight devices (primarily smart cards). By using a novel algebraic MAC based on Boneh-Boyen signatures, we achieve the most efficient proving protocol compared to existing schemes. To demonstrate the practicality of our scheme in real applications, including large-scale services such as public transportation or e-government, we present an implementation on a standard, off-the-shelf, Multos smart card. While using significantly higher security parameters than most existing implementations, we achieve performance that is more than 44% better than the current state-of-the-art implementation.

Keywords: Privacy · Anonymous credentials · Authentication · Smart cards

1 Introduction

Using cryptographic credentials, users can anonymously prove the ownership of their personal attributes, such as age, nationality, sex or ticket validity. In the recent two decades, many proposals for anonymous credential schemes have been published. Starting with the fundamental works of Chaum [23], Brands [10], Camenish and Lysyanskaya [17], until recent schemes [1,3,22,27,34], researchers try to find a scheme that fulfills all requirements on privacy, is provably secure and is so efficient that it can be implemented on constrained devices. While there are schemes that fulfill all the requirements and can be implemented on PC and smartphone platforms, existing schemes deployed on smart cards are

© IFIP International Federation for Information Processing 2019
Published by Springer Nature Switzerland AG 2019
G. Dhillon et al. (Eds.): SEC 2019, IFIP AICT 562, pp. 286–298, 2019.
https://doi.org/10.1007/978-3-030-22312-0_20

still not sufficiently fast for many applications, such as e-ticketing and eIDs. Yet, smart cards are the most appropriate platform for storing and proving personal attributes in everyday life, due to their size, security and reliability.

There are two major reasons why we lack practical implementations of anonymous credentials on smart cards. First, the complexity of asymmetric cryptographic algorithms used in anonymous credentials is quite high even for modern smart cards. Second, modern cryptographic schemes, including anonymous credentials, are mostly based on operations over an elliptic curve, while most available smart cards do not provide API for these operations. Particularly, the very popular operation of bilinear maps is still unsupported on this platform and simple operations, such as EC point scalar multiplication and addition, are significantly restricted.

In this paper, we address both these concerns: First, we propose a novel keyed-verification anonymous credential scheme that is designed to allow for smart card implementations. Our scheme has the most efficient proving algorithm to date and requires only operations that are available on existing off-the-shelf smart cards. Second, we present the implementation of our anonymous credential scheme that is 44%–72% faster than the current state-of-the-art implementation, while even providing a higher security level.

1.1 Related Work

Cryptographic anonymous credential schemes were first defined by the seminal works of Chaum [23], Brands [10] and Camenisch and Lysyanskaya [17]. The schemes were gradually improved by adding revocation protocols [15,18], using more efficient algebraic structures [19,34] and developing security models and formal proofs [16]. Idemix [21] and U-Prove [32] are the examples of the most evolved schemes aiming for a practical use. Recently, a new approach to obtain more efficient anonymous credential schemes was proposed. Chase et al. [22] argue that in many scenarios where anonymous credentials could be deployed, the issuer of the credential will also serve as the verifier. This means that the verifier possesses the issuer key, which can be leveraged to obtain more efficient anonymous credential schemes tailored to setting. They formally define these so-called *Keyed-Verification Anonymous Credentials (KVAC)* and propose two instantiations. Barki et al. [2] propose a new KVAC scheme which is currently the most efficient: Proving possession of a credential with u hidden attributes costs $u + 12$ exponentiations. Couteau and Reichle [24] construct a KVAC scheme with presentation cost $2u + 3$ exponentiations in a 2048-bit group, which is less efficient, but works in the standard model. Some of the new constructions were already implemented on the PC platform with promising results [28,34]. Yet, the implementations on the smart card platform are available only for the former schemes that are based on traditional, rather inefficient modular structures [6,30,33,38]. Furthermore, most implementations use only 1024-bit RSA groups that are considered insufficient by today's standards [37]. Implementations with higher security parameters [1–4] either need distribution of computation to another device (usually a mobile phone) or use a non-standard proprietary

API for EC operations and rely on pre-computations (which is impossible in crucial applications like e-ticketing and eID where the card is inactive and starts only for the attribute presentation). Regarding speed, the best-performing implementation of Idemix by the IRMA project [38] is able to compute the unlinkable attribute proof in at least 0,9 s, which is not convenient for time-critical applications where the proof should be presented in less than 500 ms. Currently, there is no cryptographic proposal and its implementation that would realize unlinkable anonymous credentials on the smart card platform with performance and security parameters necessary for a practical deployment.

1.2 Our Contribution

We propose a novel cryptographic scheme for anonymous attribute-based credentials that is designed primarily for smart cards. It provides all necessary privacy-protection features, i.e., the anonymity, unlinkability, untraceability and selective disclosure of attributes. The scheme is based on our original algebraic MAC that makes its proving protocol very efficient. The computational complexity of our proving protocol is the lowest from related schemes (only $u + 2$ scalar multiplications to present an attribute ownership proof) and we need only basic arithmetic operations that are already provided by existing smart cards' APIs. We present the results of the full implementation of our proving protocol that is faster by at least 44% than the state-of-the-art implementation. By reaching the time of 366 ms including overhead, which is required for proving personal attributes on a 192-bit EC security level, we argue that the anonymous credentials are finally secure and practical even for time-critical and large-scale applications like eIDs, e-ticketing and mass transportation.

2 Preliminaries

2.1 Notation

We describe (signature) proof of knowledge protocols (SPK) using the efficient notation introduced by Camenisch and Stadler [20]. The protocol for proving the knowledge of discrete logarithm of c with respect to g is denoted as SPK$\{\alpha :$ $c = g^{\alpha}\}$. The symbol ":" means "such that" and $|x|$ is the bitlength of x. The symbol $\mathcal{H}$ denotes a secure hash function. We write $a \xleftarrow{\$} A$ when a is sampled uniformly at random from A. Let GroupSetup(1^{κ}) be an efficient algorithm that generates a group $\mathbb{G} = \langle g \rangle$ of prime order q, such that $|q| = \kappa$. Let $\mathbf{e}$ denote a bilinear map.

2.2 Weak Boneh-Boyen Signature

We recall the weak Boneh-Boyen signature scheme [9], which is existentially unforgeable against a weak (non-adaptive) chosen message attack under the q-SDH assumption.

Setup: On input security parameter τ, generate a bilinear group $(q, \mathbb{G}_1, \mathbb{G}_2, \mathbb{G}_T, \mathbf{e}, g_1, g_2) \leftarrow \mathcal{G}(1^\tau)$. Take $x \xleftarrow{\$} \mathbb{Z}_q$, compute $w = g_2^x$, and output $sk = x$ as private key and $pk = (q, \mathbb{G}_1, \mathbb{G}_2, \mathbb{G}_T, g_1, g_2, \mathbf{e}, w)$ as public key.

Sign: On input message $m \in \mathbb{Z}_q$ and secret key sk, output $\sigma = g_1^{\frac{1}{x+m}}$.

Verify: On input the signature σ, message m, and public key pk, output 1 iff $\mathbf{e}(\sigma, w) \cdot \mathbf{e}(\sigma^m, g_2) = \mathbf{e}(g_1, g_2)$ holds.

2.3 Algebraic MACs

Compared to traditional Message Authentication Codes (MACs), algebraic MACs can be efficiently combined with zero knowledge proofs. In terms of security, algebraic MACs [22] are no different from traditional MACs. A MAC scheme consists of algorithms (Setup, KeyGen, MAC, Verify). Setup sets up the system parameters par that are given as implicit input to the other algorithms. KeyGen creates a new secret key, MAC(sk, m) computes a MAC on message m, and Verify is used to verify MACs. We recall the security definitions due to Dodis et al. [25] and slightly strengthened by Chase et al. [22], and require completeness and unforgeability under chosen message and verification attack (uf-cmva).

Definition 1. *A MAC scheme* (Setup, KeyGen, MAC, Verify) *is complete if the following probability is negligible in κ for all messages m:*

$$\Pr\left[\mathsf{Verify}(sk, m, \sigma) = 0 \mid par \xleftarrow{\$} \mathsf{Setup}(1^\kappa),\right.$$

$$\left.(ipar, sk) \xleftarrow{\$} \mathsf{KeyGen}(par), \sigma \xleftarrow{\$} \mathsf{MAC}(sk, m)\right].$$

Definition 2. *A MAC scheme* (Setup, KeyGen, MAC, Verify) *is $(t, \epsilon, q_{\mathsf{MAC}}, q_{\mathsf{Verify}})$-unforgeable under chosen message and verification attack if there exists no adversary $\mathcal{A}$ running in time t making at most q_{MAC} MAC queries and at most q_{Verify} Verify queries, for which the following probability is at least ϵ:*

$$\Pr\left[\mathsf{Verify}(sk, m^*, \sigma^*) = 1 \wedge m^* \notin Q \mid par \xleftarrow{\$} \mathsf{Setup}(1^\kappa),\right.$$

$$\left.(ipar, sk) \xleftarrow{\$} \mathsf{KeyGen}(par), (\sigma^*, m^*) \xleftarrow{\$} \mathcal{A}^{\mathcal{O}^{\mathsf{MAC}(sk, \cdot)}, \mathcal{O}^{\mathsf{Verify}(sk, \cdot, \cdot)}}(par, ipar)\right].$$

3 Our Algebraic MAC

This section describes our novel algebraic MAC scheme $\mathsf{MAC}_{\mathsf{wBB}}$, which is based on the weak Boneh-Boyen signature. It works in a prime order group and can MAC vectors of n messages $\vec{m} = (m_1, \ldots, m_n)$, with $m_i \in \mathbb{Z}_q^*$, using the technique due to Camenisch et al. [13] to extend the Boneh-Boyen signature to multiple messages. The scheme is composed of the following algorithms.

Setup(1^κ): Output $par = (\mathbb{G}, g, q) \leftarrow \mathsf{GroupSetup}(1^\kappa)$.
KeyGen(par): Choose $x_i \xleftarrow{\$} \mathbb{Z}_q^*$ for $i = (0, \ldots, n)$. Output secret key $sk = (x_0, \ldots, x_n)$ and public issuer parameters $ipar \leftarrow (X_0, \ldots, X_n)$ with $X_i = g^{x_i}$.

$\mathsf{MAC}(sk, \vec{m})$: Let $sk = (x_0, \ldots, x_n)$ and $\vec{m} = (m_1, \ldots, m_n)$. Compute $\sigma = g^{\frac{1}{x_0 + \sum_{i=1}^{n} m_i x_i}}$ and auxiliary information $\sigma_{x_i} \leftarrow \sigma^{x_i}$ for $i = (1, \ldots, n)$.[1] Output the authentication code $(\sigma, \sigma_{x_1}, \ldots, \sigma_{x_n})$.

$\mathsf{Verify}(sk, \vec{m}, \sigma)$: Let $sk = (x_0, \ldots, x_n)$ and $\vec{m} = (m_1, \ldots, m_n)$. Output 1 iff $g = \sigma^{x_0 + \sum_{i=1}^{n} m_i x_i}$.

Unforgeability of our MAC scheme holds under the SCDHI assumption, which is a variation of the SDDHI assumption [14].

Theorem 1. *Our MAC scheme is unforgeable, as defined in Definition 2, under the SCDHI assumption. More precisely, if n-SCDHI is (t, ϵ)-hard, then our MAC scheme is (t, ϵ)-unforgeable.*

We introduce the SCDHI problem, prove its hardness in generic groups and formally prove Theorem 1 in the full manuscript [11].

4 Keyed-Verification Anonymous Credential Scheme

We construct our keyed-verification anonymous credential (KVAC) scheme using the algebraic MAC scheme presented in Sect. 3 above. Unlike traditional anonymous attribute-based credential schemes (ABCs), the verifier needs to know the secret keys to be able to verify user's attributes in keyed-verification anonymous credential schemes. This feature is particularly convenient for scenarios where attribute issuers are the same entities as attribute verifiers. The mass transportation settings is an example of such a scenario because the transportation authority both issues and checks the tickets and passes. The KVAC scheme supports all the standard privacy-enhancing features of ABC schemes, such as anonymity, unlinkability, untraceability, and selective disclosure of attributes, and is compatible with major credential schemes [21,32] and standard revocation schemes [12,17].

4.1 Definition of Keyed-Verification Anonymous Credential Schemes

A KVAC scheme consists of algorithms (Setup, CredKeygen, Issue, Obtain, Show, ShowVerify)[2] that are executed by users and an issuer who also serves as a verifier.

Setup(1^k) takes as input the security parameter and outputs the system parameters *par*. We assume that *par* is given as implicit input to all algorithms.

---

[1] Note that the auxiliary information σ_{x_i} can be omitted as they are not required for verification. However, in our keyed verification credentials, it will turn out that adding these values will make credential presentation more efficient.

[2] Note that Chase et al. [22] define BlindIssue and BlindObtain, but as we do not show efficient algorithms for blind issuance, we omit them from the definition here. Instead, we define Obtain, which lets a user check that a credential is indeed valid using only the public issuer parameters.

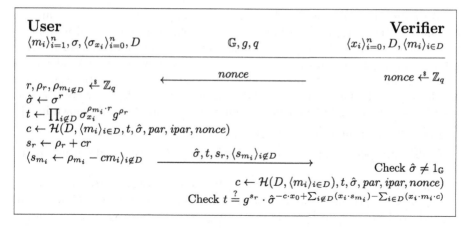

Fig. 1. Definition of the Show and ShowVerify algorithms of our KVAC scheme.

CredKeygen(par) outputs a issuer secret key sk and issuer parameters $ipar$.

Issue($sk, (m_1, \ldots, m_n)$) takes as input the issuer secret key and attribute values $(m_1, \ldots, m_n)$ and outputs a credential $cred$. The issuance of attributes must be done over a secure channel (as the attributes and private AMAC are sent between the user and issuer) and the credential should be stored on a tamper-proof device (we use a smart-card).

Obtain($ipar, cred, (m_1, \ldots, m_n)$) lets a user verify a credential by giving as input the public issuer parameters, the credential and the attribute values.

Show($ipar, cred, (m_1, \ldots, m_n), \phi$) $\leftrightarrow$ ShowVerify(sk, ϕ) is an interactive algorithm. The user runs Show on input the public issuer parameters, the credential, the attribute values and attribute predicate, and the verifier runs ShowVerify on input the issuer secret key and the attribute predicate, which will output 1 iff it accepts the credential presentation.

4.2 Our KVAC Scheme Based on MAC$_{\mathsf{wBB}}$

In this section, we present our novel KVAC scheme that uses MAC$_{\mathsf{wBB}}$ as introduced in Sect. 3. Our scheme certifies attributes in $\mathbb{Z}_q^*$ and is parametrized by n, the amount of attributes in a credential. We describe our scheme using *selective disclosure* as attribute predicates, i.e., a predicate ϕ can be seen as a set $D \subseteq \{1, \ldots, n\}$ containing the indices of the disclosed attributes and the attribute values of the disclosed attributes $\langle m_i \rangle_{i \in D}$. On a high level, we follow the approach from Chase et al. [22] and build our KVAC scheme from our algebraic MAC presented in Sect. 3 and zero knowledge proofs. One novel trick allows us to strongly improve the efficiency of our scheme. Instead of computing a standard noninteractive Schnorr-type proof of knowledge, we use the fact that the verifier knows the secret key. This allows us to omit elements that the verifier can compute by itself and saves the prover a lot of work.

We note that our Issue algorithm does not support the efficient issuance of committed attributes, i.e., the blind issuance. This feature is useful in applications where a user needs to transfer his attributes among credentials or needs to get issued attributes that are only private to him. However, we consider these scenarios rare in targeted applications such as e-ticketing, mass transportation and loyalty cards. Furthermore, if the issuance of committed attributes is necessary, it can be done by employing Paillier encryption [31], as is shown in [5].

Setup(1^k): Output $par = (\mathbb{G}, g, q) \leftarrow \mathsf{GroupSetup}(1^\kappa)$.

CredKeygen(par): Run $(sk, ipar) \leftarrow \mathsf{MAC_{wBB}.KeyGen}(par)$ and output sk and $ipar$.

Issue($sk, (m_1, \ldots, m_n)$): Run $(\sigma, \langle \sigma_{x_i} \rangle_{i=0}^n) \leftarrow \mathsf{MAC_{wBB}.MAC}(sk, (m_1, \ldots, m_n))$. Next, provide a proof that allows a user to verify the validity of the credential: $\pi \leftarrow SPK\{(x_0, \ldots, x_n) : \bigwedge_{i=0}^n \sigma_{x_i} = \sigma^{x_i} \wedge X_i = g^{x_i}\}$. Output credential $cred \leftarrow (\sigma, \langle \sigma_{x_i} \rangle_{i=0}^n, \pi)$.

Obtain($ipar, cred, (m_1, \ldots, m_n)$): Parse $ipar$ as $(X_0, \ldots, X_n)$ and parse $cred$ as $(\sigma, \langle \sigma_{x_i} \rangle_{i=0}^n, \pi)$. Check that $\sigma_{x_0} \cdot \prod_{i=1}^n \sigma_{x_i}^{m_i} = g$ and verify π with respect to $ipar$ and σ.

Show($ipar, cred, (m_1, \ldots, m_n), (D, \langle m_i \rangle_{i \in D})$): In credential presentation, we want to let the user prove possession of a valid credential with the desired attributes. On a high level, we want to prove knowledge of a weak Boneh-Boyen signature, so we can apply the efficient proof due to Arfaoui et al. [1] and Camenisch et al. [12], by extending it to support a vector of messages: Take a random $r \xleftarrow{\$} \mathbb{Z}_q^*$ and let $\hat{\sigma} \leftarrow \sigma^r$ and $\hat{\sigma}_{x_i} \leftarrow \sigma_{x_i}^r$ for $i = 0, \ldots, n$, and prove

$$SPK\{(\langle m_i \rangle_{i \notin D}, r) : \hat{\sigma}_{x_0} \prod_{i \in D} \hat{\sigma}_{x_i}^{m_i} = g^r \prod_{i \notin D} \hat{\sigma}_{x_i}^{-m_i}\}.$$

The verifier simply checks that the $\hat{\sigma}_{x_i}$ values are correctly formed and verifies the proof.

While this approach is secure and conceptually simple, it is not very efficient. We now present how we can construct a similar proof in a much more efficient manner. The key observation is that the user does not have to compute anything that the verifier, who is in possession of the issuer secret key sk, can compute. This means we can omit the computation of the $\hat{\sigma}_{x_i}$ values and define Show as follows. Randomize the credential by taking a random $r \leftarrow \mathbb{Z}_q^*$ and setting $\hat{\sigma} \leftarrow \sigma^r$. Take $\rho_r, \rho_{m_{i \notin D}} \xleftarrow{\$} \mathbb{Z}_q$ and compute

$$t = \prod_{i \notin D} \sigma_{x_i}^{\rho_{m_i} \cdot r} g^{\rho_r}, \quad c \leftarrow \mathcal{H}(D, \langle m_i \rangle_{i \in D}, t, \hat{\sigma}, par, ipar, nonce),$$

and let $s_r = \rho_r + cr, \langle s_{m_i} = \rho_{m_i} - cm_i \rangle_{i \notin D}$. Send $(\hat{\sigma}, t, s_r, \langle s_{m_i} \rangle_{i \notin D})$ to the verifier.

ShowVerify($sk, (D, \langle m_i \rangle_{i \in D})$): The verifier running ShowVerify will receive $(\hat{\sigma}, t, s_r, \langle s_{m_i} \rangle_{i \notin D})$ from the user. It recomputes

$$c \leftarrow \mathcal{H}((m_1, \ldots, m_n), (D, \langle m_i \rangle_{i \in D}), t, \hat{\sigma}, par, ipar, nonce)$$

and checks

$$t \stackrel{?}{=} g^{s_r} \cdot \hat{\sigma}^{-c \cdot x_0 + \sum_{i \notin D}(x_i \cdot s_{m_i}) - \sum_{i \in D}(x_i \cdot m_i \cdot c)}.$$

Output 1 if valid and 0 otherwise. The Show and ShowVerify algorithms are depicted in Fig. 1.

Theorem 2. *Our keyed-verification credential scheme is secure following the definition by Chase et al. [22] (ommitting the blind issuance), under the n-SCDHI assumption in the random oracle model.*

We formally prove Theorem 2 in the full manuscript [11].

4.3 Efficiency

Our Show and ShowVerify algorithms were designed to be efficient enough to run on smart cards. We avoided computing bilinear pairings due to their computational cost and the lack of support on existing smart cards. The use of the second most expensive operation, the exponentiation (or scalar multiplication of EC points respectively), is reduced to a minimum. Our proving algorithm, the part of the protocol we envision being executed on a smart card, only requires $u + 2$ exponentiations, where u is the number of undisclosed attributes.

Table 1 compares the efficiency of our Show protocol to existing KVAC schemes [3,22], well-known anonymous credential schemes U-Prove [32] and Identity Mixer [21], and a recent scheme by Ringers et al. [34]. Idemix takes place in the RSA group, meaning that the exponentiations are much more expensive than exponentiations in a prime order group. U-Prove lacks the unlinkability property. Compared to MAC$_{BB}$, our scheme requires only 2 exponentiations without hidden attributes, whereas MAC$_{BB}$ requires 12, showing that especially for a small number of undisclosed attributes, our scheme is significantly faster than MAC$_{BB}$.

Table 1. Comparison of presentation protocols of credential schemes.

	Exp. prime	Exp. RSA	Unlink.	MAC	Security
U-Prove [32]	$u + 1$	0	✗	✗	-
Idemix [21]	0	$u + 3$	✓	✗	sRSA [35]
Ringers et al. [34]	$n + u + 9$	0	✓	✗	whLRSW [39]
MAC$_{DDH}$ [22]	$6u + 12$	0	✓	✓	DDH [7]
MAC$_{GGM}$ [22]	$5u + 4$	0	✓	✓	GGM [36]
MAC$_{BB}$ [3]	$u + 12$	0	✓	✓	q-sDH [8]
NIKVAC [24]	$2u + 3$	0	✓	✓	GGM+IND-CPA
This work	$u + 2$	0	✓	✓	n-SCDHI [11]

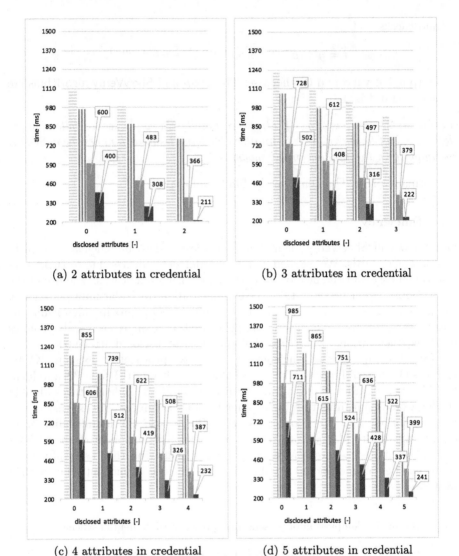

(a) 2 attributes in credential   (b) 3 attributes in credential

(c) 4 attributes in credential   (d) 5 attributes in credential

Fig. 2. Speed of our proving protocol compared to Vullers and Alpár (VA) implementation [38]. Blue - our algorithm time, orange - our total time with overhead, verticals - VA algorithm time and horizontals - VA total time with overhead. (Color figure online)

5   Implementation Results

There are many cryptographic schemes for anonymous attribute-based credentials available. Nevertheless, the smart card implementations are only very few [26,30,38] and not practically usable as they use only small insecure security parameters to be able to achieve reasonable speed. Particularly, only 1024-bit RSA or DSA groups are used. That is considered insecure for any practical deployment today.

The `Show` and `ShowVerify` algorithms of our scheme were implemented using a standard NIST P-192 curve [29] on the Multos ML3 smart card. Only standard Multos API and free public development environment (Eclipse IDE for C/C++ Developers, SmartDeck 3.0.1, MUtil 2.8) were used. For terminal application, Java BigInteger class and BouncyCastle API were used. We compare our results (blue and orange) with the state-of-the-art results of Vullers and Alpár (VA) [38] (black and white) for different numbers of attributes stored and disclosed in Fig. 2. We note that our implementation uses significantly higher security parameters (1024-bit used by Vullers and Alpár vs. 1776-bit DSA group equivalent according to [37] used by us). The algorithm time (blue) tells the time necessary to compute all algorithms on the card. The overhead time (orange) adds time necessary to do all the supporting actions, mainly establishing the communication with a reader connected to PC and transferring APDUs. All results are arithmetic means of 10 measurements in milliseconds. Compared to VA's implementation of Idemix, our implementation of all proving protocol algorithms on the card is at least 44% faster in all cases, see Fig. 2 for details.

In the case of only 2 attributes stored on the card, our scheme is by 72% faster than VA's implementation. The card needs only 211 ms to compute the ownership proof for disclosed attributes. The total time of around 360 ms necessary for the whole proof generation on the card including communication with and computations on a terminal (standard PC, Core i7 2.4 GHz, 8 GB RAM) makes the implementation suitable also for time-critical applications like public transportation and ticketing. We also evaluated our scheme using an embedded device (Raspberry Pi 3) instead of the PC as a terminal. Even in that case the total time including overhead was below 450 ms. Based on our benchmarks, we expect that increasing security parameters to the 256-bit EC level would cost acceptable 15%–20% in performance.

Our implementation is artificially limited to 10 attributes per a user, but the smart card's available memory resources (approx. 1.75 KB RAM and 7.5 KB usable EEPROM) would allow storing upto 50 attributes on a single card.

6 Conclusion

Practical anonymous credential schemes are only very few, with implementations on smart cards either too slow or providing insufficient security levels. Our approach to address this problem was twofold: (1) to propose a novel cryptographic scheme that is more efficient than all comparable schemes and formally prove its security; and (2) to develop a software implementation that is significantly faster than existing implementations, although they use lower security parameters. By achieving these results, we hope that we get privacy-enhanced authentication closer to practical applications.

Our future steps, besides further optimization, are the integration with a suitable revocation scheme (e.g., [12]) and implementation and benchmarks on higher security levels, hopefully on a wider range of smart cards, if they become available on the market.

Acknowledgment. This paper is supported in part by European Union's Horizon 2020 research and innovation programme under grant agreement No 830892, project SPARTA, the Ministry of Industry and Trade grant # FV20354 and the National Sustainability Program under grant LO1401. For the research, infrastructure of the SIX Center was used.

References

1. Arfaoui, G., Lalande, J.F., Traoré, J., Desmoulins, N., Berthomé, P., Gharout, S.: A practical set-membership proof for privacy-preserving NFC mobile ticketing. In: PoPETs, pp. 25–45 (2015)
2. Barki, A., Brunet, S., Desmoulins, N., Gambs, S., Gharout, S., Traoré, J.: Private eCash in practice (short paper). In: Grossklags, J., Preneel, B. (eds.) FC 2016. LNCS, vol. 9603, pp. 99–109. Springer, Heidelberg (2017). https://doi.org/10.1007/978-3-662-54970-4_6
3. Barki, A., Brunet, S., Desmoulins, N., Traoré, J.: Improved algebraic MACs and practical keyed-verification anonymous credentials. In: Avanzi, R., Heys, H. (eds.) SAC 2016. LNCS, vol. 10532, pp. 360–380. Springer, Cham (2017). https://doi.org/10.1007/978-3-319-69453-5_20
4. Barki, A., Desmoulins, N., Gharout, S., Traoré, J.: Anonymous attestations made practical. In: ACM WiSec 2017 Proceedings, pp. 87–98 (2017)
5. Belenkiy, M., Camenisch, J., Chase, M., Kohlweiss, M., Lysyanskaya, A., Shacham, H.: Randomizable proofs and delegatable anonymous credentials. In: Halevi, S. (ed.) CRYPTO 2009. LNCS, vol. 5677, pp. 108–125. Springer, Heidelberg (2009). https://doi.org/10.1007/978-3-642-03356-8_7
6. Bichsel, P., Camenisch, J., Groß, T., Shoup, V.: Anonymous credentials on a standard java card. In: ACM CCS 2009 Proceedings, pp. 600–610 (2009)
7. Boneh, D.: The decision Diffie-Hellman problem. In: Buhler, J.P. (ed.) ANTS 1998. LNCS, vol. 1423, pp. 48–63. Springer, Heidelberg (1998). https://doi.org/10.1007/BFb0054851
8. Boneh, D., Boyen, X.: Short signatures without random oracles. In: Cachin, C., Camenisch, J.L. (eds.) EUROCRYPT 2004. LNCS, vol. 3027, pp. 56–73. Springer, Heidelberg (2004). https://doi.org/10.1007/978-3-540-24676-3_4
9. Boneh, D., Boyen, X.: Short signatures without random oracles and the SDH assumption in bilinear groups. J. Cryptol. **21**, 149–177 (2008)
10. Brands, S.A.: Rethinking Public Key Infrastructures and Digital Certificates: Building in Privacy (2000)
11. Camenisch, J., Drijvers, M., Dzurenda, P., Hajny, J.: Fast keyed-verification anonymous credentials on standard smart cards. Cryptology ePrint Archive, Report 2019 (2019). https://eprint.iacr.org/2019/
12. Camenisch, J., Drijvers, M., Hajny, J.: Scalable revocation scheme for anonymous credentials based on n-times unlinkable proofs. In: ACM CCS WPES 2016 Proceedings, pp. 123–133 (2016)
13. Camenisch, J., Dubovitskaya, M., Neven, G.: Oblivious transfer with access control. In: ACM CCS 2009 Proceedings, pp. 131–140 (2009)

14. Camenisch, J., Hohenberger, S., Kohlweiss, M., Lysyanskaya, A., Meyerovich, M.: How to win the clonewars: efficient periodic n-times anonymous authentication. In: ACM CCS 2006 Proceedings, pp. 201–210 (2006)
15. Camenisch, J., Kohlweiss, M., Soriente, C.: Solving revocation with efficient update of anonymous credentials. In: Garay, J.A., De Prisco, R. (eds.) SCN 2010. LNCS, vol. 6280, pp. 454–471. Springer, Heidelberg (2010). https://doi.org/10.1007/978-3-642-15317-4_28
16. Camenisch, J., Krenn, S., Lehmann, A., Mikkelsen, G.L., Neven, G., Pedersen, M.Ø.: Scientific comparison of ABC protocols (2014)
17. Camenisch, J., Lysyanskaya, A.: An efficient system for non-transferable anonymous credentials with optional anonymity revocation. In: Pfitzmann, B. (ed.) EUROCRYPT 2001. LNCS, vol. 2045, pp. 93–118. Springer, Heidelberg (2001). https://doi.org/10.1007/3-540-44987-6_7
18. Camenisch, J., Lysyanskaya, A.: Dynamic accumulators and application to efficient revocation of anonymous credentials. In: Yung, M. (ed.) CRYPTO 2002. LNCS, vol. 2442, pp. 61–76. Springer, Heidelberg (2002). https://doi.org/10.1007/3-540-45708-9_5
19. Camenisch, J., Neven, G., Rückert, M.: Fully anonymous attribute tokens from lattices. In: Visconti, I., De Prisco, R. (eds.) SCN 2012. LNCS, vol. 7485, pp. 57–75. Springer, Heidelberg (2012). https://doi.org/10.1007/978-3-642-32928-9_4
20. Camenisch, J., Stadler, M.: Efficient group signature schemes for large groups. In: Kaliski, B.S. (ed.) CRYPTO 1997. LNCS, vol. 1294, pp. 410–424. Springer, Heidelberg (1997). https://doi.org/10.1007/BFb0052252
21. Camenisch, J., Van Herreweghen, E.: Design and implementation of the idemix anonymous credential system. In: ACM CCS 2002 Proceedings, pp. 21–30 (2002)
22. Chase, M., Meiklejohn, S., Zaverucha, G.: Algebraic MACs and keyed-verification anonymous credentials. In: ACM SIGSAC 2014 Proceedings, pp. 1205–1216 (2014)
23. Chaum, D.: Security without identification: transaction systems to make big brother obsolete. Commun. ACM **28**, 1030–1044 (1985)
24. Couteau, G., Reichle, M.: Non-interactive keyed-verification anonymous credentials. Cryptology ePrint Archive, Report 2019/117 (2019). https://eprint.iacr.org/2019/117
25. Dodis, Y., Kiltz, E., Pietrzak, K., Wichs, D.: Message authentication, revisited. In: Pointcheval, D., Johansson, T. (eds.) EUROCRYPT 2012. LNCS, vol. 7237, pp. 355–374. Springer, Heidelberg (2012). https://doi.org/10.1007/978-3-642-29011-4_22
26. Hajny, J., Malina, L.: Unlinkable attribute-based credentials with practical revocation on smart-cards. In: Mangard, S. (ed.) CARDIS 2012. LNCS, vol. 7771, pp. 62–76. Springer, Heidelberg (2013). https://doi.org/10.1007/978-3-642-37288-9_5
27. Hinterwälder, G., Riek, F., Paar, C.: Efficient E-cash with attributes on MULTOS smartcards. In: Mangard, S., Schaumont, P. (eds.) RFIDSec 2015. LNCS, vol. 9440, pp. 141–155. Springer, Cham (2015). https://doi.org/10.1007/978-3-319-24837-0_9
28. Isaakidis, M., Halpin, H., Danezis, G.: UnlimitID: privacy-preserving federated identity management using algebraic MACs. In: ACM CCS WPES 2016 Proceedings, pp. 139–142 (2016)
29. Kerry, C.F., Secretary, A., Director, C.R.: FIPS PUB 186-4 Federal Information Processing Standards Publication: Digital Signature Standard (DSS) (2013)
30. Mostowski, W., Vullers, P.: Efficient U-prove implementation for anonymous credentials on smart cards. In: Rajarajan, M., Piper, F., Wang, H., Kesidis, G. (eds.) SecureComm 2011. LNICST, vol. 96, pp. 243–260. Springer, Heidelberg (2012). https://doi.org/10.1007/978-3-642-31909-9_14

31. Paillier, P.: Public-key cryptosystems based on composite degree residuosity classes. In: Stern, J. (ed.) EUROCRYPT 1999. LNCS, vol. 1592, pp. 223–238. Springer, Heidelberg (1999). https://doi.org/10.1007/3-540-48910-X_16

32. Paquin, C.: U-Prove cryptographic specification v1.1. Technical report, Microsoft Corporation (2011)

33. de la Piedra, A., Hoepman, J.-H., Vullers, P.: Towards a full-featured implementation of attribute based credentials on smart cards. In: Gritzalis, D., Kiayias, A., Askoxylakis, I. (eds.) CANS 2014. LNCS, vol. 8813, pp. 270–289. Springer, Cham (2014). https://doi.org/10.1007/978-3-319-12280-9_18

34. Ringers, S., Verheul, E., Hoepman, J.-H.: An efficient self-blindable attribute-based credential scheme. In: Kiayias, A. (ed.) FC 2017. LNCS, vol. 10322, pp. 3–20. Springer, Cham (2017). https://doi.org/10.1007/978-3-319-70972-7_1

35. Rivest, R.L., Kaliski, B.: RSA problem. In: van Tilborg, H.C.A. (ed.) Encyclopedia of Cryptography and Security, pp. 532–536. Springer, New York (2005). https://doi.org/10.1007/0-387-23483-7_363

36. Shoup, V.: Lower bounds for discrete logarithms and related problems. In: Fumy, W. (ed.) EUROCRYPT 1997. LNCS, vol. 1233, pp. 256–266. Springer, Heidelberg (1997). https://doi.org/10.1007/3-540-69053-0_18

37. Smart, N.: Yearly report on algorithms and keysizes. Katholieke Universiteit Leuven, Technical report (2012)

38. Vullers, P., Alpár, G.: Efficient selective disclosure on smart cards using idemix. In: Fischer-Hübner, S., de Leeuw, E., Mitchell, C. (eds.) IDMAN 2013. IFIP AICT, vol. 396, pp. 53–67. Springer, Heidelberg (2013). https://doi.org/10.1007/978-3-642-37282-7_5

39. Wei, V.K., Yuen, T.H.: More short signatures without random oracles (2005). https://eprint.iacr.org/2005/463

BlockTag: Design and Applications of a Tagging System for Blockchain Analysis

Yazan Boshmaf[1(✉)], Husam Al Jawaheri[2], and Mashael Al Sabah[1]

[1] Qatar Computing Research Institute, Doha, Qatar
yboshmaf@hbku.edu.qa
[2] Qatar University, Doha, Qatar

Abstract. Annotating blockchains with auxiliary data is useful for many applications. For example, criminal investigation of darknet marketplaces, such as Silk Road and Agora, typically involves linking Bitcoin addresses, from which money is sent or received, to user accounts and web activities. We present BlockTag, an open-source tagging system for blockchains that facilitates such tasks. We describe BlockTag's design and demonstrate its capabilities through a real-world deployment of three applications in the context of privacy research and law enforcement.

Keywords: Blockchain · Tagging · Bitcoin · Privacy ·
Law enforcement

1 Introduction

Public blockchains contain data that describe various financial transactions. As of December 2018, Bitcoin's blockchain amounted to 18.5 GB of raw data and is growing rapidly. Such data is crucial for understanding different aspects of cryptocurrencies, including their privacy properties and market dynamics. Blockchain analysis systems, such as BlockSci [16], have enabled "blockchain science" by addressing three pain points: Poor performance, limited capabilities, and cumbersome programming interfaces. These systems, however, are focused on analyzing on-chain data and are not designed to incorporate off-chain data into their analysis pipeline. This limitation makes it difficult to use existing blockchain analysis systems for tasks that require linking off/on-chain data and searching for vulnerabilities or clues, which are common in privacy research and law enforcement.

We present BlockTag: An open-source tagging system for blockchain analysis. BlockTag uses vertical crawlers to annotate on-chain data with customizable, off-chain tags. In BlockTag, a tag is a mapping between a block, a transaction, or an address identifier and external auxiliary data. For example, the system can tag a Bitcoin address with the Twitter user account of its likely owner. BlockTag also provides a novel query interface for linking and searching. For example, BlockTag provides best-effort responses to high-level queries used in e-crime investigations, such as "which Twitter user accounts paid $\geq$ ฿10.0 to Silk Road in 2014."

We designed BlockTag based on the observation that blockchain analysis systems transform raw blockchain data into a stripped-down, simple data structure which can fit in or map to OS memory. Therefore, on-chain data that is not part of basic transaction information, such as hashes, scripts, and off-chain auxiliary data, cannot be part of this data structure and must have their own mappings. This naturally leads to a layered system architecture where a tagging layer sits on top of an analysis layer with a well-defined and extendable cross-layer interface.

In our implementation, we used BlockSci for analysis as it is much faster than its contenders. For BlockTag, we developed four vertical crawlers that annotate Bitcoin addresses with three types of tags: User tags representing BitcoinTalk and Twitter user accounts, service tags representing Tor hidden service providers, and text tags representing user-generated Blockchain.info labels. We extended BlockSci's analysis library and implemented a novel query engine to link, search, and aggregate off/on-chain Bitcoin data in a SQL-like syntax.

We deployed BlockTag in January 2018 for three months on a single, locally-hosted, machine. As of March 2018, the crawlers have ingested about 5B tweets, 2.2M BitcoinTalk user profiles, 1.5K Tor onion pages, and 30K Blockchain.info labels. This has resulted in 45K user, 88 service, and 29K text tags. In addition to BlockTag, our *contributions* include the following findings from three real-world applications that demonstrate BlockTag's capabilities:

(1) Linking: We showed how to deanonymize Tor hidden service user by linking their Bitcoin payments to their social network accounts. In total, we were able to link 125 user accounts to 20 service providers, which included illegal and controversial ones, such as Silk Road and The Pirate Bay. Such deanonymization is possible because of Bitcoin's pseudo-anonymous privacy model and the lack of retroactive operational security, as originally highlighted by Nakamoto [21]. From a law enforcement perspective, BlockTag offers a valuable capability that is useful in e-crime investigations. In particular, showing a link between a user account on a website and illegal activities on darknet marketplaces could be used to secure a subpoena and collect more information about the user from the website [25].

(2) Market economics: We analyzed the market of Tor hidden services by calculating their "balance sheets." We found that WikiLeaks is the highest receiver of payments in terms of volume, with about 26.4K transactions. In terms of the total value of incoming payments, however, Silk Road tops the list with more than ฿29.6K received on its address. We also found that total value of incoming and outgoing payments of service addresses are nearly the same, meaning they have nearly-zero balances. This suggests that service providers do not keep their bitcoins on the addresses on which they receive payments, but distribute them to other addresses. From transaction dates of service addresses, we found that all but three of the top-10 revenue making service providers are active in 2018.

(3) Forensics: We performed an exploratory investigation of MMM: One of the world's largest Ponzi schemes. In total, we were able to link 24.2K users and

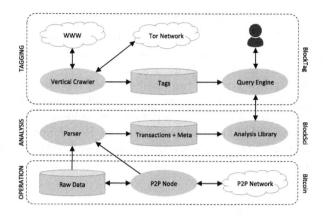

Fig. 1. Layered blockchain system architecture.

202 labels to MMM and its affiliates using BlockTag's full-text search capabilities. We found that all of the linked users are BitcoinTalk users who are mostly male, 20–40 years old, and are located worldwide in more than 80 countries. Moreover, we found that only 313 of these users have logged in to the forum at least once a day and made one or more activities, such as writing posts. After further analysis, we found that all of the linked user accounts were created as part of the "MMM Extra" scheme, which promises "up to 100% return per month for performing simple daily tasks that take 5–15 min." We also used BlockTag to retrieve and model MMM transactions as a directed graph, consisting of 14.3K addresses and 32.K transactions. We found that two of the top-10 ranked addresses, in terms of their PageRank, have been flagged on BitcoinTalk as known scammer addresses. As of December 1, these addresses have received more than ₿2M combined.

2   Design and Architecture

BlockTag's design follows a layered system architecture. As depicted in Fig. 1, each layer in the stack is responsible for a separate set of tasks and can interact with other layers through programmable interfaces. We present a high-level view of BlockTag's design, and leave the details in the technical report [7].

Tags. In BlockTag, a tag is a mapping between a block, a transaction, or an address identifier and a list of JSON-serializable objects. Each object specifies the type, the source, and other information representing auxiliary data describing the tagged identifier. As raw blockchain data is stored in a format that is efficient for validating transactions and ensuring immutability, the data must be parsed and transformed into a simple data structure that is efficient for analysis. For example, BlockSci uses a memory-mapped data structure to represent core transaction data as a graph. All other transaction data, such as hashes and scripts, are stored separately as mappings that are loaded when needed. Block-Tag follows this design choice, and uses a persistent key-value database, such as RocksDB [12], with an in-memory cache in order to store and manage blockchain tags, as they can grow arbitrarily large in size.

BlockTag defines four types of tags, namely user, service, text, and custom tags. A user tag represents a user account on an online social network, such as BitcoinTalk and Twitter. A service tag represents an online service provider, such as Tor hidden services like Silk Road and The Pirate Bay. A text tag represents a user-generated textual label, such as address labels submitted to Blockchain.info. A custom tag can hold arbitrary data, including other tags, and is usually used by analysts to create tags manually.

Vertical Crawlers. In BlockTag, a vertical crawler is used to scrape a data source, typically an HTML website or a RESTful API, in order to automatically create block, transaction, or address tags of a particular type using a website-specific parser. A crawler can be configured to run according to a crontab-like schedule, and to bootstrap on the first run with previously crawled raw HTML/JSON data, which can also be used to initialize blockchain tags.

For example, BitcoinTalk, the most popular Bitcoin forum with more than 2.2M users and 42.2M posts, is a good data source to collect public Bitcoin addresses and their associated user accounts. Behind the scene, a BitcoinTalk user crawler downloads user account pages through a URL that is unique for each user account. In addition to a BitcoinTalk user crawler, BlockTag implements a Twitter user crawler that consumes Twitter's API, a Tor hidden service crawler that scrapes onion landing pages of Ahmia-indexed service providers, and a Blockchain.info text crawler that scrapes textual labels that are self-signed by address owners or submitted by arbitrary users. By default, the vertical crawlers create Bitcoin address tags, but can be configured to scrape auxiliary data of other cryptocurrencies, including Litecoin, Namecoin, and Zcash.

Query Engine. BlockTag query engine is inspired from NoSQL document databases, such as MongoDB [9], where queries are specified using a JSON-like structure. Selecting, grouping, and aggregating transactions is provided through a simple query interface. To write a query, the analyst starts with specifying block, transaction, or address properties to which the results should match. BlockTag treats each property as having an implicit boolean AND, but also supports boolean OR queries using a special operator. In addition to exact matches, BlockTag has operators for string matching, numerical comparisons, etc. The analyst can also specify the properties by which the results are grouped. Finally, the analyst can specify which properties to return per result. While this query interface is suitable for many tasks, BlockTag's Python package also exposes lower-level functionality to analysts who have tasks with more sophisticated requirements.

One important capability of BlockTag's query engine is address clustering [18], which can be configured to operate on a particular source, namely inputs, outputs, or both, using one of the supported clustering methods. Address

Table 1. Summary of created tags.

Source	Type	# addresses	
		Original	Clustering
BitcoinTalk	User	40,970	19,213,141
Twitter	User	4,183	623,189
Tor Network	Service	88	–
Blockchain.info	Text	29,643	–

clustering expands the set of Bitcoin addresses that are mapped to a unique user, service, or text tag through a technique called closure analysis. As a result, this allows the analyst to identify more links between different tags by considering a larger number of transactions in the blockchain.

BlockTag supports multiple address clustering methods. The first method is based on the heuristic proposed by Meiklejohn et al. [18], which works as follows: If a transaction has addresses A and B as inputs, then A and B belong to the same cluster. The rationale behind this heuristic is that such addresses are highly likely to be controlled by the same entity. While efficient, this method can result in large clusters that include addresses which belong to different entities, due to mixing services, exchanges, mining pools and CoinJoin transactions. In order to tackle this issue, BlockTag implements a novel minimal clustering method that prematurely terminates the original clustering method before the clusters grow to their maximum size. Minimal clustering includes a final trimming phase to find clusters that share at least one address and consequently merges them, after which they are conditionally removed if their size exceeds a defined limit, which defaults to cluster size > 1 (i.e., unconditional removal of merged clusters). Doing so ensures that the clusters are mutually-exclusive and likely to belong to separate entities, but also means the clusters are smaller than usual, reducing the chance of linking different tags as a result.

3 Real-World Deployment

We deployed BlockTag on a single machine from January 1–March 21, 2018.[1] The machine was running Ubuntu v16.04.4 LTS, Bitcoin Core v0.16.0, and BlockSci v0.5.0 on two 2 GHz quad-core CPUs, 128 GB of system memory, and 2 TB of NAS storage. We used BlockTag to tag Bitcoin's blockchain at the address level. As of March 2018, the crawlers have ingested nearly 5B tweets, 2.2M BitcoinTalk profiles, 1.5K Tor onion pages, and 30K Blockchian.info labels, resulting in 45K user, 88 service, and 29K text tags. We used a previously collected dataset consisting of 4.8B tweets, which were posted in 2014, to bootstrap Twitter user tags. Moreover, for the first application where we link users to services, we configured address clustering for inputs from user tags using the minimal clustering method. We summarize the created tags in Table 1.

---

[1] For research ethics considerations, please refer to the technical report [7].

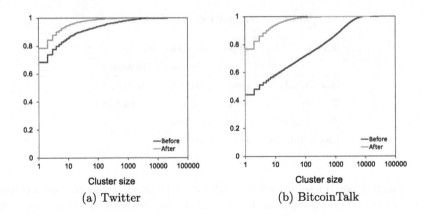

Fig. 2. CDFs of minimal clustering's cluster size before and after trimming

Figure 2 shows the CDFs of the cluster size for BitcoinTalk and Twitter user tags, before and after the trimming phase of minimal clustering. As illustrated in the figure, there is a significant drop in the size of clusters after trimming; the average size of a cluster decreased from 75 addresses to 7 for Twitter users, and from 452 addresses to 6 for BitcoinTalk users. The standard deviation also decreased from 606 to 67 and from 1194 to 114, respectively. This suggests that cluster sizes are getting closer to the mean. In fact, more than 90% of the users in both sources have 10 addresses or less in their clusters after trimming. The figure also suggests that more BitcoinTalk users have larger cluster size than Twitter users, as shown by the difference in their before/after distributions.

To cross-validate minimal clustering, we used WalletExplorer: An online service that uses a similar approach to find and tag clusters based on aggregated information from the web. We crawled cluster information from WalletExplorer for both user tag sources. Overall, we found that our closure analysis coincides with that obtained from WalletExplorer. All clusters that had less than 700 addresses were untagged on WalletExplorer, which means it is likely that these are user clusters, not services. When we used this number as a limit for the trimming, the percentage of clusters with size 700 or less changed from 83% to 99.95% for BitcoinTalk users and from 97.63% to 99.75% for Twitter users.

4 Applications

4.1 Linking Users to Services

In e-crime investigations of Tor hidden services, analysts often try to link cryptocurrency transactions to user accounts and activities. This can start with a known transaction that is part of a crime, such as a Bitcoin payment to buy drugs on Silk Road. Alternatively, a wider search criteria can be used to understand the landscape of activities of illegal services, such as finding service providers

Table 2. Top-10 linked service providers.

Name	# linked users		
	Twitter	BitcoinTalk	Total
WikiLeaks	11	35	46
Silk Road	4	18	22
Internet Archives	3	13	16
Snowden Defense Fund	3	8	11
The Pirate Bay	3	7	10
DarkWallet	9	1	10
ProtonMail	1	7	8
Darknet Mixer	1	2	3
Liberty Hackers	0	2	2
CryptoLocker Ransomware	1	0	1

that receive the most payments. Either way, the analysts need to link users to services, which is a core feature of effective blockchain analysis.

BlockTag was able to link 28 Twitter user accounts to 14 service providers via 167 transactions and 97 BitcoinTalk user accounts to 20 service providers via 115 transactions. Some of these users were linked to multiple service providers. In total, 125 users were linked to 20 services. The results suggest that although Twitter users are smaller in number compared to BitcoinTalk users, they are more active and have a larger number of transactions with services. In fact, some of these users are "returning customers," as they have performed multiple transactions with the same service provider.

From services perspective, Table 2 lists the top-10 service providers ranked by how many users were linked to them. The list is topped by WikiLeaks, which is a service that publishes secret information provided by anonymous sources, with 46 linked users. This is followed by Silk Road, the famous darknet marketplace, with 22 linked users whose spent coins have been seized by the FBI. Although the Silk Road address was seized, it still appears in transactions until recently. However, based on further analysis, we found that a number of transactions were performed prior to the seizure. Ranked fifth, The Pirate Bay, which is known for infringing IP and copyright laws by facilitating the distribution of protected digital content, was linked to 10 users. As the linked users have accounts with various personally identifiable information (PII), their identities could be deanonymized. We next focus on two case studies that illustrate this threat.

Actionable Links. Purchasing products and services from darknet marketplaces is generally considered illegal and calls for legal action. Some of the 22 users who are linked to Silk Road through transactions with seized coins shared enough PII to completely deanonymoize their identity. For example, one user is a teenager from the U.S. The user has been a registered BitcoinTalk member since 2013, and

Table 3. Balance sheet of top-10 service providers ranked by incoming coins.

Name	Volume (# txs)	Flow of money (฿)		Lifetime (dd/mm/yyyy)		
		Incoming	Outgoing	First tx	Last tx	# days
Silk Road	1,242	29,676.99	29,658.80	02/10/2013	19/03/2018	1,628
WikiLeaks	26,399	4,043.00	4,040.74	15/06/2011	21/03/2018	2,470
VEscudero Escrow Service	192	842.42	842.42	27/05/2012	20/08/2017	1,910
Internet Archives	2,957	775.86	746.89	06/09/2013	21/03/2018	1,656
Freenet Project	280	691.87	687.62	23/02/2011	16/03/2018	2,577
Snowden Defense Fund	1,722	218.95	218.95	11/08/2013	18/03/2018	1,680
ProtonMail	3,096	208.40	208.36	17/06/2014	18/03/2018	1,369
Ahmia Search Engine	1,423	176.51	176.50	27/03/2013	06/03/2018	1,652
DarkWallet	983	114.62	97.40	16/04/2014	02/11/2016	931
The Pirate Bay	1,214	76.80	76.80	29/05/2013	21/08/2017	1,544

has a transaction with Silk Road in 2013, the takedown year. The corresponding user account points to his personal website, which contains links to his user profiles on Facebook, Twitter, and Youtube. Even if users do not share PII or use fake identities on their accounts, simply having an account on social networks is enough to track them online, or even secure a subpoena to collect identifiable information, such as login IP addresses. For example, three out of the 18 BitcoinTalk users recently logged in to the forum.

A Matter of Jurisdiction. One of the users who are linked to The Pirate Bay is a middle-aged man from Sweden. The Pirate Bay was founded by a Swedish organization called Piratbyrån. Furthermore, the original founders of the website were found guilty in the Swedish court for copyright infringement activities. Since then, the website has been changing its domain constantly, and eventually operated as a Tor hidden service. Consequently, having such a link to The Pirate Bay through recent transactions in Sweden can lead to legal investigation, at least, and potentially be incriminating.

4.2   Market Economics

Keeping track of market statistics of Tor hidden services is useful for identifying thriving services, measuring the impact of law enforcement, and prioritizing e-crime investigations. As such, an analyst may start with calculating a financial "balance sheet" for service providers, which includes the number of transactions with which a service is involved (i.e., volume), the amount of coins a service has received or sent (i.e., money flow), and the difference between the timestamps of the last and first transactions (i.e., operation lifetime). Table 3 shows the balance sheet of the top-10 service providers ranked by incoming coins.

Volume. While the number of created service tags is small, the corresponding service providers have been involved in a relatively large number of transactions.

For example, WikiLeaks tops the list with 26.4K transactions. The Darknet Mixer, which did not make it to the top-10 list in Table 3, has a volume of 22.1K transactions that is greater than the remaining services combined. One explanation for this popularity is that users are actually aware of the possibility of linking, and try to use mixing services in order to make traceability more difficult and improve their anonymity.

Money Flow. One interesting observation is that service providers have a nearly zero balance, which means almost the same amount of coins comes in and goes out of their addresses. This indicates that the coins is likely distributed to other addresses and is not kept on payment-receiving addresses. One explanation for this behavior is that by distributing coins among multiple addresses, a service provider can reduce coin traceability. Moreover, service providers still need to distribute their revenues among owners and sellers. Among all service providers listed in Table 3, Silk Road stands out with an income of ₿29.6K.

Lifetime. The services vary in their lifetime, ranging from two to seven years of operation. The first transaction date indicates the date on which the service provider started receiving payments through the tagged addresses. Looking at last transaction dates, all but three services are still active in 2018. For example, Silk Road has been receiving money since October 2013, even after the address has been seized by the FBI and its coins auctioned for sale in June, 2014. However, a large number of post-seizure transactions appear to be novelty tips.

4.3   Forensics

Organizations responsible for consumer protection, such as trade commission agencies and financial regulatory authorities, have a mandate to research and identify fraud cases involving cryptocurrencies, including unlawful initial coin offerings and Ponzi schemes. Given the popularity of Ponzi schemes in Bitcoin [26,27], we focus on this type of fraud and show how BlockTag can help analysts flag users who are likely victims or operators of such schemes.

A Ponzi scheme, also known as a high yield investment program, is a fraudulent financial activity promising unusually high returns on investment, and is named after a famous fraudster, Charles Ponzi, from the 1920s. The scheme is designed in such a way that only early investors will get benefits and once the sustainability of the scheme is at risk the majority of shareholders will lose the money they invested [1]. Among various Ponzi schemes in Bitcoin, MMM is considered one of the largest schemes that is hard to detect solely based on blockchain transaction analysis [2], highlighting the need for a systematic integration of auxiliary data into blockchain analysis. As such, an analyst can start the investigation with BlockTag using a full-text search query of keywords associated with MMM scheme, such as its name, without requiring prior knowledge of who is involved in the scheme or how it works.

BlockTag's search returned 24.2K user accounts, all of which are BitcoinTalk users, and 202 Blockchain.info text labels. For BitcoinTalk user accounts, the

Table 4. Top-10 frequent MMM labels.

Label	Frequency
mmm universe.help	46
mmm global	13
bonus from mmm universe.help	9
mmm indonesia	6
mmm nusantara	4
mmm china	2
mmm india	2
mmm indonesia	2
mmm philippines	2
mmm russia	2

full-text search matched the website property of an account, which contained a URL pointing to the user's profile on MMM website. As for Blockchain.info text tags, the search matched the self-signed label property, which contained "mmm" substring, as summarized in Table 4. We next analyze the user accounts looking for clues related to MMM operation.

User Demographics. Out of 24.2K users, 52.86%, 18.31%, and 12.48% shared their gender, age, and geo-location information, respectively. Based on this data, we found that the users are mostly male (75.44%), between 20–40 years old (average $= 32$), and are located worldwide in more than 80 different countries. However, 70.69% of the users were located in only five countries, namely Indonesia, China, India, South Africa, and Thailand. Interestingly, most of these countries have a corresponding MMM label, as listed partially in Table 4.

Forum Activity. Using activity-related properties of user accounts, we found that 99.44% of the users registered on the forum between August 2015–March 2016. Moreover, 98.21% of the users made their last activity on the forum during the same period. This suggest that users have short-lived accounts. In fact, we found that 94.25% of the users were active for 30 days or less, and that 78.45% of users were dormant, meaning they were active for less than a day after registration. This also suggests that most of the users are not engaged with the forum. Indeed, only 313 users made at least one activity, and even for these users, they never engaged with the forum for more than once a day, on average. After manually inspecting the accounts on the website, we found that most of them were created as part of its "MMM Extra" scheme, which promises "up to 100% return per month for performing simple daily tasks that take 5–15 min," such as promoting MMM on social networks. This was evident from the accounts' signatures, which the crawler did not parse, that included messages such as "MMM Extra is the right step towards the goal" and "MMM participants get up to 100% per month."

Table 5. Properties of MMM transaction graphs where n is the order, m is the size, $\bar{C}$ is the average clustering coefficient, d is the diameter, and r is the radius.

Type		n	m	LSCC		$\bar{C}$	Triangles		Distance	
Input	Output			n	m		#	% closed	d	r
User	User	14,227	31,819	5,850	17,498	0.11	6,566	0.08	17	7
User	Label	129	125	1	0	0.00	0	0.00	0	0
Label	User	64	45	1	0	0.00	0	0.00	0	0
Label	Label	61	508	20	246	0.64	943	61.04	3	2
Any	Any	14,319	32,497	5,934	18,128	0.11	7,576	0.09	17	7

Financial Operation. We can invistigate how MMM scheme operates financially through transaction graph analysis [23]. In this analysis, Bitcoin transactions are modeled as a weighted, directed graph where nodes represent addresses, edges represent transactions, and weights represent information about transactions, such as input/output values and dates. Analyzing the topological properties of this graph can provide insights into which addresses are important and how the money flows. For example, having a few "influential" nodes and a small clustering coefficient suggest that most of the money funnels through these nodes and does not flow back to others, which are indicative of a Ponzi operation [2,26,27]. In BlockTag, an analyst can easily model case-specific transaction graphs by linking tags based on some search criteria.

We modeled and analyzed five transaction graphs, one for every combination of tag types, as summarized in Table 5. The MMM transaction graph includes addresses of any type, and consisted of 14.3K addresses (i.e., order) and 32.5K transactions (i.e., size). This graph is also sparsely connected, as suggested by the small-sized largest strongly connected component (LSCC), low clustering, and long distance measures. Moreover, it consists of two subgraphs, the user→user subgraph, which is also sparsely connected, and the label→label subgraph, which is dense and small. Even though the two subgraphs are loosely connected through only 170 edges, an order of magnitude more money has flown from users to labels than the reverse direction.

To find influential nodes in the graph, we computed their PageRank, where weights represented input address values of transactions. All of the top-10 ranked nodes were located in the user→user subgraph, which mapped to unique Bit-coinTalk users. After manually inspecting the corresponding accounts, we found that the first and the third users have been reported as scammers on BitcoinTalk for operating fraudulent services, namely Dr.BTC and OreMine.org. While the first user has received a total of ฿426.7K on her address, the third has received a total of ฿1.8M on his address that is associated with Huobi, an exchange service, suggesting that the user has exchanged the received coins.

5   Discussion

Limitations. BlockTag's main limitation is the validity of its tags, since they are created automatically by crawlers from open, public data sources. This limitation is part of a larger problem that is common with Internet content providers, such as Google and Facebook, especially when content is generated mostly by users [17,28]. In general, the validity issue is especially important for user identities, as fraudsters can always create fake accounts in order to hide their real identity [13]. While doing so improves their anonymity, law enforcement agencies can use the links found through BlockTag to secure a subpoena in order to collect more information about suspects from website operators [25].

Work In-progress. We are designing BlockSearch, an open-source Google-like searching layer that sits on top of BlockTag. BlockSearch allows analysts to search blockchains for useful information in plain English and in real-time, without having to go through the hassle of performing low-level queries using BlockTag. The system also provides in a dashboard for analysts that displays real-time results of important queries, such as the ones we used in the paper. Based on feedback from trade commission agencies and financial regulatory authorities, such capabilities are extremely helpful to protect customers, comply with know you customer (KYC) and anti-money laundering (AML) laws, and draft new, investor-friendly cryptocurrency regulations.

Future Work. In order to address the main limitation of BlockTag, we plan to define confidence scores for tag sources. The scores can be computed using various "truth discovery" algorithms [10], which are generally based on the intuition that the more sources confirm a tag the more confidence is assigned to it.

   BlockTag is modular by design. This means we can easily enhance or add new capabilities. As such, we plan to implement more vertical crawlers for services such as WalletExplorer, ChainAlysis, BitcoinWhosWho, and Reddit. We also plan to support more clustering methods and develop a systematic way to automatically tag clusters, in addition to blocks, transactions, and addresses, based on label propagation algorithms [15].

6   Related Work

Analysis Systems. Blockchain analysis systems parse and analyze raw transaction data for many applications. Recently, Kalodner et al. proposed BlockSci [16], an open-source, scalable blockchain analysis system that supports various blockchains and analysis tasks. BlockSci incorporates an in-memory, analytical database, which makes it several hundred times faster than its contenders. While there is a minimal support for tagging in its programming interface, BlockSci is designed for analysis of core blockchain data. At the cost of performance, annotation and tagging can be integrated into the analysis pipeline through a centralized, transactional database. For example, Spagnuolo et al.

proposed BitIodine [24], an open-source blockchain analysis system that supports tagging through address labels. However, BitIodine, relies on Neo4j [19], a general-purpose graph database that is not designed for blockchain data and its append-only nature, which makes it inefficient for common blockchain analysis tasks, such as address linking. In contrast, BlockTag is the first open-source tagging system that fills this role.

Linking. The impact of Bitcoin address linking on user anonymity and privacy has been known for a while now [11,14,18,22]. Reid and Harrigan [22] showed that passive analysis of public Bitcoin information can lead to a serious information leakage. They constructed two graphs representing transactions and users from Bitcoin's blockchain data and annotated the graphs with auxiliary data, such as user accounts from BitcoinTalk and Twitter. The authors used visual content discovery and flow analysis techniques to investigate Bitcoin theft. Alternatively, Fleder et al. [14] explored the level of anonymity in the Bitcoin network. The authors annotated addresses in the transaction graph with user accounts collected from BitcoinTalk in order to show that users can be linked to transactions through their public Bitcoin addresses. These studies show the value of using public data sources for Bitcoin privacy research and law enforcement, which is our goal behind designing BlockTag.

Tor and Darknet Markets. Tor hidden services have become a breeding ground for darknet marketplaces, such as Silk Road and Agora, which offer illicit merchandise and services [5,20]. Moore and Rid [20] studied how hidden services are used in practice, and noted that Bitcoin was the dominant choice for accepting payments. Although multiple studies [14,18] showed that Bitcoin transactions could be linked to identities, Bitcoin remains the most popular digital currency on the Dark Web [8], and many users choose to use it despite its false sense of anonymity. Recent research explored the intersection between Bitcoin and Tor privacy [3,4], and found that legitimate hidden service users and providers are one class of Bitcoin users whose anonymity is particularly important. Moreover, Biryukov et al. [5] found that hidden services devoted to anonymity, security, human rights, and freedom of speech are as popular as illegal services. While BlockTag makes it possible to link users to such services, we designed it to help analysts understand the privacy threats and identify malicious actors.

Forensics. Previous research showed that cryptocurrencies, Bitcoin in particular, have a thriving market for fraudulent services, such as fake wallets, fake mining pools, and Ponzi schemes [6,26]. Recently, Bartoletti et al. [2] proposed a data mining approach to detect Bitcoin addresses that are involved in Ponzi schemes. The authors manually collected and labeled Bitcoin addresses from public data sources, defined a set of features, and trained multiple classifiers using supervised machine learning. The best classifier correctly labelling 31 addresses out of 32 with 1% false positives. Interestingly, MMM was excluded because it had a complex scheme. In concept, BlockTag complements such techniques by providing an efficient and easy way to collect and explore data that is relevant to the investigation. This data can be then analyzed using different machine learning and graph algorithmic techniques with the help of existing tools [27].

7   Conclusion

State-of-the-art blockchain analysis systems, such as BlockSci, while efficient, are not designed to annotate and analyze auxiliary blockchain data systematically. We presented BlockTag, an open-source tagging system for blockchains. We used BlockTag to uncover privacy issues with using Bitcoin in Tor hidden services, and to flag Bitcoin addresses that are likely to be part of a large Ponzi scheme.

References

1. Artzrouni, M.: The mathematics of Ponzi schemes. Math. Soc. Sci. **58**(2), 190–201 (2009)
2. Bartoletti, M., Pes, B., Serusi, S.: Data mining for detecting Bitcoin Ponzi schemes. arXiv preprint arXiv:1803.00646 (2018)
3. Biryukov, A., Khovratovich, D., Pustogarov, I.: Deanonymisation of clients in Bitcoin P2P network. In: Proceedings of the 2014 ACM SIGSAC Conference on Computer and Communications Security, pp. 15–29. ACM (2014)
4. Biryukov, A., Pustogarov, I.: Bitcoin over tor isn't a good idea. In: 2015 IEEE Symposium on Security and Privacy (SP), pp. 122–134. IEEE (2015)
5. Biryukov, A., Pustogarov, I., Thill, F., Weinmann, R.P.: Content and popularity analysis of tor hidden services. In: 2014 IEEE 34th International Conference on Distributed Computing Systems Workshops (ICDCSW), pp. 188–193. IEEE (2014)
6. Bohr, J., Bashir, M.: Who uses Bitcoin? An exploration of the Bitcoin community. In: 2014 Twelfth Annual Conference on Privacy, Security and Trust (PST), pp. 94–101. IEEE (2014)
7. Boshmaf, Y., Jawaheri, H.A., Sabah, M.A.: BlockTag: design and applications of a tagging system for blockchain analysis. arXiv preprint arXiv:1809.06044 (2018)
8. del Castillo, M.: Bitcoin remains most popular digital currency on dark web. https://bit.ly/2UOZNS6 (2016). Accessed 01 July 2018
9. Chodorow, K.: MongoDB: The Definitive Guide: Powerful and Scalable Data Storage. O'Reilly Media Inc., Sebastopol (2013)
10. Dong, X.L., Berti-Equille, L., Srivastava, D.: Integrating conflicting data: the role of source dependence. Proc. VLDB Endow. **2**(1), 550–561 (2009)
11. DuPont, J., Squicciarini, A.C.: Toward de-anonymizing Bitcoin by mapping users location. In: Proceedings of the 5th ACM Conference on Data and Application Security and Privacy, pp. 139–141. ACM (2015)
12. Facebook: RocksDB: an embeddable persistent key-value store for fast storage (2014). https://rocksdb.org. Accessed 01 July 2018
13. Ferrara, E., Varol, O., Davis, C., Menczer, F., Flammini, A.: The rise of social bots. Commun. ACM **59**(7), 96–104 (2016)
14. Fleder, M., Kester, M.S., Pillai, S.: Bitcoin transaction graph analysis. arXiv preprint arXiv:1502.01657 (2015)
15. Gregory, S.: Finding overlapping communities in networks by label propagation. New J. Phys. **12**(10), 103018 (2010)

16. Kalodner, H., Goldfeder, S., Chator, A., Möser, M., Narayanan, A.: BlockSci: design and applications of a blockchain analysis platform. arXiv preprint arXiv:1709.02489 (2017)
17. Li, Y., et al.: A survey on truth discovery. ACM SIGKDD Explor. Newsl. **17**(2), 1–16 (2016)
18. Meiklejohn, S., et al.: A fistful of bitcoins: characterizing payments among men with no names. In: Proceedings of the 2013 Conference on Internet Measurement Conference, pp. 127–140. ACM (2013)
19. Miller, J.J.: Graph database applications and concepts with Neo4j. In: Proceedings of the Southern Association for Information Systems Conference, Atlanta, GA, USA, vol. 2324, p. 36 (2013)
20. Moore, D., Rid, T.: Cryptopolitik and the Darknet. Survival **58**(1), 7–38 (2016)
21. Nakamoto, S.: Bitcoin: a peer-to-peer electronic cash system (2008). Bitcoin.org
22. Reid, F., Harrigan, M.: An analysis of anonymity in the Bitcoin system. In: Altshuler, Y., Elovici, Y., Cremers, A., Aharony, N., Pentland, A. (eds.) Security and Privacy in Social Networks, pp. 197–223. Springer, New York (2013). https://doi.org/10.1007/978-1-4614-4139-7_10
23. Ron, D., Shamir, A.: Quantitative analysis of the full Bitcoin transaction graph. In: Sadeghi, A.-R. (ed.) FC 2013. LNCS, vol. 7859, pp. 6–24. Springer, Heidelberg (2013). https://doi.org/10.1007/978-3-642-39884-1_2
24. Spagnuolo, M., Maggi, F., Zanero, S.: BitIodine: extracting intelligence from the Bitcoin network. In: Christin, N., Safavi-Naini, R. (eds.) FC 2014. LNCS, vol. 8437, pp. 457–468. Springer, Heidelberg (2014). https://doi.org/10.1007/978-3-662-45472-5_29
25. Theymos: DPR subpoena (2014). https://bitcointalk.org/index.php?topic=881488.0. Accessed 01 July 2018
26. Vasek, M., Moore, T.: There's no free lunch, even using Bitcoin: tracking the popularity and profits of virtual currency scams. In: Böhme, R., Okamoto, T. (eds.) FC 2015. LNCS, vol. 8975, pp. 44–61. Springer, Heidelberg (2015). https://doi.org/10.1007/978-3-662-47854-7_4
27. Vasek, M., Moore, T.: Analyzing the Bitcoin Ponzi scheme ecosystem. In: Zohar, A., et al. (eds.) FC 2018. LNCS, vol. 10958, pp. 101–112. Springer, Heidelberg (2019). https://doi.org/10.1007/978-3-662-58820-8_8
28. Yin, X., Han, J., Philip, S.Y.: Truth discovery with multiple conflicting information providers on the web. IEEE Trans. Knowl. Data Eng. **20**(6), 796–808 (2008)

Forward Secure Identity-Based Signature Scheme with RSA

Hankyung Ko[1], Gweonho Jeong[1], Jongho Kim[2], Jihye Kim[2(✉)],
and Hyunok Oh[1(✉)]

[1] Hanyang University, Seoul, Korea
{hankyungko,jkho1229,hoh}@hanyang.ac.kr
[2] Kookmin University, Seoul, Korea
{bi04208,jihyek}@kookmin.ac.kr

Abstract. A forward secure identity based signature scheme (FSIBS)
provides forward secrecy of secret keys. In order to mitigate the damage
when keys are leaked, it is desirable to evolve all the secret keys, i.e., *both*
the user keys and the master key. In this paper, we propose a new RSA-
based FSIBS scheme which requires constant size keys and generates
constant size signatures. The experimental results show that it takes
3 ms to generate a signature in the proposed scheme while it takes 75 ms
in the existing pairing based approach. The proposed scheme is provably
secure under the factoring assumption in the random oracle model.

Keywords: Forward security · Digital signature · ID based ·
Private key generator · RSA

1 Introduction

The identity based signatures, so called IBS, are digital signatures where an
identifier, such as e-mail address etc., is used as a public key. In the IBS system,
a private key generator (PKG) publishes a public parameter and issues a secret
signing key to an identified user using its master secret key. Because the IBS system utilizes publicly known information such as an identifier as its verification
key, it allows any party to verify signatures without the explicit authenticated
procedure of the verification key. On the other hand, the forward secure signature is a way to mitigate the damage caused by key exposure. A forward secure
signature scheme divides the lifetime into several time periods, and uses a different key at each time period. With this idea, even if a signature key is exposed at
a specific point of time, all signatures which are generated before the exposure
time can be kept valid. In order to have advantages of both the IBS system
and the forward secure signature system, several researches [6,12,17–20] have
been conducted to add forward security to the signing key issued by a private
key generator (PKG). However, forward secrecy in such forward secure ID based
signature schemes is limited and incomplete because only the user private keys

© IFIP International Federation for Information Processing 2019
Published by Springer Nature Switzerland AG 2019
G. Dhillon et al. (Eds.): SEC 2019, IFIP AICT 562, pp. 314–327, 2019.
https://doi.org/10.1007/978-3-030-22312-0_22

evolve each time period; Once the master secret key is compromised, all signatures made by any user under this system cannot be considered valid even if the signing keys evolve properly. Mainly, these attacks can be done by the powers, buying off the system managers. In order to minimize the risk from the leakage of the master secret key, the PKG master key as well as the user's signing key also needs to satisfy forward security. This paper simply notates FSIBS *with* forward-secure PKG as FSIBS hereafter.

The notion of FSIBS was considered first in [15] and the scheme was based on an elliptic curve pairing function. The proposed scheme in [15] satisfies the forward security of both the master secret key and the users' signing keys. That is, even if a master secret key of time period t is exposed, all signatures generated before t are not invalidated. Although the scheme in [15] efficiently generates a signature, its application is limited due to the non-constant size signing keys. Considering the forward secrecy is widely applied including the IoT environment [9], providing more options for FSIBS is desirable. Given the only pairing-based scheme, it may be difficult to cover the resource-constrained IoT devices due to its complex pairing operation.

In this paper, we propose a new forward secure identity based signature scheme (FSIBS) in the RSA setting. Our scheme is constructed by extending the forward secure RSA based signature scheme in [1] into an identity-based scheme. The proposed scheme also allows the master key update for the forward secrecy of the master key. To update the keys, the scheme does not require any interaction between the user and the PKG, once an initial signing key is delivered. The proposed scheme is secure under the factoring assumption in the random oracle model.

Table 1. Comparison of our scheme and [15]

		Our scheme	[15]
Size	Master Secret Key	$O(1)$	$O(\log^2 T)$
	User Signing Key	$O(1)$	$O(\log^2 T)$
	Verification Key	$O(1)$	$\log T + 2l + 5$
	Signature	$O(1)$	$O(1)$
Computation	PKGKeyGen	$O(lTk^2)$	$O((a+e) \cdot \log^2 T + p)$
	KeyIssue	$O(lTk^2)$ (opt. $O(k^3)$)	$O(a \cdot l \log T + (a+e) \cdot \log^2 T)$
	MSKUpdate	$O(lk^2)$	$O((a+e) \cdot \log^2 T)$
	UKUpdate	$O(lk^2)$	$O((a+e) \cdot \log^2 T)$
	Sign	$O(lTk^2)$ (opt. $O(k^3)$)	$O((\log T + 2l) \cdot a + 3 \cdot e)$
	Verify	$O(lTk^2)$	$O(5p + (\log T + 2l) \cdot a)$

Table 1 summarizes the comparison between the proposed scheme and [15]. In Table 1, T is the maximum number of periods, l is the bit length of the hash output, k is the bit length in RSA, a is multiplication time in bilinear group, p is pairing time, and e is exponentiation time in bilinear group. As shown in Table 1, in the proposed scheme, the sizes of all keys such as a master secret key, a user signing key and a verification key, and a signature are constant while they are not in [15]. The costs of KeyIssue, Update, and Sign algorithms in the optimized proposal are independent of the maximum number of period T. According to our experiment, overall, our (optimized) scheme is faster than [15], except in the Verify algorithm. In particular, for $T = 2^{15}$, generating a signature requests only 3 ms with 2048-bit k of RSA, while the scheme in [15] requires 75 ms with the 224-bit ECC key.

This paper is organized as follows. We begin by discussing the related works in Sect. 2. In Sect. 3, we define the notion of forward-secure ID based signature scheme with forward-secure key generation, the background assumption that our scheme is based on, and the formal security model. We construct our signature scheme in Sect. 4. After that, we describe the security proof in Sect. 5. In Sect. 6, we extend our proposed scheme to improve the signing performance. Moreover, experimental results are shown in Sect. 7. Finally, we conclude in Sect. 8.

2 Related Work

The concept of forward security was first proposed by Anderson [2], and Bellare and Miner [3] made the formal definition of forward secure digital signatures (FSS). [3] proposed two forward secure signature constructions. One of them is a generic construction which can be derived by any signature scheme, of which complexity is at least $O(\log T)$-factor times of the original scheme, where T is the number of total time periods. The other is a variation on the Fiat-Shamir signature scheme [7] for a forward secure version that has constant-size signatures and takes $O(T)$-time for signing and verification. Next, Abdalla and Reyzin proposed a FSS scheme with a short public key [1]. However, its key generation, signing and verification are slow. Itkis and Reyzin suggested a FSS scheme with efficient signing and verification based on Guillous-Quisquater signatures [8], however, it costs more update time. On the other hands, Kozlov and Reyzin suggested a FSS scheme with fast update, but its signing and verification takes longer. So far, time complexity of some algorithm components for any suggested FSS constructions is depends on the total number of period T.

Krawczyk proposed a generic FSS construction with a constant-size private key, but with $O(T)$-size storage (possibly non-private) [11]. An efficient generic construction with unlimited time periods [13] is suggested by Malkin, Micciancio, and Miner using Merkletrees [14].

Boyen et al. [4] proposed the concept of forward secure signatures with untrusted updates for additional protection of the private key. Its private key is encrypted by the second factor (i.e., user password), and the key update proceeds with encryption. They construct an efficient scheme with constant-size signatures and provide the security reduction based on the Bilinear Diffie-Hellman Inversion assumption (BDHI) without random oracles.

Liu et al. proposed the forward-secure ID-based signature scheme without providing specific security definition and proof [12]. Yu et al. [18] formalized the security definition and proposed a scheme with a formal security proof. Meanwhile, Ebri et al. [6] proposed an efficient generic construction of forward-secure ID-based signature. When it is instantiated with Schnorr signatures [16], their construction provides an efficient scheme in the random oracle model. Yu et al. [19] made an extension of forward-secure ID-based signature with untrusted updates. Zhang et al. [20] proposed the two forms of lattice-based constructions under lattice assumption, in the random oracle model and in the standard model. Presently, Wei et al. [17] proposed an efficient revocation forward-secure ID-based signature to provide forward security and backward security upon key exposure. However, user key update requires interaction with PKG for each period to support revocation. The update algorithm run by the PKG generates an updated private key for each unrevoked user at each period. However, the master secret key of PKG is not updated.

All of the above [6,12,17–20] consider only the security of the user's private keys. Oh et al. [15] proposed first and only FSIBS system with forward secure PKG in a bilinear group setting. The scheme is constructed as a three-dimensional hierarchical ID based signature scheme; the first dimension is related with periods, the second dimension with identifiers, and the third dimension with messages. The hierarchical ID based signatures are taken in a double manner; first for issuing users' signing keys, and second for signing. To implement [15], no extra interaction between PKG and users is required, if the signing key is issued properly.

3 Preliminaries

We define the model for a FSIBS scheme with forward secure private key generator as in [15] in this section. And we define the security model of our scheme and describe its underlying assumption.

3.1 FSIBS Scheme

A FSIBS scheme with forward secure private key generator consists of six algorithms $\Sigma_{FSIBS} = ($PKGKeyGen, MSKUpdate, KeyIssue, UKUpdate, Sign, Verify$)$ as following:

- PKGKeyGen(k, T): This algorithm takes security parameter k and the maximum number of time periods T, and outputs the public verification key VK and the PKG master key MSK_t. The PGK master key MSK_t includes the current time period t which is initialized to 1.
- MSKUpdate(MSK_t): This algorithm takes the PKG master key at current time period t as an input, and outputs a new PKG master key MSK_{t+1}. Once a new PKG master key is computed, the previous PKG master key is removed. If $t + 1 \geq T$, then the old key is removed and no new keys are created.
- KeyIssue(MSK_t, ID): This algorithm takes the PKG master key at current time period t and an identifier ID. It outputs a secret signing key $SK_{t,ID}$ for the specific identifier ID. The signing key also includes the time period t.
- UKUpdate$(SK_{t,ID})$: This algorithm takes in the signing key of an identifier ID at current time period t and outputs a new signing key $SK_{t+1,ID}$. Once a new signing key is created, the previous signing key is removed. If $t + 1 \geq T$, then the old key is removed and no new keys are created.

- Sign($SK_{t,ID}$, M): This algorithm takes in the signing key of an identifier ID at current time period t, and a message M. It outputs a signature σ for time period t.
- Verify(ID, σ, M, VK, t): This algorithm takes an identifier ID, a signature σ, a message M, a verification key VK, and a time period t. It outputs either valid or invalid.

Definition 1. *A FSIBS scheme is perfectly correct if*

$$
\Pr \left[\begin{array}{l} \text{Verify}(ID, \text{Sign}(SK_{t,ID}, M), M, VK, t) = 1 | \\ SK_{i,ID} \leftarrow \text{KeyIssue}(MSK_i, ID) \vee \text{UKUpdate}(SK_{i-1,ID}) \\ \wedge (MSK_1, VK) \leftarrow \text{PKGKeyGen}(k, T) \\ \wedge MSK_j \leftarrow \text{MSKUpdate}^j(MSK_1) \end{array} \right] = 1
$$

3.2   Security Model

We use a game between an adversary $\mathcal{A}$ and a challenger $\mathcal{C}$ to define the security. The game captures the notion of PKG master key forward security, users' signing key forward security and the traditional unforgeability with security parameter k, maximum time period T, and negligible function ϵ. The game proceeds as following:

[**Setup phase**] The time period t is set to 1. Challenger $\mathcal{C}$ generates verification key VK and master key MSK_1 through PKGKeyGen, and give VK to adversary $\mathcal{A}$.

[**Interactive query phase**] In this phase the adversary $\mathcal{A}$ is allowed to adaptively query the following oracles.

KeyIssue: The adversary $\mathcal{A}$ can choose a specific ID and ask the challenger $\mathcal{C}$ for the signing key for ID of current time period t. The challenger $\mathcal{C}$ returns the signing key $SK_{t,ID}$ for ID of current time period t to adversary $\mathcal{A}$.

Update: The adversary $\mathcal{A}$ can request the challenger $\mathcal{C}$ to execute the MSKUpdate algorithm and UKUpdate algorithm.

Sign: The adversary $\mathcal{A}$ can choose a random message M and a user identifier ID and ask the challenger to sign a message M for ID on the current time period t. Then, it receives a signature σ from the challenger.

Hash: The adversary $\mathcal{A}$ can make queries to the random oracle and get the corresponding random values.

[**PKG corrupt phase**] The adversary $\mathcal{A}$ requests the PKG master key of the current time period t' from the challenger $\mathcal{C}$. The challenger $\mathcal{C}$ returns the master key $MSK_{t'}$. Note that the signing key of any identifier at the current time period and after can be generated using $MSK_{t'}$.

[**Final forgery phase**] The adversary $\mathcal{A}$ produces a fake signature which consists of a time period, user's ID, message and signature tuple $(t^*, ID^*, M^*, \sigma^*)$. The adversary is successful if $t^* < t'$, the signature is valid for time t^*, and the adversary had not queried for a signature on M^* and ID^* at the time period t^*.

We define the advantage of an algorithm $\mathcal{A}$ attacking scheme Σ by the probability that $\mathcal{A}$ is successful in the game, and denote the advantage by $Adv_{\mathcal{A},\Sigma}^{\mathsf{FSIBS}}$. We say that a PKG forward secure scheme Σ is $(\tau, q_{key}, q_{sign}, q_{hash}, \epsilon)$-secure if for all adversaries $\mathcal{A}$ running in time τ, making q_{key} key issue query, q_{sign} sign query and q_{hash} hash query, if:

$$Adv_{\mathcal{A},\Sigma}^{\mathsf{FSIBS}}(k, T) \leq \epsilon(k).$$

3.3 Factoring Assumption

Let $\mathcal{A}$ be an adversary for the problem of factoring Blum integers; an integer $n = pq$, where p, q are distinct odd primes and $p \equiv q \equiv 3 \pmod 4$, is called a Blum integer. We define the following experiment.

Experiment Factoring$_{\mathcal{A}}(k)$

    Randomly choose two $k/2$ bit primes p and q, s.t.:

    $p \equiv q \equiv 3 \pmod{4}$

    $N \leftarrow pq$

    $(p', q') \leftarrow \mathcal{A}(N)$

    If $p'q' = N$ and $p' \neq 1$ and $q' \neq 1$ then return 1 else return 0

Let $Adv_{\mathcal{A}}^{FAC}(k)$ denote the probability that experiment Factoring$_{\mathcal{A}}(k)$ returns 1. We say that the factoring assumption holds if for all PPT algorithm $\mathcal{A}$ with negligible function ϵ,

$$Adv_{\mathcal{A}}^{FAC}(k) \leq \epsilon(k).$$

3.4 Multiple Forking

Multiple forking (MF) is an extension of general forking to accommodate nested oracle replay attacks [5]. The modularity of the MF Lemma allows one to abstract out the probabilistic analysis of the rewinding process from the actual simulation in the security argument. The MF algorithm $\mathcal{M}_{\mathcal{Y},n}(x)$ associated to $\mathcal{Y}$ and n is defined in [5].

Lemma 1. *Let $\mathcal{G}$ be a randomized algorithm that takes no input and returns a string. Let $\mathcal{Y}$ be a randomized algorithm that on input a string x and elements $s_1, ..., s_q \in \mathbb{S}$ returns a triple (I, J, σ) consisting of two integers $0 \leq J < I \leq q$ and a string σ. Let*

$$mfrk := Pr[(b = 1) \mid x \xleftarrow{\$} \mathcal{G}; (b, \{\sigma_0, ..., \sigma_n\}) \xleftarrow{\$} \mathcal{M}_{\mathcal{Y},n}(x)] \text{ and}$$

$$acc := Pr[(I \geq 1) \wedge (J \geq 1)] \mid x \xleftarrow{\$} \mathcal{G}; \{s_1, ..., s_q\} \xleftarrow{U} \mathbb{S}; (I, J, \sigma) \xleftarrow{\$} \mathcal{Y}(x, s_1, ..., s_q)]$$

then

$$mfrk \geq acc \cdot \left(\frac{acc^n}{q^{2n}} - \frac{(n+1)(n+3)}{8|\mathbb{S}|} \right),$$

where q is the sum of the upper bound on the queries to the random oracles involved and n is, loosely speaking, the number of forking.

4   The Proposed Scheme

In this section, we construct a forward-secure ID based digital signature scheme with forward-secure key generation, based on the factoring assumption. Our construction is based on a forward secure signature scheme in [1] and extends it into the identity based schemes. The proposed scheme can be considered as a recursive signature scheme; The first signature is generated with the master key to issue the user's signing key and the second signature is performed with the signing key to sign messages. This doubly recursive design should be carefully handled so that the update of the master key is correctly and independently synchronized with the update of the user's signing key. Finally, verifying the final signature must validate the authenticity of identifier, message, and the period. Given a maximum time period T, a security parameter k, and a maximum hash length l, The detailed construction of our FSIBS scheme is described in Algorithm 1.

Algorithm 1. Factoring based FSIBS

PKGKeyGen(k, T, l):

    Generated random primes p, q s.t:

        $p \equiv q \equiv 3 \pmod 4$

        $2^{k-1} \leq (p-1)(q-1)$

        $pq < 2^k$

    $N \leftarrow pq$

    $S \xleftarrow{\$} \mathbb{Z}_N^*$

    $msk_1 \leftarrow S^{2^{3l}} \bmod N$

    $U \leftarrow 1/S^{2^{3l(T+1)}} \bmod N$

    $MSK_1 \leftarrow (N, T, 1, msk_1)$

    $VK \leftarrow (N, U, T)$

    return (MSK_1, VK)

KeyIssue(MSK_i, ID):

    parse MSK_i as (N, T, i, msk_i)

    $R \xleftarrow{\$} \mathbb{Z}_N^*$

    $Y \leftarrow R^{2^{3l(T+1-i)}} \bmod N$

    $sk_{i,ID} \leftarrow R \cdot (msk_i)^{H_1(Y \| ID)} \bmod N$

    $SK_{i,ID} \leftarrow (N, T, Y, sk_{i,ID})$

    return $SK_{i,ID}$

MSKUpdate(MSK_j):

    parse MSK_j as (N, T, j, msk_j)

    if $j = T$ then

        $MSK_{j+1} \leftarrow \perp$

    else

        $msk_{j+1} \leftarrow (msk_j)^{2^{3l}} \bmod N$

        $MSK_{j+1} \leftarrow (N, T, j+1, msk_{j+1})$

    return MSK_{j+1}

UKUpdate($SK_{j,ID}$):

    parse $SK_{j,ID}$ as $(N, T, Y, sk_{j,ID})$

    if $j = T$ then

        $SK_{j+1,ID} \leftarrow \perp$

    else

        $sk_{j+1,ID} \leftarrow (sk_{j,ID})^{2^{3l}} \bmod N$

        $SK_{j+1,ID} \leftarrow (N, T, Y, sk_{j+1,ID})$

    return $SK_{j+1,ID}$

Sign($SK_{j,ID}, M, j$):

    parse $SK_{j,ID}$ as $(N, T, Y, sk_{j,ID})$

    $R' \xleftarrow{\$} \mathbb{Z}_N^*$

    $Y' \leftarrow (R')^{2^{3l(T+1-j)}} \bmod N$

    $\sigma_j \leftarrow R' \cdot (sk_{j,ID})^{H_2(Y \| Y' \| j \| M)} \bmod N$

    return (σ_j, Y', Y)

Verify($ID, M, (\sigma_i, Y', Y), VK, i$):

    parse VK as (N, U, T)

    $h_1 \leftarrow H_1(Y \| ID)$

    $h_2 \leftarrow H_2(Y \| Y' \| i \| M)$

    if $\sigma_i^{2^{3l(T+1-i)}} \cdot U^{h_1 h_2} = Y' \cdot Y^{h_2}$

    $\pmod N$

        return 1

    else

        return 0

For the following correctness description, we assume that the user signing key is generated as $SK_{j,ID} = (N, T, Y, sk_{j,ID} = R \cdot (msk_j)^{h_1})$ for some time period j where $j < T$. Then, notice that the user signing key at time period i for $i > j$ evolves into $SK_{i,ID} = (N, T, Y, sk_{i,ID} = (R \cdot (msk_j)^{H_1(Y\|ID)})^{2^{3l(i-j)}} \mod N)$ by the UKUpdate algorithm.

Correctness: We show that our scheme is correct by computing the equation below:

$$
\begin{aligned}
\sigma_i^{2^{3l(T+1-i)}} &= (R' \cdot (sk_{i,ID})^{h_2})^{2^{3l(T+1-i)}} \\
&= R'^{2^{3l(T+1-i)}} (R^{2^{3l(i-j)}} (msk_j^{2^{3l(i-j)}})^{h_1})^{h_2 2^{3l(T+1-i)}} \\
&= R'^{2^{3l(T+1-i)}} (R^{2^{3l(T+1-j)}} (msk_j^{2^{3l(T+1-j)}})^{h_1})^{h_2} \\
&= R'^{2^{3l(T+1-i)}} (R^{2^{3l(T+1-j)}} (S^{2^{3lj}})^{h_1 2^{3l(T+1-j)}})^{h_2} \\
&= R'^{2^{3l(T+1-i)}} (R^{2^{3l(T+1-j)}})^{h_2} (S^{2^{3l(T+1)}})^{h_1 h_2} \\
&= Y'(Y)^{h_2} (1/U)^{h_1 h_2}
\end{aligned}
$$

where $h_1 = H_1(Y \| ID)$, $h_2 = H_2(Y \| Y' \| i \| M)$.

5   Security Proof

Let k and l be two security parameters. Let p and q be primes and $N = pq$ be a k-bit integer (Since $p \equiv q \equiv 3 \pmod 4$, N is a Blum integer). Let Q denote the set of non-zero quadratic residues modulo N. Note that for $x \in Q$, exactly one of its four square roots is also in Q.

Lemma 2. *Given $\alpha \neq 0, \lambda > 0, v \in Q$ and $X \in \mathbb{Z}_N^*$ such that $v^\alpha \equiv X^{2^\lambda}$ (mod N) and $\alpha < 2^\lambda$, one can easily compute y such that $v \equiv y^2$ (mod N).*

Proof. Let $\alpha = 2^\gamma \beta$ where β is odd. Note that $\lambda > \gamma$. Let $\beta = 2\delta + 1$. Then $(v^{2\delta+1})^{2^\gamma} \equiv v^\alpha \equiv X^{2^\lambda} \pmod N$, so $v^{2\delta+1} \equiv X^{2^{\lambda-\gamma}} \pmod N$. Note that it is allowed to take roots of degree 2^γ since both sides are in Q. Let $y = X^{2^{\lambda-\gamma-1}}/v^\delta \mod N$. Then $y^2 \equiv X^{2^{\lambda-\gamma}}/v^{2\delta} \equiv v \pmod N$. Note that since $\alpha < 2^\lambda, \lambda - \gamma - 1 \geq 0$. ∎

Theorem 1. *Let $\mathcal{A}$ be an adversary attacking Σ_{FSIBS} with running in time τ, making q_{key} number of key issue queries, q_{sign} number of sign queries and q_{hash} number of hash queries. If the advantage $Adv_{\mathcal{A},\Sigma}^{FSIBS}$ is ϵ, then there exists an algorithm $\mathcal{B}$ that succeeds in solving the factoring problem in expected time at most τ' with probability at least ϵ', where $\tau' = 4\tau + O(k^2 lT + k^3)$ and $\epsilon' = \frac{\epsilon}{2T} \cdot \left(\frac{\epsilon^3}{T^3 \cdot q^6} - \frac{28}{2^{k+3}} \right)$ with $q = q_{hash} + q_{sign} + q_{key}$.*

Proof. Given $\mathcal{A}$, we show how to construct an algorithm $\mathcal{B}$ of solving the factoring problem with success probability $\frac{\epsilon}{2T} \cdot \left(\frac{\epsilon^3}{T^3 \cdot q^6} - \frac{28}{2^{k+3}} \right)$ approximately in time $4\tau + O(k^2 lT + k^3)$. To factor its input N, $\mathcal{B}$ will select a random $x \in \mathbb{Z}_N^*$, compute

$v \equiv x^2 \bmod N$, and attempt to use the adversary $\mathcal{A}$ to find a square root y of v. Because v has four square roots and x is random, with probability $1/2$ we have that $x \not\equiv \pm y \pmod N$ and, then $\mathcal{B}$ will be able to find a factor of N by computing the gcd of $x - y$ and N. To do so, we utilize the forking lemma technique in the following. Before start of the game, $\mathcal{B}$ guesses the time period b', $1 < b' \leq T$ that $\mathcal{A}$ will make a forgery.

[**Setup phase**] $\mathcal{B}$ selects a random value $x \in \mathbb{Z}_N^*$, compute $v \equiv x^2 \bmod N$ and sets $msk_{b'} = v$. Then, $\mathcal{B}$ sets $U \equiv 1/msk_{b'}^{2^{3l(T+1-b')}} \bmod N$ and gives $VK = (N, U, T)$ to $\mathcal{A}$. $\mathcal{B}$ maintains two hash tables to respond to the hash queries to H_1 and H_2.

[**Interactive query phase**] In this phase, $\mathcal{A}$ is allowed to query four types of oracles **KeyIssue, Update, Sign,** and **Hash** as follows.

KeyIssue query: The adversary $\mathcal{A}$ chooses a specific ID and a time period i, and sends them to $\mathcal{B}$. If $i < b'$ then $\mathcal{B}$ selects a random $Z \in Q$ and chooses random string $h_1 \in \{0,1\}^l$. Y is computed as $U^{h_1} Z^{2^{3l(T+1-i)}}$. If $H_1(Y, ID)$ is defined in the H_1 table then fails and aborts; otherwise, sets $H_1(Y, ID)$ as h_1 and adds the entry (Y, ID, h_1) into the H_1 table. Set $SK_{i,ID} = Z$. After that, $\mathcal{B}$ returns the signing key $SK_{i,ID} = (N, T, Y, sk_{i,ID})$ for ID of current time period i to adversary $\mathcal{A}$. Consider $i \geq b'$. Since $\mathcal{B}$ has $msk_{b'} = v$, it can obtain msk_i by executing MSKUpdate algorithm. $\mathcal{B}$ returns the singing key $SK_{i,ID}$ simply by applying the KeyIssue algorithm. $\mathcal{B}$ maintains a KeyIssue table and updates the entry $(SK_{i,ID}, i, ID)$ into the table.

Update query: The adversary $\mathcal{A}$ requests the challenger $\mathcal{C}$ to execute the MSKUpdate algorithm and UKUpdate algorithm. If $i \leq b'$ then nothing is performed. Otherwise MSKUpdate and UKUpdate is called.

Sign query: $\mathcal{A}$ chooses a random message M and a user identifier ID, and asks $\mathcal{B}$ to sign a message M for ID on the current time period i. If $i \geq b'$, $\mathcal{B}$ can generate any signing key as above. Otherwise, $\mathcal{B}$ first checks whether the ID and i entry exist in the KeyIssue table. If so, $\mathcal{B}$ generates a signature using the private key. If the ID and i entry does not exist in the table, $\mathcal{B}$ generates the key in a manner similar to the **KeyIssue query** phase. $\mathcal{B}$ selects random $\sigma_i \in \mathbb{Z}_N^*$ and $h_2 \in \{0,1\}^l$. $\mathcal{B}$ selects a random Z and h_1 again, computes $Y = Z^{2^{3l(T+1-i)}} \cdot U^{h_1}$, and adds them into H_1 and uses them. Then, Y' is computed as $\sigma_j^{2^{3l(T+1-t)}} \cdot U^{h_1 \cdot h_2} / Y^{h_2}$. $\mathcal{B}$, finally, returns (σ', Y, Y') to $\mathcal{A}$.

Hash query: The adversary $\mathcal{A}$ can make queries to the random oracle and get the corresponding random values.

[**PKG corrupt phase**] Upon the request of the PKG master secret key, $\mathcal{B}$ marks the current time period as b. If $b < b'$, $\mathcal{B}$ aborts. Otherwise, since $\mathcal{B}$ has $msk_{b'}$, $\mathcal{B}$ can compute $msk_b \leftarrow v^{2^{3l(b-b')}}$ and return it to $\mathcal{A}$.

[**Final forgery phase**] In order to compute the factors of blum integer N, we utilize double forking here. As usual, $\mathcal{B}$ runs $\mathcal{A}$ twice with the same random tape, then $H_1(Y \parallel ID^*)$ would be h_1 at the first case, and $\overline{h_1}$ for the second case. In each case, $\mathcal{B}$ runs $\mathcal{A}$ twice, so that each of them outputs different h_2's this time.

In this phase, $\mathcal{A}$ gives a forgery signature (σ_j, Y'^*, Y^*) of identifier ID^* and message M^* to $\mathcal{B}$, where $j < b'$. As we described before, $H_1(Y^* \parallel ID^*)$ returns different h_1 and $\overline{h_1}$ for each case. First of all, when h_1 was returned, $H_2(Y^* \parallel Y'^* \parallel j \parallel M^*)$ is h_2, and $\mathcal{B}$ is given the forgery signature (σ_j, Y'^*, Y^*) from $\mathcal{A}$. So, the Eq. (1) is valid.

$$(\sigma_j)^{2^{3l(T+i-j)}} \cdot U^{h_1 \cdot h_2} = Y'^* \cdot Y^{*h_2} \tag{1}$$

$\mathcal{B}$, then, rewinds with the same random tape, $H_2(Y^* \parallel Y'^* \parallel j \parallel M^*)$ is h_2^* and get the different forgery signature (σ_j^*, Y'^*, Y^*) from $\mathcal{A}$. We obtain the Eq. (2) at this time.

$$(\sigma_j^*)^{2^{3l(T+i-j)}} \cdot U^{h_1 \cdot h_2^*} = Y'^* \cdot Y^{*h_2^*} \tag{2}$$

With dividing (1) by (2), Eq. (3) comes out.

$$(\sigma_j/\sigma_j^*)^{2^{3l(T+1-j)}} = (Y^*/U^{h_1})^{h_2 - h_2^*} \tag{3}$$

$\mathcal{B}$ resets the nesting fork, $H_1(Y^* \parallel ID^*)$ is $\overline{h_1}$ in this time. In this round, we can get two Eqs. (4) and (5), and Eq. (6) is computed by (4)/(5).

$$\overline{(\sigma_{j'})}^{2^{3l(T+1-j')}} \cdot U^{\overline{h_1} \cdot \overline{h_2}} = \overline{Y'^*} \cdot Y^{*\overline{h_2}} \tag{4}$$

$$\overline{(\sigma_{j'}^*)}^{2^{3l(T+1-j')}} \cdot U^{\overline{h_1} \cdot \overline{h_2^*}} = \overline{Y'^*} \cdot Y^{*\overline{h_2^*}} \tag{5}$$

$$(\overline{\sigma_{j'}}/\overline{\sigma_{j'}^*})^{2^{3l(T+1-j')}} = (Y^*/U^{\overline{h_1}})^{\overline{h_2} - \overline{h_2^*}} \tag{6}$$

By calculating $(3)^{(\overline{h_2} - \overline{h_2^*})}/(6)^{(h_2 - h_2^*)}$ and replacing U with $1/v^{2^{3l(T+1-b')}}$, the following Eq. (7) can be derived. We assume without loss of generality that $j \geq j'$.

$$\frac{(\sigma_j/\sigma_j^*)^{(\overline{h_2} - \overline{h_2^*}) \cdot 2^{3l(b'-j)}}}{(\overline{\sigma_{j'}}/\overline{\sigma_{j'}^*})^{(h_2 - h_2^*) \cdot 2^{3l(b'-j')}}} = v^{(h_1 - \overline{h_1})(h_2 - h_2^*)(\overline{h_2} - \overline{h_2^*})} \tag{7}$$

By Lemma 2 in, if we set $\alpha = (h_1 - \overline{h_1})(h_2 - h_2^*)(\overline{h_2} - \overline{h_2^*})$, $X = \frac{Z}{\overline{Z}^{2^{3l(j-j')}}}$, $\lambda = 3l(b' - j)$, we can easily get the square root of v, where $Z = (\sigma_j/\sigma_j^*)^{(\overline{h_2} - \overline{h_2^*})}$ and $\overline{Z} = (\overline{\sigma_{j'}}/\overline{\sigma_{j'}^*})^{(h_2 - h_2^*)}$.

Because v has four square roots and x is random, with probability $1/2$ we have that $x \not\equiv \pm y \pmod{N}$ and, hence $\mathcal{B}$ will be able to find a factor of N by computing the gcd of $x - y$ and N.

To succeed the forgery, i should be $j < i \leq t'$. So, the probability loss is at most $\frac{1}{T}$, and this is used as acc in Lemma 1. According to Lemma 1, the final probability that $\mathcal{B}$ succeeds becomes at least $\frac{\epsilon}{2T} \cdot \left(\frac{\epsilon^3}{T^3 \cdot q^6} - \frac{28}{2^{k+3}} \right)$ where $n = 3$, and $q = q_{hash} + q_{sign} + q_{key}$.

6   Optimization of Performance Improvement

The key issuing and singing in our scheme require computing Y and Y' such that $Y \leftarrow R^{2^{3l(T+1-j)}} \bmod N$ and $Y' \leftarrow (R')^{2^{3l(T+1-i)}} \bmod N$ for randomly chosen numbers R, and R', of which time complexity is proportional to the maximum period T. Fast-AR scheme [10] accelerates the signing algorithm without sacrificing the other factors. In the scheme, instead of squaring R to compute Y, it chooses an exponent r randomly and then computes $R \leftarrow g^r$ and $Y \leftarrow X^r$ where X is precomputed and $X = g^{2^{l(T+1)}}$ (mod N).

We can apply the Fast-AR scheme to the proposed FSIBS. In the KeyIssue algorithm, we choose a random exponent r to compute R and Y such that $R \leftarrow g^r$ and $Y \leftarrow X^r$. In the Sign algorithm, similarly we choose an exponent r' randomly and compute R' and Y' such that $R' \leftarrow g^{r'}$ and $Y' \leftarrow X^{r'}$. Note that X is precomputed in advance. The verification algorithm is revised similar to the Fast-AR [10]. Then the time complexities of the KeyIssue and the Sign algorithms are independent of the maximum period T. In the experiment we compare the optimized FSIBS with the original FSIBS.

7   Experiment

In this section, we implement the proposed FSIBS and the optimized FSIBS(opt-FSIBS) in Sect. 6 to validate the practicality. The FSIBS and opt-FSIBS are implemented using openssl library in C. For comparison, we also implement the pairing based scheme [15] using PBC library in C. The experiment is performed on Intel i7-8700K 3.7 GHz machine with 24 GB RAM using Ubuntu 18.04.

Table 2. Comparison of FSIBS, opt-FSIBS, and [15]

	FSIBS	opt-FSIBS	[15]
PKGKeyGen [s]	0.070354	0.065916	2.026031
KeyIssue [s]	24.142792	0.003250	1.991108
Sign [s]	24.112566	0.003366	0.075265
Verify [s]	24.169596	24.017984	0.061528
MSKUpdate [s]	0.000275	0.000360	1.869376
UKUpdate [s]	0.000269	0.000275	1.871108
MSK [Byte]	264	264	2644
VK [Byte]	260	260	5440
SK [Byte]	520	520	3060
Signature [Byte]	388	388	104

Table 2 shows the execution time and the key sizes when the security parameter k is 2048 for FSIBS and opt-FSIBS and $k = 224$ for [15]. The maximum time period T is 2^{15} and the hash length l is 160. The execution times of FSIBS and

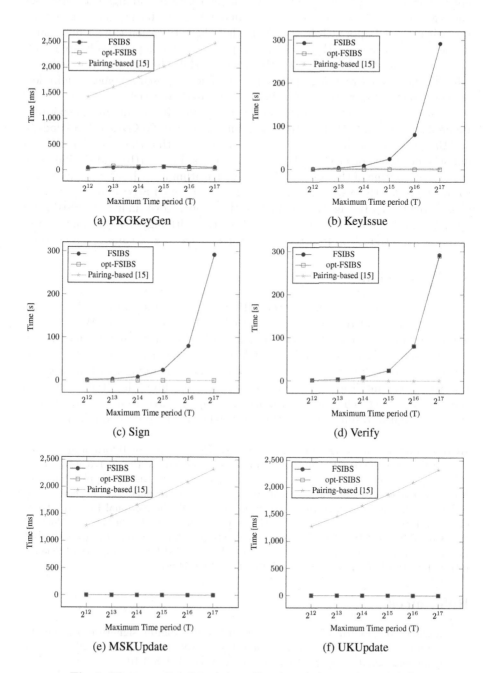

Fig. 1. The execution time by varying the maximum time periods

opt-FSIBS are almost equivalent except for the KeyIssue and Sign time. KeyIssue and Sign times decrease to 3 ms from 24 s in opt-FSIBS which is much faster than the pairing based approach in [15]. Note that the verification time is quite large in the proposed scheme. However, the proposed scheme is practically applicable when the verification is not frequently occurred while the sign is performed periodically such as a streaming application [9]. The proposed schemes require less memory space to store keys than the pairing based approach.

Figure 1 illustrates the execution time by varying the maximum time periods. Figure 1(a), (e), and (f) show the execution times of PKGKeyGen, MSKUpdate, and UKUpdate algorithms in each scheme. They show that the execution times of our schemes are shorter than those of [15] always. Figure 1(b) and (c) illustrate the execution time of the KeyIssue and Sign algorithm in each scheme. According to these graphs, the KeyIssue time and Sign time of opt-FSIBS and [15] are both short enough. Lastly, Fig. 1(d) shows that the Verify algorithm in both FSIBS and opt-FSIBS takes much longer than [15]. Note that although Verify is much slower in our scheme than [15], it is relatively insignificant in a streaming application.

8   Conclusion

This paper proposes a forward-secure ID-based digital signature scheme with forward secure private key generator based on RSA assumption. We define its notion and provide practical constructions and its security proof under the factoring assumption in the random oracle model. The signing algorithm in the proposed scheme is fast enough to be applied in a streaming real-time application with constant size keys and constant size signatures while the verification time is quite long. The proposed scheme is implemented in a real system and the experimental results validate the practicality of the proposed scheme.

Acknowledgement. This work was supported by Institute for Information & communications Technology Promotion (IITP) grant funded by the Korea government (MSIT) (No. 2016-6-00599, A Study on Functional Signature and Its Applications and No. 2017-0-00661, Prevention of video image privacy infringement and authentication technique), by Basic Science Research Program through the National Research Foundation of Korea (NRF) funded by the Ministry of Education (No. 2017R1A2B4009903 and No. 2016R1D1A1B03934545), and by Basic Research Laboratory Program through the National Research Foundation of Korea (NRF) funded by the Ministry of Science, ICT & Future Planning (MSIP) (No. 2017R1A4A1015498).

References

1. Abdalla, M., Reyzin, L.: A new forward-secure digital signature scheme. In: Okamoto, T. (ed.) ASIACRYPT 2000. LNCS, vol. 1976, pp. 116–129. Springer, Heidelberg (2000). https://doi.org/10.1007/3-540-44448-3_10
2. Anderson, R.: Two remarks on public-key cryptology - invited lecture. In: The Fourth ACM Conference on Computer and Communications Security (CCS) (1997)
3. Bellare, M., Miner, S.K.: A forward-secure digital signature scheme. In: Wiener, M. (ed.) CRYPTO 1999. LNCS, vol. 1666, pp. 431–448. Springer, Heidelberg (1999). https://doi.org/10.1007/3-540-48405-1_28

4. Boyen, X., Shacham, H., Shen, E., Waters, B.: Forward-secure signatures with untrusted update. In: Juels, A., Wright, R.N., di Vimercati, S.D. (ed.) Proceedings of the 13th ACM Conference on Computer and Communications Security, CCS 2006, Alexandria, VA, USA, 30 October–3 November 2006, pp. 191–200. ACM (2006)

5. Chatterjee, S., Kamath, C.: A closer look at multiple forking: leveraging (in)dependence for a tighter bound. Algorithmica **74**(4), 1321–1362 (2016)

6. Ebri, N.A., Baek, J., Shoufan, A., Vu, Q.H.: Forward-secure identity-based signature: new generic constructions and their applications. JoWUA **4**(1), 32–54 (2013)

7. Fiat, A., Shamir, A.: How to prove yourself: practical solutions to identification and signature problems. In: Odlyzko, A.M. (ed.) CRYPTO 1986. LNCS, vol. 263, pp. 186–194. Springer, Heidelberg (1987). https://doi.org/10.1007/3-540-47721-7_12

8. Itkis, G., Reyzin, L.: Forward-secure signatures with optimal signing and verifying. In: Kilian, J. (ed.) CRYPTO 2001. LNCS, vol. 2139, pp. 332–354. Springer, Heidelberg (2001). https://doi.org/10.1007/3-540-44647-8_20

9. Kim, J., Lee, S., Yoon, J., Ko, H., Kim, S., Oh, H.: PASS: privacy aware secure signature scheme for surveillance systems. In: 2017 IEEE Symposium on Advanced Video and Signal-Based Surveillance (AVSS). IEEE (2017)

10. Kim, J., Oh, H.: Forward-secure digital signature schemes with optimal computation and storage of signers. In: De Capitani di Vimercati, S., Martinelli, F. (eds.) SEC 2017. IAICT, vol. 502, pp. 523–537. Springer, Cham (2017). https://doi.org/10.1007/978-3-319-58469-0_35

11. Krawczyk, H.: Simple forward-secure signatures from any signature scheme. In: CCS 2000, Proceedings of the 7th ACM Conference on Computer and Communications Security, Athens, Greece, 1–4 November 2000, pp. 108–115 (2000)

12. Liu, Y., Yin, X., Qiu, L.: ID-based forward-secure signature scheme from the bilinear pairings. In: Yu, F., Luo, Q., Chen, Y., Chen, Z., (eds.) Proceedings of The International Symposium on Electronic Commerce and Security, ISECS 2008, Guangzhou, China, 3–5 August 2008, pp. 179–183. IEEE Computer Society (2008)

13. Malkin, T., Micciancio, D., Miner, S.: Efficient generic forward-secure signatures with an unbounded number of time periods. In: Knudsen, L.R. (ed.) EUROCRYPT 2002. LNCS, vol. 2332, pp. 400–417. Springer, Heidelberg (2002). https://doi.org/10.1007/3-540-46035-7_27

14. Merkle, R.C.: A digital signature based on a conventional encryption function. In: Pomerance, C. (ed.) CRYPTO 1987. LNCS, vol. 293, pp. 369–378. Springer, Heidelberg (1988). https://doi.org/10.1007/3-540-48184-2_32

15. Oh, H., Kim, J., Shin, J.S.: Forward-secure ID based digital signature scheme with forward-secure private key generator. Inf. Sci. **454–455**, 96–109 (2018)

16. Schnorr, C.P.: Efficient identification and signatures for smart cards. In: Quisquater, J.-J., Vandewalle, J. (eds.) EUROCRYPT 1989. LNCS, vol. 434, pp. 688–689. Springer, Heidelberg (1990). https://doi.org/10.1007/3-540-46885-4_68

17. Wei, J., Liu, W., Hu, X.: Forward-secure identity-based signature with efficient revocation. Int. J. Comput. Math. **94**(7), 1390–1411 (2017)

18. Yu, J., Hao, R., Kong, F., Cheng, X., Fan, J., Chen, Y.: Forward-secure identity-based signature: security notions and construction. Inf. Sci. **181**(3), 648–660 (2011)

19. Yu, J., Xia, H., Zhao, H., Hao, R., Fu, Z., Cheng, X.: Forward-secure identity-based signature scheme in untrusted update environments. Wirel. Pers. Commun. **86**(3), 1467–1491 (2016)

20. Zhang, X., Xu, C., Jin, C., Xie, R.: Efficient forward secure identity-based shorter signature from lattice. Comput. Electr. Eng. **40**(6), 1963–1971 (2014)

Integrity

Integrity

On the Effectiveness of Control-Flow Integrity Against Modern Attack Techniques

Sarwar Sayeed$^{(\boxtimes)}$ (ID) and Hector Marco-Gisbert (ID)

University of the West of Scotland, Paisley, UK
{sarwar.sayeed,hector.marco}@uws.ac.uk

Abstract. Memory error vulnerabilities are still widely exploited by attackers despite the various protections developed. Attackers have adopted new strategies to successfully exploit well-known memory errors bypassing mature protection techniques such us the NX, SSP, and ASLR. Those attacks compromise the execution flow to gain control over the target successfully.

Control-flow Integrity (CFI) is a protection technique that aims to eradicate memory error exploitation by ensuring that the instruction pointer (IP) of a running program cannot be controlled by a malicious attacker. In this paper, we assess the effectiveness of 14 CFI techniques against the most popular exploitation techniques including code reuse attacks, return-to-user, return-to-libc and replay attacks.

Surveys are conducted to classify those 14 CFI techniques based on the security robustness and implementation feasibility. Our study indicates that the majority of the CFI techniques are primarily focused on restricting indirect branch instructions and cannot prevent all forms of vulnerability exploitation. Moreover, we show that the overhead and implementation requirement make some CFI techniques impractical. We conclude that the effort required to have those techniques in real systems, the high overhead, and also the partial attack coverage is discouraging the industry from adopting CFI protections.

Keywords: CFI Protection Techniques · CFI attacks

1 Introduction

Cyber Security is a changing platform, where new defense advances are being evolved every moment to cope with the ongoing challenges. Due to continuous changes in the attacking methods, a protection technique often remains outdated and having required to come up with something more advanced. From the past few decades, code-injection attack was most significant to corrupt the control-flow of a program. To meet such attacking challenges, various strong protection techniques were introduced by security developers. However, in-time the attackers have advanced their ability to corrupt control-flow in a more efficient way;

© IFIP International Federation for Information Processing 2019
Published by Springer Nature Switzerland AG 2019
G. Dhillon et al. (Eds.): SEC 2019, IFIP AICT 562, pp. 331–344, 2019.
https://doi.org/10.1007/978-3-030-22312-0_23

hence, it was imperative to introduce another protection technique which would mitigate such leading threats.

CFI was first initiated by Microsoft in 2005 to obstruct the control-flow exploitation challenges. CFI is a security policy which can be implemented to mitigate various levels of severe attacks that mainly occur to corrupt the control-flow of a program. To accomplish an attack, an adversary goes through various attacking stages where obtaining control over the IP is the very first step of the vulnerability exploitation. A compromised control-flow may lead to various exploitation techniques, such as Code-reuse attacks (CRA), Code injection, return-to-libc. Protection techniques such as Stack Smashing Protector (SSP), Address Space Layout Randomization (ASLR) and Non-executable (NX) bit are some mechanisms that are present in all modern systems, but unfortunately, recent attacking techniques are improved to bypass all those protection mechanisms [23].

The main contributions of this paper are:

- We analyze 14 CFI techniques revealing their main features and weaknesses.
- We show the competence between hardware and software based CFI implementation.
- We conduct a survey to classify the CFI techniques based on attacks that can bypass them.
- We summarize the performance overhead revealing which of the CFI techniques are prohibitive because of the overhead introduced.

The paper is organized as follows; Sect. 2 is the background section which discusses control-flow transfers, attacks, and integrity. Section 3 defines various attack vectors that subvert the control-flow of a program. In Sect. 4, we review 14 CFI techniques and point out the limitations associated with each technique. Section 5 involves analyzing software and hardware-based CFI techniques and produces a solution by debating which techniques are more protective against control-flow attacks. In addition to that, the impact of performance overhead is also discussed. Finally the concluding Sect. 6, which summarizes the findings and discusses future work.

2   Background

To comprehend the complete control-flow mechanism this section describes types of control-flow transfers and attacks adhering to them. CFI method and its effectiveness towards control-flow attacks are also discussed.

2.1   Control Flow Transfers

Control-flow transfers can be direct or indirect. Direct control-flow transfer comprises read-only permissions; hence, these types of transfers are more secure and protection is implemented by the memory management unit (MMU). Whereas indirect control-flow transfer relies on run-time information; such as register or memory values. In a control-flow attack, attackers tend to divert the indirect control-flow to their chosen location to perform arbitrary code execution [32].

2.2 Control Flow Attacks

A control-flow attack is a run-time exploitation technique which is performed during a run-time state of a program [13]. In this attacking technique, the adversary gets hold of the instruction pointer to divert the execution flow by exploiting an application's weakness. It is also used to overwrite the buffer in the stack. Two major classes of control-flow attacks are mainly performed by random attackers; Code injection and Code-reuse attack.

2.3 Control Flow Integrity

Control-Flow Integrity is a defense policy intending to restrict unintended control flow transfers to unauthorized locations [31]. It is able to defend against various types of attacks whose primary intention is to redirect the execution flow elsewhere. Many CFI techniques [4,6,24,27–29,35,38] have been proposed over the past few years. However, they were not fully adopted due to practical challenges and significant limitations.

3 Threats

In this section, we present several attacking techniques utilized by adversaries to subvert the control-flow. The nature of exploitation strategies and their ability to perform the attacks are pointed out.

3.1 Code Reuse Attack

Code-reuse attack is an attacking technique which relies on reusing the existing code [5]. CRA exploitation occurs using codes that are already present in the target's application. For instance, the very first step of CRA begins by exploiting a vulnerability in an application that runs in the targeted system. Once the vulnerability is discovered, then the target machine can be exploited by malicious input.

3.2 Code Injection

Code injection involves injecting and executing malicious code in the memory address space [30]. The exploitation can be achieved by providing malicious payload as input and then get processed by the program. Code injection occurs when a program bug handles untrusted data. For instance, if a program does not perform bounds checking of the given input, then an adversary might provide large data than the actual limit resulting in possible buffer corruption.

3.3 Disclosure Attack

Disclosure attack endorses an attacker to uncover sensitive information, which may include source code, stack information, passwords or database information [19]. This attack can be exploited by authenticating users confidential information and then apply such information to perform further attacks.

3.4   Return-to-User

Return-to-user (ret2usr) overwrites kernel data with user address space [21]. To conduct this attack vector, an adversary gets hold of the return address, dispatch tables, and function pointers to perform arbitrary code execution. The ultimate cause involves hijacking the kernel level control-flow to redirect towards the userspace code.

3.5   Return-to-Libc

Return-to-libc occurs by jumping to the function address and allocating arguments [34]. The adversary does not require to inject payload to exploit the target. It overwrites the instruction pointer with the address pointing to the Global Offset Table (GOT), which contains pointers to glibc library functions.

3.6   Replay Attack

In this attack, an adversary copies series of data between two users and takes advantage of the event by communicating with one or both parties [25]. The adversary aims to eavesdrop the exchange of messages or aware of the message rule from earlier communications between users. Correctly encrypted message, sent by attackers, is considered as legit request and the necessary task is performed accordingly.

4   CFI Protection Techniques and Limitations

In this section, we discuss the most relevant CFI techniques that can be used to prevent control-flow hijacking. In our discussion, we also point out the limitations associated with individual techniques. Table 1 shows the enforced mechanisms in each CFI and Table 2 represents the essential characteristics related to each technique.

4.1   CFI Principles, Implementations, and Applications (CFI)

CFI was proposed by Abadi et al. [1,2] and it is the first CFI proposal for CFI implementation. They have implemented inlined CFI for windows on the x86 platform. Their work suggests that a Control flow graph (CFG) be obtained before program execution. The CFG monitors runtime behavior; therefore, any inconsistency in the program results in CFI exception to be called and application to terminate. However, Davi et al. [14] point out three main limitations of this technique. First, the source code is not always available. Second, binaries lack the required debug information and finally, it causes high execution overhead because of dynamic rewriting and run-time checks. It is also unable to determine if the function returns to the current call site [9].

4.2   CCFI: Cryptographically Enforced CFI (CCFI)

CCFI possesses new pointer arrangements, which can not be imposed with static approaches [24]. It comprises two prime attributes. First, it recategorizes function pointers at runtime to boost typecasting. Second, it restricts swapping of two valid pointers which consist of the same type. Nevertheless, CCFI comprises an average overhead of 52% on all benchmarks. CCFI is vulnerable to replay attacks [28]. It also fails to identify structure pointers. It is possible to disrupt the control flow by altering the current pointer with the old pointer. CCFI mainly focuses on defending the user level program and does not include kernel level security [24].

4.3   CFI for COTS Binaries (binCFI)

In this technique, CFI is applied to stripped binaries on x86/Linux architecture. It involves implementing CFI to the shared libraries; for instance glibc [38]. binCFI focuses on overcoming the drawbacks which are highlighted by the static analysis technique. According to Niu et al. [29], bin-CFI permits a function to return to every viable return addresses; hence, the accuracy of this CFI is fragile to Return-Oriented programming (ROP) based attacks.

4.4   Practical CFI and Randomization for Binary Executables (CCFIR)

CCFIR gathers the legit target of indirect branch instructions and places them randomly in a "Springboard Section" [37]. CCFIR restricts indirect branch instructions and permits them only towards white-list destinations. The average execution overhead of CCFIR is 3.6% and can be a maximum of 8.6%. It can be challenging to disassemble a PE file properly. CCFIR utilizes three ID's for each branch instruction, excluding the shadow stack. A ROP chain can be built to subvert CCFIR [27].

4.5   Hardware (CFI) for an IT Ecosystem (HW-CFI)

HW-CFI is a security proposal by NSA information assurance [4]. They put forward two notional features to enhance CFI. One of the features recommends implementing CFG to hardware. The other feature protects the dynamic control-flow by a protected shadow stack. This CFI proposal is a notional CFI design and might not be compatible with all system architecture. Though shadow stack can be an excellent option but monitoring shadow stack explicitly might not often be possible.

4.6   Per-Input CFI (PICFI)

PICFI imposes computed CFG to each input. It is certainly difficult to consider all inputs of an application and CFG for each of the inputs. Therefore, PICFI

Table 1. Mechanism that is enforced in CFI techniques

CFI techniques	CFI enforcement
CFI	Inlined CFI
CCFI	Dynamic Analysis
binCFI	Static Binary Rewriting
CCFIR	Binary Rewriting
HW-CFI	Landing Point
PICFI	Static Analysis
KCofi	SVA Compiler Instrumentation
Kernel CFI	Retrofitting Approach
IFCC	Dynamic Analysis
CFB	Precise Static CFI
SAFEDISPATCH	Static Analysis
C-Guard	Dynamic Instrumentation
RAP	Type Based
O-CFI	Static Rewriting

runs an application with empty CFG and lets the program to discover the CFG by itself [29]. PICFI consists overall run-time overhead as low as 3.2%. PICFI statically computes CFG to determine the edges that will be added on run-time and implements DEP to defend against code injection. However, statically computed CFG does not produce a proper result, and various experiments prove that DEP is by-passable.

4.7 KCoFI: Complete CFI for Commodity Operating System Kernels (KCoFI)

KCoFI ensures protection for commodity operating systems from attacks, such as, ret2usr, code segment modification [12]. KCoFI performs its tasks in-between the stack and the processor. KCoFI includes a conventional label-based approach to deal with indirect branch instructions. KCoFI consists of about 27% overhead on transferred file with the size between 1 KB and 8 KB and on smaller files it consists average overhead of 23%. Though KCoFI fulfills all the requirements of managing event handling based on SVA but the outcome of this CFI enforcement is too expensive, over 100% [17].

4.8 Fine-Grained CFI for Kernel Software (Kernel CFI)

Kernel CFI implements retrofitting approach entirely to FreeBSD, the MINIX microkernel system, MINIX's user-space server and partially to BitVisor hypervisor [17]. It follows two main approaches to CFI implementation. The average performance overhead ranges from 51%/25% for MINIX and 12%/17% for

Table 2. Key features of CFI techniques.

CFI techniques	Based on		Compiler modified	Shadow stack	CFG	Label	Coarse grained	Fine grained	Backward edge
	HW	SW							
CFI [1,2]		✓		✓	✓	✓	✓		✓
CCFI [24]	✓	✓	✓					✓	✓
binCFI [38]		✓			✓	✓			
CCFIR [37]		✓			✓	✓			✓
HW-CFI [4]	✓	✓	✓	✓	✓	✓		✓	✓
PICFI [29]		✓	✓		✓			✓	✓
KCoFI [12]		✓	✓		✓	✓			✓
Kernel CFI [17]		✓			✓			✓	✓
IFCC [35]		✓	✓		✓			✓	
CFB [6]		✓		✓	✓			✓	✓
SAFEDISPATCH [20]		✓	✓				✓		
C-Guard [26]		✓	✓		✓		✓		
RAP [18]		✓	✓		✓			✓	✓
O-CFI [27]	✓	✓					✓	✓	✓

FreeBSD. Though Kernel CFI cuts down the indirect transfer up to 70%; however, there still a chance remains for indirect branch instructions to transfer control to an unintended destination.

4.9   Enforcing Forward-Edge CFI in GCC & LLVM (IFCC)

Indirect Function-Call Checks (IFCC), a CFI transformation mechanism, is imposed over LLVM. IFCC introduces a dynamic tool which can be used to analyze CFI and locate forward edge CFI vulnerabilities [35]. The implementation mainly concentrates on three compiler-based mechanisms. It consists 1% to 8.7% performance overhead measured on SPECCPU2006 benchmark. Nevertheless, IFCC fails to protect against control Jujutsu attack, a fine-grained attacking technique which aims to execute malicious payload [16].

4.10   Control-Flow Bending: On the Effectiveness of CFI (CFB)

CFB comprises static CFI implementation [6]. It is based on non-control-data attacks. For instance; if arguments are overwritten directly, then that is considered as a data-only attack as it did not require to invade the control-flow for such operation, but if the overwritten data is non-control-data, then it has affected the control-flow. CFB implements fully-precise static CFG which can be undecidable [16]. CFB also violates certain functions at a high level and execution of such functions likely to alter the return address and corrupt control-flow [29].

4.11   SAFEDISPATCH: Securing C++ Virtual Calls from Memory Corruption Attacks (SAFEDISPATCH)

SAFEDISPATCH defenses against vtable hijacking. It examines C++ programs statically and carries out a run-time check to ensure that the control-flow at virtual method call sites is not hijacked by attackers [20]. SAFEDISPATCH is an enhanced C++ compiler and is built based on Clang++/LLVM. The run time overhead of SAFEDISPATCH is quite low as 2.1% and having a memory overhead of 7.5%. However, all the compiler based fine-grained illustration undergoes common problems: Shared libraries; as recent programs frequently use shared library or dynamic loaded libraries [28]. SAFEDISPATCH is also unable to protect binaries; hence, making them vulnerable to ROP based attacks [27].

4.12   Control Flow Guard (C-Guard)

Control Flow Guard is a highly implemented security mechanism developed by Microsoft to defense against memory error vulnerabilities [26]. It enhances high restrictions so that arbitrary codes cannot be executed through vulnerabilities such as memory buffer overflow. However, it is unable to verify when a function returns to some unauthorized destination [18].

4.13   Reuse Attack Protector (RAP)

RAP is imposed on GCC compiler as a plugin; therefore, developers do not have to use a reformed compiler to utilize RAP [18]. RAP has a commercial version, which comes with two prime defense mechanisms to protect against control-flow attacks. However, RAP's implemented approach is very much similar to the traditional label-based approach. Label based CFI suffers from security issues as a function could return to any call site. RAP does not have a solid protection against ret2usr attacks [22].

4.14   Opaque CFI (O-CFI)

O-CFI comprises binary software randomization for CFI enforcement [27]. It protects legacy binaries without even accessing the source code. CFI checks are done on Intel x86/x64 memory protection extensions (MPX), which is hardware driven. It consists of a performance overhead of only 4.7%. O-CFI comprises static binary phase; therefore, it fails to protect codes that are generated dynamically. It is also incompatible with the Windows Component Object Model(COM). Moreover, MPX is much slower than software-based implementation and not protective against memory based errors.

5   Analysis

In this section, we assess the effectiveness of software and hardware-based CFI techniques against the threats presented in Sect. 3. The assessment involves surveying on the security experiments that are done by various research groups

Table 3. State of the art of the attacks bypassing CFI

CFI Tech.	Code reuse	Code-injection	Disclosure	Return2user	Return2libc	Replay
CFI	[7]		[7]			
CCFI						[28]
binCFI	[27]					
CCFIR			[14]			
HW CFI	[8]					
PICFI	[15]					
KCofi	[7]		[7]			
Kernel CFI	[16,33]					
IFCC	[16]					
CFB		[36]				
SafeDispatch	[14]					
C-Guard	[10]	[3]				
RAP				[22]		
O-CFI	[11]					

and then put them together to define the flaws. We establish the outcome of our evaluation by suggesting an optimal security solution and also discuss the impact of performance overhead towards CFI implementation.

5.1 Software-Based CFI

Software-based CFI enforcement primarily focuses on program instructions, which are corrupted by indirect branch instructions.

Table 3 shows that CFI [1] does not comprise strong protection and according to Chen et al. [7], it is possible to execute ROP based attacks while this technique is deployed. bin-CFI does not comprise shadow stack policy and uses ID/label for each branch instructions. Mohan et al. [27] illustrates that the most recent experiments prove that label based approaches are also vulnerable to ROP based attacks. Though CCFIR enhances a security policy that restricts indirect branch instructions to predefined functions; however, it misuses external library call dispatching policy in Linux and also causes boundless direct calls to critical functions in windows libraries which can be exploited. Moreover, the springboard section can be exploited by disclosure attacks [14]. PICFI lacks security, and the control-flow can be compromised by performing three distinguish attacking stages illustrated by [15]. KCoFI and Kernel CFI are kernel based CFI techniques. KCoFI depends only on the source code; therefore, ensuring minimal protection to binaries. It also does not provide stack protection; hence, it is exploitable by various memory error vulnerabilities, such as CRA and memory disclosure [7]. Beside that, Kernel CFI is able to build a minimal challenge to defend against ROP based attacks [33]. Table 3 also shows that IFCC is unable to mitigate control-flow exploitation and can be by-passable by control-flow attack [16]. Though CFB enhances strong CFI enforcement by imposing shadow stack. However, it is evidenced that it can be by-passable by CFG-Aware Attack [36]. SAFEDISPATCH is a compiler based CFI enforcement focuses on securing indirect calls to virtual methods in C++, and it can also be subverted by CRA such as ROP [14]. Control Flow Guard fails to protect against indirect jumps. Moreover, It is fully by-passable by Back To The Epilogue (BATE) attack [3]. RAP makes it very hard for a ROP chain to be built up; however, it is unable to provide security against ret2usr attacks [22].

5.2   Hardware-Based CFI

Hardware-based CFI enforcement requires the system to have hardware-based components in place for deployment. Our assessment involves 3 CFI techniques, which are implemented in both hardware and software. Hardware implementation is an expensive option as it might not be compatible with the running system; hence, it may require to transform the whole system. CCFI requires AES-NI implementation besides compiler fulfillment. Experiments suggest that AES-NI can be exploited by replay attack [28]. HW-CFI comprises shadow stack to protect backward edge and landing point instructions for indirect branch transfers. In the context of security, this CFI technique will fail to mitigate indirect branch transfers if implemented. An adversary will be able to direct forward edge to any landing point instruction causing control-flow corruption [8]. Memory Protection Extention (MPX) is adopted by O-CFI; however, MPX is not a quick approach and hits 4x slow down compared to the software approach. O-CFI can also be exploited by function-reuse attacks [11].

5.3   Optimal Protection

We present that all the 14 CFI techniques comprise major limitations; hence, they are very much prone to be compromised. Table 3 does not give any indication on how much effort is required to subvert the individual CFI; however, it reveals the exploitation method, by providing a reference, related to particular CFI technique.

Based on our analysis, we are able to identify two software-based CFI techniques, which are more practical and realistic for industry deployment.

CCFIR, a coarse-grained approach, does not rely on weak implementations such as CFG or shadow stack. Since it involves binary instrumentation; hence, it does not need source code or debug information. It also protects backward edges by allowing them only towards a white-list destination. CCFIR avoids most of the weak implementation classes, which are used in most software-based CFI techniques.

RAP, a fine-grained approach, has already been adopted by the industry. A modified compiler is not required to utilize RAP. RAP enhances security by ensuring that a function is called from a designated place and returned to that

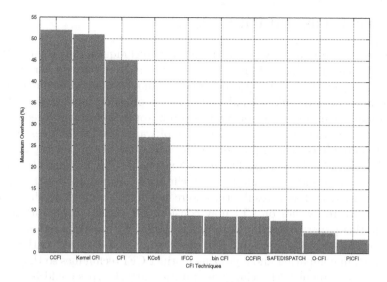

Fig. 1. Performance overhead of major CFI techniques

specific function. RAP instruments Linux kernel at compile time to implement strict CFI at runtime and assures that code pointer are not corrupted by the adversary.

Though CCFIR and RAP are not a completely secure CFI enforcement technique; however, they are able to restrict control-flow hijacking to an extensive level.

5.4 Performance Overhead

Overhead plays a significant part in CFI implementation. Figure 1 presents the maximum performance overhead of 10 CFI techniques. Distinct CFI techniques use different platforms to measure execution, performance, and space overhead. Figure 1 shows that CFI, KCofi, Kernel CFI cause the most overhead ranging from 27%–51%. CFI enforcement with such amount of overhead is not accepted and must receive a denial. bin-CFI consists considerable amount of overhead. Hardware-based enforcement, CCFI, comprises 52% overhead raising the question if hardware implementation is worth enough beside software implementation. However, O-CFI another hardware-based implementation comprises only 4.7% overhead. Compiler implemented CFI approaches, such as PICFI, IFCC, SAFEDISPATCH comprise very low overhead too. CCFIR also consists of very low overhead, 8.6%.

We evidence that an advanced CFI technique with high overhead may not be accepted since, besides integrity, performance is an important factor.

6   Conclusions and Future Work

In this paper, we have surveyed 14 major CFI techniques. It is clear that each technique comprises severe limitations and can be subverted by various attack vectors. It is identified that software-based techniques are more secure compared to hardware-based techniques and also based on practical implication. They are easy to implement and does not require an improper architectural requirement. Based on our findings, we have upheld two software-based techniques, assuming that they provide enhanced protection in-terms of security. The impact of high overheads has also been brought out in regards to CFI implementation. Our survey has established that most CFI techniques are dis-functional to provide proper security, and as a result, they were not fully acquired by the industry. Hence, future researches on CFI may consider overcoming the limitations discussed in this paper to develop a more advanced CFI implementation.

For our future work, we would like to develop a standardized method, which can be used in comparing and analyzing distinct CFI techniques so that particular information about CFI techniques and their attack types could be obtained.

References

1. Abadi, M., Budiu, M., Erlingsson, U., Ligatti, J.: Control-flow integrity. In: Proceedings of the 12th ACM Conference on Computer and Communications Security, CCS 2005, pp. 340–353. ACM, New York (2005). https://doi.org/10.1145/1102120.1102165
2. Abadi, M., Budiu, M., Erlingsson, U., Ligatti, J.: Control-flow integrity principles, implementations, and applications. ACM Trans. Inf. Syst. Secur. **13**(1), 4:1–4:40 (2009). https://doi.org/10.1145/1609956.1609960
3. Biondo, A., Conti, M., Lain, D.: Back to the epilogue: evading control flow guard via unaligned targets. In: Network and Distributed Systems Security (NDSS) Symposium 2018 (2018). https://doi.org/10.14722/ndss.2018.23318
4. NSA Information Assurance: Hardware control flow integrity CFI for an IT ecosystem. NSA, April 2015
5. Bletsch, T., Jiang, X., Freeh, V.W., Liang, Z.: Jump-oriented programming: a new class of code-reuse attack. In: Proceedings of the 6th ACM Symposium on Information, Computer and Communications Security, ASIACCS 2011, pp. 30–40. ACM, New York (2011). https://doi.org/10.1145/1966913.1966919
6. Carlini, N., Barresi, A., Payer, M., Wagner, D., Gross, T.R.: Control-flow bending: on the effectiveness of control-flow integrity. In: Proceedings of the 24th USENIX Conference on Security Symposium, SEC 2015, pp. 161–176. USENIX Association, Berkeley (2015). http://dl.acm.org/citation.cfm?id=2831143.2831154
7. Chen, X., Slowinska, A., Andriesse, D., Bos, H., Giuffrida, C.: StackArmor: comprehensive protection from stack-based memory error vulnerabilities for binaries. In: NDSS 2015. Internet Society, San Diego (2015). https://doi.org/10.14722/ndss.2015.23248
8. Christoulakis, N., Christou, G., Athanasopoulos, E., Ioannidis, S.: HCFI: hardware-enforced control-flow integrity. In: Proceedings of the Sixth ACM Conference on Data and Application Security and Privacy, CODASPY 2016, pp. 38–49. ACM, New York (2016). https://doi.org/10.1145/2857705.2857722

9. de Clercq, R., Verbauwhede, I.: A survey of hardware-based control flow integrity (CFI). CoRR abs/1706.07257 (2017). http://arxiv.org/abs/1706.07257

10. Power of Community: Windows 10 Control Flow Guard Internals (2014). http://www.powerofcommunity.net/poc2014/mj0011.pdf. Accessed 15 Jan 2018

11. Crane, S.J., et al.: It's a TRaP: table randomization and protection against function-reuse attacks. In: Proceedings of the 22nd ACM SIGSAC Conference on Computer and Communications Security, CCS 2015, pp. 243–255. ACM, New York (2015). https://doi.org/10.1145/2810103.2813682

12. Criswell, J., Dautenhahn, N., Adve, V.: KCoFI: complete control-flow integrity for commodity operating system kernels. In: Proceedings of the 2014 IEEE Symposium on Security and Privacy, SP 2014, pp. 292–307. IEEE Computer Society, Washington, DC (2014). https://doi.org/10.1109/SP.2014.26

13. Davi, L., Sadeghi, A.-R.: Building Secure Defenses Against Code-Reuse Attacks. SCS. Springer, Cham (2015). https://doi.org/10.1007/978-3-319-25546-0

14. Davi, L., Sadeghi, A.R., Lehmann, D., Monrose, F.: Stitching the gadgets: on the ineffectiveness of coarse-grained control-flow integrity protection. In: Proceedings of the 23rd USENIX Conference on Security Symposium, SEC 2014, pp. 401–416. USENIX Association, Berkeley (2014). http://dl.acm.org/citation.cfm?id=2671225.2671251

15. Ding, R., Qian, C., Song, C., Harris, B., Kim, T., Lee, W.: Efficient protection of path-sensitive control security. In: 26th USENIX Security Symposium (USENIX Security 2017), pp. 131–148. USENIX Association, Vancouver (2017). https://www.usenix.org/conference/usenixsecurity17/technical-sessions/presentation/ding

16. Evans, I., et al.: Control jujutsu: on the weaknesses of fine-grained control flow integrity. In: Proceedings of the 22nd ACM SIGSAC Conference on Computer and Communications Security, CCS 2015, pp. 901–913. ACM, New York (2015). https://doi.org/10.1145/2810103.2813646

17. Ge, X., Talele, N., Payer, M., Jaeger, T.: Fine-grained control-flow integrity for kernel software. In: Proceedings of the IEEE European Symposium on Security and Privacy, pp. 179–194, March 2016

18. grsecurity: How Does RAP Works. https://grsecurity.net/rap_faq.php. Accessed 3 Feb 2018

19. Guan, L., Lin, J., Luo, B., Jing, J., Wang, J.: Protecting private keys against memory disclosure attacks using hardware transactional memory. In: Proceedings of the 2015 IEEE Symposium on Security and Privacy, SP 2015, pp. 3–19. IEEE Computer Society, Washington, DC (2015). https://doi.org/10.1109/SP.2015.8

20. Jang, D., Tatlock, Z., Lerner, S.: SafeDispatch: securing C++ virtual calls from memory corruption attacks. In: NDSS 2014. Internet Society, San Diego, February 2014. http://dx.doi.org/doi-info-to-be-provided-late

21. Kemerlis, V.P., Polychronakis, M., Keromytis, A.D.: Ret2dir: rethinking kernel isolation. In: Proceedings of the 23rd USENIX Conference on Security Symposium, SEC 2014, pp. 957–972. USENIX Association, Berkeley (2014). http://dl.acm.org/citation.cfm?id=2671225.2671286

22. Li, J., Tong, X., Zhang, F., Ma, J.: Fine-CFI: fine-grained control-flow integrity for operating system kernels. IEEE Trans. Inf. Forensics Secur. **13**(6), 1535–1550 (2018). https://doi.org/10.1109/TIFS.2018.2797932

23. Marco-Gisbert, H., Ripoll, I.: On the effectiveness of NX, SSP, RenewSSP, and ASLR against stack buffer overflows. In: NCA, pp. 145–152. IEEE Computer Society (2014)

24. Mashtizadeh, A.J., Bittau, A., Boneh, D., Mazières, D.: CCFI: cryptographically enforced control flow integrity. In: Proceedings of the 22nd ACM SIGSAC Conference on Computer and Communications Security, CCS 2015, pp. 941–951. ACM, New York (2015). https://doi.org/10.1145/2810103.2813676

25. Microsoft: Replay Attacks (2017). https://docs.microsoft.com/en-us/dotnet/framework/wcf/feature-details/replay-attacks. Assessed May 2018

26. Microsoft.com: Control Flow Guard (2013). https://courses.cs.washington.edu/courses/cse484/14au/reading/25-years-vulnerabilities.pdf. Accessed 29 Mar 2018

27. Mohan, V., Larsen, P., Brunthaler, S., Hamlen, K.W., Franz, M.: Opaque control flow integrity. In: 22nd Annual Network and Distributed System Security Symposium, NDSS 2015, San Diego, California, USA, 8–11 February 2015 (2015). https://www.ndss-symposium.org/ndss2015/opaque-control-flow-integrity

28. Muench, M., Pagani, F., Shoshitaishvili, Y., Kruegel, C., Vigna, G., Balzarotti, D.: Taming transactions: towards hardware-assisted control flow integrity using transactional memory. In: Monrose, F., Dacier, M., Blanc, G., Garcia-Alfaro, J. (eds.) RAID 2016. LNCS, vol. 9854, pp. 24–48. Springer, Cham (2016). https://doi.org/10.1007/978-3-319-45719-2_2

29. Niu, B., Tan, G.: Per-input control-flow integrity. In: Proceedings of the 22nd ACM SIGSAC Conference on Computer and Communications Security, CCS 2015, pp. 914–926. ACM, New York (2015). https://doi.org/10.1145/2810103.2813644

30. OWASP: Code Injection (2013). https://www.owasp.org/index.php/Code_Injection. Accessed 28 Sept 2017

31. Pappas, V.: Defending Against Return-Oriented Programming (2015). https://www.cs.columbia.edu/~angelos/Papers/theses/vpappas_thesis.pdf. Accessed 21 Feb 2018

32. Payer, M.: Control-Flow Integrity: An Introduction (2016). https://nebelwelt.net/blog/20160913-ControlFlowIntegrity.html. Accessed 21 April 2018

33. Pomonis, M., Petsios, T., Keromytis, A.D., Polychronakis, M., Kemerlis, V.P.: kr^x: comprehensive kernel protection against just-in-time code reuse. In: EuroSys, pp. 420–436. ACM (2017)

34. Shellblade.net: Performing a ret2libc Attack (2018). https://www.shellblade.net/docs/ret2libc.pdf. Accessed 25 May 2017

35. Tice, C., et al.: Enforcing forward-edge control-flow integrity in GCC & LLVM. In: Proceedings of the 23rd USENIX Conference on Security Symposium, SEC 2014, pp. 941–955. USENIX Association, Berkeley (2014). http://dl.acm.org/citation.cfm?id=2671225.2671285

36. van der Veen, V., et al.: Practical context-sensitive CFI. In: Proceedings of the 22nd ACM SIGSAC Conference on Computer and Communications Security, CCS 2015, pp. 927–940. ACM, New York (2015). https://doi.org/10.1145/2810103.2813673

37. Zhang, C., et al.: Practical control flow integrity and randomization for binary executables. In: Proceedings of the 2013 IEEE Symposium on Security and Privacy, SP 2013, pp. 559–573. IEEE Computer Society, Washington, DC (2013). https://doi.org/10.1109/SP.2013.44

38. Zhang, M., Sekar, R.: Control flow integrity for cots binaries. In: Proceedings of the 22nd USENIX Conference on Security, SEC 2013, pp. 337–352. USENIX Association, Berkeley (2013). http://dl.acm.org/citation.cfm?id=2534766.2534796

Automatically Proving Purpose Limitation in Software Architectures

Kai Bavendiek[1]([envelope]), Tobias Mueller[2], Florian Wittner[3], Thea Schwaneberg[4], Christian-Alexander Behrendt[4], Wolfgang Schulz[3], Hannes Federrath[2], and Sibylle Schupp[1]

[1] Hamburg University of Technology (TUHH), Hamburg, Germany
`kai.bavendiek@tuhh.de`
[2] University of Hamburg (UHH), Hamburg, Germany
[3] Hans-Bredow-Institut for Media Research (HBI), Hamburg, Germany
[4] University Medical Center Hamburg-Eppendorf (UKE), Hamburg, Germany

Abstract. The principle of purpose limitation is one of the corner stones in the European General Data Protection Regulation. Automatically verifying whether a software architecture is capable of collecting, storing, or otherwise processing data without a predefined, precise, and valid purpose, and more importantly, whether the software architecture allows for re-purposing the data, greatly helps designers, makers, auditors, and customers of software. In our case study, we model the architecture of an existing medical register that follows a rigid Privacy by Design approach and assess its capability to process data only for the defined purposes. We demonstrate the process by verifying one instance that satisfies purpose limitation and two that are at least critical cases. We detect a violation scenario where data belonging to a purpose-specific consent are passed on for a different and maybe even incompatible purpose.

Keywords: Medical register · GDPR · Purpose limitation · Compliance · Data protection · Privacy verification · Software architectures

1   Introduction

Purpose limitation is a very relevant concept in the medical context where the adverse effects of misused data are arguably perceived more strongly and the requirement for privacy is not necessarily concerned with data ownership but much more with access and use of data [20]. The loss of confidentiality arises when the entity holding the patient's confidence conveys private information to another, unauthorised party. While it is hard to find documentations of medical studies that breached that confidence, "novel protocols for achieving confidentiality and security are urgently needed by the data mining community" [6]. Correspondingly, the GDPR recognises health data as particularly sensitive and in need of special protection by including it in Art. 9. On the other hand, medical

G. Dhillon et al. (Eds.): SEC 2019, IFIP AICT 562, pp. 345–358, 2019.
https://doi.org/10.1007/978-3-030-22312-0_24

registers are socially and politically accepted and must give access to researchers to serve their purpose.

The current practice of granting researchers access to medical data involves a complex and time-consuming process in which several boards and review committees are involved to make decisions. For example, the German Centre for Cardiovascular Research (DZHK) generously provides access to the data they have collected for several studies.[1] The process to get access to that data involves at least eight parties and has a lower bound of 10 weeks processing time, some of which is spent on determining whether privacy-related preconditions are met. A more automated process arguably increases the trust placed in the software system by developers, authorities, and users. Additionally, operators of a (medical) data collection platform will appreciate a proof of the system in which it is impossible to illegitimately collect or process data. Examples of the types of data collected by a medical platform to facilitate machine-assisted gathering of insights are age, sex, emergency department visits, office visits, comorbidity index, dyslipidemia, hypertension, cardiovascular disease, retinopathy, or end-stage renal disease [7]. Such data is typically collected for a broad and very general purpose in order to enable research to find associations of data that were not considered relevant in that context. The GDPR provides an exception[2] for collection of medical data without a specific purpose in such registers. However, participants of such a study might appreciate a proof of the limitation of the purposes of the data they provide.

CAPVerDE is a tool for designing architectures with privacy properties and generating above mentioned proofs. It uses formal methods to verify a given statement, such as "the register has access to personal data." We use CAPVerDE for modelling the architecture of GermanVasc, a real-world medical register for symptomatic peripheral arterial diseases [5]. GermanVasc follows a Privacy by Design approach. It encrypts the data as early as possible, implements access control through encryption [19], and limits the access to data to the minimum. We show how GermanVasc limits exposure of data based on the purpose researchers ask for obtaining access to the data. We demonstrate our privacy verification in the context of GermanVasc and show that it is practical.

More succinctly, the contributions of this papers are:

- a syntax and method for automatically verifying purpose limitation in the context of a software architecture,
- a software architecture for a real-world medical register which follows a Privacy by Design approach,
- and a case study in the context of a medical register demonstrating that automatically proving correct purpose limitation is feasible and practical.

This paper is structured as follows: In Sect. 2 we give a brief legal background on the GDPR, before we present related work in Sect. 3. Section 4 describes a medical register and Sect. 5 a privacy verification logic. In Sect. 6 we present the

---

[1] Cf. https://dzhk.de/en/research/clinical-research/use-and-access/.

[2] More on purpose limitation in the GDPR can be found in Sect. 2.

case study which we verify for purpose limitation violation in Sect. 7. We discuss purpose limitation as well as our verification in Sect. 8 and conclude the paper in Sect. 9.

2   Legal Prerequisites

The concept of the European General Data Protection Regulation (GDPR) demands every act of processing – be it collecting, storing, accessing, transmitting etc. – concerning personal data to be based on a legal ground in order to be lawful and legitimate. While Art. 6 (1) offers six different legal grounds, Art. 9 essentially limits it down to explicit consent given by the affected person (or "data subject") where especially sensitive data like genetic or biometric data or data concerning a person's health are concerned. Consent, like any other legal ground, then permits acts of processing for one or more specific purposes, usually set down in a privacy statement given to the data subject beforehand. The so-called principle of purpose limitation according to Art. 5 (1) (b) – "one of the stable bedrock principles in European data protection law" [11] – argues that any further acts of processing are generally only covered by this initial legal ground as far as they are necessary for the purpose(s) initially laid down. While this could potentially be bypassed by formulating a very broad initial purpose that could cover most future and unforeseen processing acts, the ideals of informed consent according to Art. 4 No. 11 and a specific and explicit purpose according to Art. 5 (1) (b) set strict boundaries.

Further processing for new and different purposes would consequently need a new legal basis, e.g. a new and adapted declaration of consent. However, exemptions from this strict limitation are possible. Art. 5 (1) (b), 6 (4) (a) state that an initial legal basis also covers acts of processing for those new purposes that are compatible with the initial purpose(s). Usually a complex matter of case-to-case consideration, the question of compatibleness in some cases, for example regarding acts of processing for scientific research and statistical purposes, is answered by the GDPR itself: according to Art. 5, they should "not be considered to be incompatible with the initial purposes." This means that, the implementation of appropriate technical and organisational measures pursuant to Art. 89 (1), e.g. pseudonymisation, provided, no new consent (or other legal ground) has to be sought in these cases [21]. It is, however, important to note that this statutory privilege only covers those acts of processing for statistical purposes where the statistical results themselves are not personalised and are not used for future measures or decisions against specific persons, cf. recital 162.

3   Related Work

The need for preserving "privacy and security of human data" in the context of mining data from medical registers has already been acknowledged in 2002 [8]. It has also been subject of discussion in the context of genetic data which has the potential that future technologies can reveal much more about a patient than current technologies [16]. Consequently, storing medical data in a secure way has been subject of research [1,9,13]. While some of these works describe potential solutions, they did not actually implement any of them, let alone run an actual medical register with these techniques or cater for their particular needs with giving access to yet unknown researchers. Neither do they discuss measures for limiting the purpose of acquired data or provide automatic proofs.

The importance of designing software in such a way that it respects Privacy by Design principles has been acknowledged several times [12,14,17]. In this work, we present a tool that can prove a system's capability to restrict the use of data for other purposes than they were collected for.

While to the best of our knowledge approaches of verifying purpose limitation in software architectures do not exist, previous work also used purpose hierarchies to formalize privacy. Fischer-Hübner and Ott already formalized purpose binding and necessity of data processing based on an access-control scheme [10]. Their approach considered different object classes, tasks, and purposes to determine the legality of data processing in the context of an enterprise. Karjoth, Schunter and Waidner used P3P[3]-like policies for an enterprise privacy management that is also based on access-control [15]. Their paper makes use of a purpose hierarchy to determine valid sub-purposes. Ashley et al. proposed the Enterprise Privacy Authorization Language (EPAL), which is a formal enterprise privacy policy language that also considers data types, hierarchies, and purposes. EPAL uses an XML-based syntax that aims at formalizing policies to achieve a machine-readable format. The above approaches all differ from the proposal of this paper because of their focus on enterprise-internal policies, while our goal is to describe and evaluate an "ecosystem" of potentially multiple actors (inter-enterprise) [2]. While the mentioned papers go in more detail about the different types of data processing, our description is more on a high-level without explicit details. This paper not only proposes a formal language but also presents a verification logic and a tool.

4 GermanVasc

A medical register following the Privacy by Design approach is GermanVasc [5]. It collects data about patients suffering from vascular diseases with the aim to assess the quality of treatments. To that end, the platform collects data about patients and their medical history. To increase quantity and diversity data it is designed to cater for over 50 study centres with up to 500 000 patients. This decentralised and distributed collection of data creates a challenge for the register [4].

In order to implement Privacy by Design the personal data of the patients are encrypted in a way that only allows the collecting entity, i.e. the hospital, to access the data. In particular, neither other hospitals nor the register itself have access to the key required for accessing the encrypted content. Similarly, medical data is stored encrypted with the encryption key stored on a separate

---

[3] P3P is an outdated website policy protocol with B2C focus.

Table 1. Reduced syntax of architecture language

A	$::= \{R\}$	
R	$::= Has_i(\tilde{X})$	$\| Rec_{i,j}(\mathcal{P}, \{\tilde{X}\})$
	$\| Compute_i(\tilde{X} = T)$	$\| Dep_i(\tilde{Y}, \{\tilde{X}^1, ..., \tilde{X}^n\})$

medium, such that even a stolen database does not reveal any data, except for the metadata the database management system requires, e.g. row IDs. Taking the Privacy by Design approach further, researchers only get access to a subset of the available medical data after a request has been granted by the register. In order for the register to make a decision about the requested access it can take the purpose into the account, which the researchers used when making the request. The register then needs to decide whether the researcher can get access to the data.

5   CAPVerDE

CAPVerDE is an acronym for Computer-Aided Privacy Verification and Design Engineering and refers to a project that includes a formal description and verification framework for privacy properties in software architectures as well as a tool that automatically performs this verification. For this paper we have enhanced the CAPVerDE formal verification language by adding syntax and semantics for defining purpose limitation. In the following sections we will focus on our additions rather than the basics, which are explained in [3], and hence give only a brief description of the necessary syntax and new semantics.

5.1   Syntax

The architecture language is the formal description of software systems' architectures with a focus on data flow and information flow between different components, e.g. representing actors of the system. An architecture consists of a set of relations that represent the data flow and information flow. Table 1 shows the relevant syntax of the architecture language. Again, we refer to [3] for more details.

The relation $Has_i(\tilde{X})$ models an entry point for the datum $\tilde{X}$ into the system, e.g. a measurement via a sensor of the component C_i. The new relation $Rec_{i,j}(\mathcal{P}, \{\tilde{X}\})$ represents the transmission of data with a purpose attached, also called a purpose-receive. Component C_i receives a set of variables $\{\tilde{X}\}$ from component C_j for the explicit purpose $\mathcal{P}$. $Compute_i(\tilde{X} = T)$ represents the ability of component C_i to compute a new variable $\tilde{X}$ from a term. Apart from the relations modelling the explicit flow, there are also relations that model the implicit data and information flow of a system. The dependence relation $Dep_i(\tilde{Y}, \{\tilde{X}^1, ..., \tilde{X}^n\})$ represents the ability of component C_i to obtain variable $\tilde{Y}$ when in possession of a set of variables $\{\tilde{X}^1, ..., \tilde{X}^n\}$.

The property logic, next, is the formal language to express properties that an architecture should satisfy. While the architecture language describes what is, the property logic describes what should be. Currently, it supports four different types of properties: the data a component can access, the knowledge it can gain, the data it shares with a third party, and the data it stores. In this paper we extend the properties by a fifth type that regards purpose violation. We only describe this new property. The property $notPurp_i$ represents the fact that component C_i does not comply with its purpose limitation. That is, it violates the property to only use purpose-restricted data for a compatible purpose and to only pass said data on to components with a compatible purpose. If an architecture satisfies this property, this means a purpose limitation violation.

5.2   Semantics

The semantics are based on states of components and events and traces. Additionally, a purpose role hierarchy and the mapping of a labelling function are necessary for our extension. The inverse labelling function $\mathcal{L}^{-1}$: $String \rightarrow Var^n$, $\mathcal{P} \mapsto \{\tilde{X}\}$ maps from purposes to sets of variables. The purpose hierarchy is derived from user input and explicitly states the partial order relations of all purposes of an architecture. To track the purposes attached to variables, we introduce the purpose state $State_P$. It assigns a purpose to each variable and is defined as follows: $State_P = (Var \rightarrow Purp)$, where $Purp$ is a purpose role that is a label for sets of variables. The initial state of the purpose state is $(X : \bot \mid \forall X \in Var)$. Each variable gets assigned $\bot$ that symbolises the bottom of the purpose role hierarchy with $\mathcal{L}^{-1}(\bot) = \emptyset$.

In this paper we only show our purpose-limitation-related changes to CAP-VerDE. When receiving variables with a purpose attached, this purpose role is stored in the purpose state of the receiving component. Purposes do not get lost when variables are passed on or altered. For example, if a component computes a new variable, the intersection of all variables' purposes is the purpose of the new variable. This is the conservative approach to prevent problems like aliasing. When a variable that has a purpose attached is received without a specific purpose, the original purpose is passed on to the receiving component. When a component deletes a variable from its local storage, the corresponding purpose is reset.

The semantics of the original properties including soundness and completeness proofs can be found in the mentioned general paper [3]. Here we present the semantics of the purpose limitation property $notPurp_i$:

$$A \in S(notPurp_i) \Leftrightarrow \exists \sigma \in \mathcal{S}(A), \exists \tilde{X} \in \sigma_i^v, \tilde{X} \notin \mathcal{L}^{-1}(\sigma_i^p(\tilde{X})) \vee$$
$$(\sigma_i^o(\tilde{X}) = j \wedge \sigma_i^p(\tilde{X}) \not\sqsubseteq \sigma_j^p(\tilde{X}))$$

This expresses that an architecture A satisfies the property $notPurp_i$ iff a state σ of the architecture exists in which the component C_i has at least one variable $\tilde{X}$ that either does not match its purpose or was received via an incompatible purpose. S is the semantics function and $\mathcal{S}$ denotes the set of all possible

$$\mathbf{I}\wedge \frac{A \vdash \phi_1 \quad A \vdash \phi_2}{A \vdash \phi_1 \wedge \phi_2} \qquad \mathbf{I}\neg \frac{A \nvdash \phi}{A \vdash \neg\phi} \qquad \mathbf{P1} \frac{Rec_{i,j}(\mathcal{P}, E) \in A \quad \exists \tilde{X} \in E, \tilde{X} \notin \mathcal{L}^{-1}(\mathcal{P})}{A \vdash notPurp_i}$$

$$\mathbf{P2} \frac{Rec_{i,j}(\mathcal{P}, E) \in A \quad Rec_{k,i}(\mathcal{R}, F) \in A \quad \exists \tilde{X} \in E, \exists \tilde{Y} \in F, Dep_i(\tilde{Y}, Z), \tilde{X} \in Z \quad \mathcal{R} \not\sqsubseteq \mathcal{P}}{A \vdash notPurp_i}$$

Fig. 1. Verification rules regarding the purpose limitation property

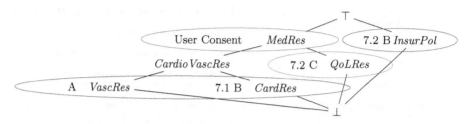

Fig. 2. Purpose hierarchy in a lattice structure

states of an architecture. σ_i^v, σ_i^p, and σ_i^o are the variable state, the purpose state and the origin state, respectively.

In the following we present verification rules that the CAPVerDE tool uses to verify properties in the context of an architecture. These rules are in the form of inference rules and a relevant excerpt is depicted in Fig. 1. Rule **P1** expresses that if component C_i receives variables with a purpose and not all the variables are covered by said purpose, the property $notPurp_i$ holds (illegitimate receiving). Rule **P2** states that if component C_i receives variables with a purpose $\mathcal{P}$ and then passes on part of these variables (or derivations) with a purpose $\mathcal{R}$ and $\mathcal{R}$ is not a sub-purpose of $\mathcal{P}$, then the property $notPurp_i$ holds (incompatible purpose). Rules **I∧** and **I¬** regard the conjunction and negation of properties, respectively.

6 Case Study

In this section we introduce a case study which regards two cases: The positive case with valid use of the data and the negative case with illegitimate use of the data. We will use CAPVerDe to check for purpose limitation violations in the next section.

Table 2. Mapping of purposes and data types as variables

Purpose	MedRes	CardioVascRes	QoLRes	VascRes	CardRes	Profiling
Variables	$\{mD\}$	$\{emD\}$	$\{depression, salary,$ $erectiledysfunction\}$	$\{bloodpressure,$ $bloodcellcount\}$	$\{bloodpressure,$ $cholestorellevel\}$	$\{pmD\}$

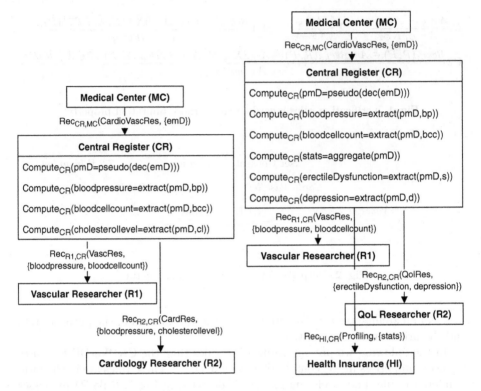

Fig. 3. Architecture for positive case   **Fig. 4.** Architecture for negative case

Our medical data register case study is based on the real-world register GermanVasc as described in Sect. 4. Our focus is on the purpose limitation aspect, therefore we omit a detailed description of the macro view and present only the relation between the register and the surrounding actors. Figure 3 shows a graphical representation of this view of the architecture following the syntax described in Sect. 5. We have merged some components of the architecture which are not relevant for our purpose limitation case study such that it leaves only four to five actors.[4]

(1) The Positive Case. A medical center has the encrypted medical data it acquired from treating its patients. Because it wants to make this data available for further research, it acquires the patients' consent pursuant to Art. 6 (1) (a), 9 (2) (a) GDPR for transmitting the encrypted data to a central register for the purpose of cardiovascular research by researchers following this purpose. Architecturally phrased this means that the medical center *MC* sends the encrypted medical data *emD* with the attached research purpose *CardioVascRes* to the

---

[4] We make the full description of the architecture available under: https://www.tuhh. de/sts/research/data-protection-machine-learning/capverde.html.

central register CR. The purpose hierarchy presented in Fig. 2 demonstrates how sub-purposes may be included in a more general purpose role. The gray area marks the entry level of consent the user has given. Table 2 shows the relation between the purpose roles and the corresponding data types, i.e. the variables that can be received. The highlighted areas will be relevant in the following section. For this example the variable emD only contains the necessary information for the two purposes of vascular research and cardiology research. The central register can decrypt and pseudonymise the medical data to obtain the pseudonymised medical data pmD. From this the data for the vascular disease researcher $R1$ and the cardiological diseases researcher $R2$ can be extracted. The researcher with the purpose of vascular diseases $VascRes$ receives the $bloodcellcount$ (abbreviated as BCC) and the $bloodpressure$ (abbreviated as BP), while the other researcher with the purpose of cardiology research $CardRes$ receives information about the $bloodpressure$ and the $cholesterollevel$ (abbreviated as CL).

In addition to the explicit relations of an architecture, the software designer has to model the dependence relations. For this architecture we have $Dep_{CR}(pmD, \{emD\})$, $Dep_{CR}(BP, \{pmD\})$, $Dep_{CR}(BCC, \{pmD\})$, and $Dep_{CR}(CL, \{pmD\})$. Additionally, the architecture includes all dependence relations deduced from transitivity, i.e. $Dep_i(\tilde{Y}, \{\tilde{X}\}) \wedge Dep_i(\tilde{Z}, \{\tilde{Y}\}) \implies Dep_i(\tilde{Z}, \{\tilde{X}\})$, and also reflexivity, i.e. $Dep_i(\tilde{X}, \{\tilde{X}\})$, for all components C_i.

(2) The Negative Case. The slightly altered architecture depicted in Fig. 4 shows a different scenario with two new actors. The second researcher is replaced by a quality of life (QoL) researcher and a health insurance actor comes into play. The central register now shares the statistical data $stats$ with the insurance company HI attached with the purpose of policy profiling $Profiling$. The medical data of the research register is used for improving the quality of the insurance predictions in order to adapt the conditions for individual insurance members depending on how the predictions compare to their characteristics. The QoL researcher receives information about $depression$ (abbreviated as D) and $erectiledysfunction$ (abbreviated as ED) about the patients with the purpose of QoL research $QoLRes$. The purpose role hierarchy in Fig. 2 stays unchanged and similarly Table 2 is still valid for the second architecture. However, the variable emD now also contains information about depression and erectile dysfunction.

For the second architecture we have the following additional dependence relations: $Dep_{CR}(ED, \{pmD\})$, $Dep_{CR}(D, \{pmD\})$, and $Dep_{CR}(stats, \{pmD\})$. Again, we assume transitivity and reflexivity.

7   Verification

In the following subsection we demonstrate how to verify whether the described architecture satisfies the purpose limitation properties. The property described in Sect. 5 checks two aspects: first, whether the purpose-receive relation itself was valid and second, whether the "purpose-chain" was valid. In the presented syntax the property that expresses that an actor has violated its purpose limitation can be expressed as $notPurp_i$. The verification is done automatically by CAPVerDE and we demonstrate the steps of this verification process by tracing the algorithm in the following sub-sections.

Dual-Researcher Example. In our first case of the architecture depicted in Fig. 3 we want to verify whether any of the three actors, namely the central register and the two researchers, has violated the principle of purpose limitation by processing the medical data for a purpose that is neither covered by nor compatible with the purpose declared in the patients' original consent. Therefore, we want to verify the property: $\neg notPurp_{CR} \wedge \neg notPurp_{R1} \wedge \neg notPurp_{R2}$.

To do this CAPVerDE pattern-matches to apply the rules presented in Fig. 1 and backtracks if necessary. The verifier tries to apply the axioms presented in Fig. 1. As the property is a conjunction of properties, it applies the conjunction rule **I∧** twice. Therefore, the verifier checks all three sub-properties in turn and only if all three hold, the conjunction property holds. As all sub-properties are negations, it applies the negation rule **I¬**. Hence, the negated properties must not hold for the whole property to hold. Starting with the first (negated) sub-property $notPurp_{CR}$ CAPVerDE tries to apply the corresponding rules **P1** and **P2**. If either one is applicable, the property holds. By looking at rule **P1**, it can pattern-match the purpose-receive with the relation $Rec_{CR,MC}(CardioVascRes, \{emD\})$. Thus, the verifier needs to check whether emD is not part of the purpose variables of $CardioVascRes$. A look at Table 2 shows that encrypted medical data are valid for this purpose. Hence, rule **P1** is not applicable. If we look at the second rule (**P2**), CAPVerDE can pattern-match the first purpose-receive, again, with the relation $Rec_{CR,MC}(CardioVascRes, \{emD\})$. There are two options for matching the second one, with either $Rec_{R1,CR}(VascRes, \{BP, BCC\})$ or $Rec_{R2,CR}(CardRes, \{BP, CL\})$. The verifier then considers the purpose-receive between the register and the vascular researcher first. In this case we have $E = \{emD\}$ and $F = \{BP, BCC\}$ as well as the purposes $\mathcal{P} = MedRes$ and $\mathcal{R} = VascRes$. CAPVerDE can deduce that $\tilde{X} = emD$. There are, in fact, the two transitive dependence relations $Dep_{CR}(BP, \{emD\})$ and $Dep_{CR}(BCC, \{emD\})$. Therefore, the verifier needs to check whether the partial order relation $VascRes \not\sqsubseteq CardioVascRes$ holds. A look at Fig. 2 shows that the latter purpose is annotated as **A**. The relation does not hold because $CardioVascRes$ is an upper bound for $VascRes$. Thus, rule **P2** does not apply for the first researcher branch. Let us now take a look at the purpose-receive between the register and the cardiology researcher. $\mathcal{R}$ changes to $CardRes$ and F changes to $\{BP, CL\}$. The dependence relation $Dep_{CR}(CL, \{emD\})$ exists, so the verifier has to check the partial order constraint $CardRes \not\sqsubseteq CardioVascRes$. Figure 2 depicts $CardRes$ as case **7.1 B**. This relation does not hold neither, because $CardioVascRes$ is also an upper bound for $CardRes$. Therefore, rule **P2** is not applicable for the second researcher branch, neither. Hence, CAPVerDE can deduce that the sub-property $notPurp_{CR}$ does not hold and that thus its negation does.

The verification of the remaining sub-properties works in a similar fashion and is, therefore, omitted. The two properties $notPurp_{R1}$ and $notPurp_{R2}$ also do not hold. As, thus, all three (negated) sub-properties of the conjunction hold, the conjunction property also does. Therefore, CAPVerDE has successfully verified the full property $\neg notPurp_{CR} \wedge \neg notPurp_{R1} \wedge \neg notPurp_{R2}$ that states that none of the three actors violate their purpose limitation.

Health Insurance Example. In our second case of the architecture, depicted in Fig. 4, the verifier proceeds in a very similar way. We want to verify that the three actors, namely the central register, the researcher, and the insurance, do not violate their respective purpose limitation property. Hence, the property $\neg notPurp_{CR} \wedge \neg notPurp_{R1} \wedge \neg notPurp_{R2} \wedge \neg notPurp_{HI}$ must hold for this.

We omit the verification of conjunctions and negations and only demonstrate the verification of the four sub-properties: $notPurp_{CR}$, $notPurp_{R1}$, $notPurp_{R2}$, and $notPurp_{HI}$. As neither the branch of the Vascular Researcher has changed nor have the purpose role hierarchy and the connected data types, we carry over the result from the previous verification. Thus, we omit this part of the verification. We also focus on the purpose-chain checks and therefore only present the verification of $notPurp_{CR}$ as we have shown the approach in the previous example. CAPVerDE checks the property for the register: $notPurp_{CR}$. The verification of the first rule is the same as in the previous example. Thus, we omit it and focus on rule **P2**. For the pattern matching of the outgoing purpose-receives we only consider the two new branches. The health insurance branch gives us $\mathcal{P} = Cardio VascRes$ and $\mathcal{R} = Profiling$ and $E = \{emD\}$ and $F = \{stats\}$. The dependence relation $Dep_{CR}(stats, \{emD\})$ shows that the verifier has to check the purpose hierarchy. *Profiling* is highlighted in Fig. 2 in red and annotated as **7.2 B**. The partial order relation *Profiling* $\not\sqsubseteq$ *Cardio VascRes* does hold because, as Fig. 2 shows, the join of the two purposes is $\top$. Therefore, rule **P2** is applicable and the property $notPurp_{CR}$ holds in this architecture. This seems to be a case of purpose limitation violation because profiling is not covered by the purpose defined in the patients' consent. As this is already a violation, the verifier would stop here. Therefore, we omit the verification of the QoL branch. In our chosen purpose hierarchy CAPVerDE also detects a violation in this branch (cf. Fig. 2 annotated as **7.2 C**).

Therefore, our tool detects this case as a violation of purpose limitation due to the way we modelled the purposes and their compatibility. While these acts of processing still serve the overall purpose of research for furthering the medical treatment of certain illnesses, the patients gave consent for the narrower purpose of cardiovascular research. Therefore, the principle of purpose limitation seems to be, again, not preserved. However, in both cases the principle of purpose limitation could still be preserved. As our model only checks if the new purpose is still covered by the initial purpose, it does not make a statement about the question of compatibility. Further processing for the purposes of quality-of-life research and profiling could, at least potentially, be covered by the assumed-compatibility clause in Art. 5 (1) (b). Although our approach does not assume such compatibility, one could argue that the "parent" of the two purposes is medical research and therefore the purpose *QoLRes* is not strictly incompatible with the initial purpose of the patient.

8 Discussion

Our contribution is a method and an extension to a tool for effectively applying a rigorous Privacy by Design approach manifested in the limitation of purpose of the collected data. As CAPVerDE is open-source software and the approach very generic, it is possible to apply the same method to other contexts. One limitation of the presented approach to automatically prove purpose limitation is the need for a correct model. If the model does not accurately reflect the data flows within the software, the verification cannot provide reliable information. Similarly, while the verifier can automatically answer questions, it is essential that the correct questions be asked. We have suggested a set of classifications and formulae for the specific domain of a medical register. This set is, obviously, not directly usable for other contexts but we envision that it can serve as a template for other domains.

Also we have to discuss the obvious trade-off between innovation and purpose limitation. This discussion has been had many times regarding the conflict between purpose limitation and big data analytics in general. While the GDPR does offer some relief in this regard with the compatibility clauses in Art. 5 (1) (b) it remains to be seen if this is enough to allow sufficient innovation. One reason to doubt that stems from the fact that due to several flexibility clauses like Art. 89 (2) regarding processing for (medical) research and statistical purposes, the details are up to the member states and their respective national laws. This fragmentation makes innovation difficult and contradicts the purpose of unification otherwise pursued by the GDPR [18].

From the technical aspect, our presented approach considers an inflexible purpose role hierarchy that prevents, for instance, new associations in the medical field that would attract further research. If we wanted to allow a new association, the purpose roles have to be re-evaluated and updated frequently. Here, the question of compatibility would need to be checked on two levels: firstly, regarding the pre-formulated assumptions of compatibility in Art. 5 (1) (b). And secondly, where none of the explicit purposes described there apply, a freehand check taking into account and balancing the legitimate interests of both controller and data subject(s) as stated by Art. 6 (4). While this would arguably be an ambitious goal for an approach like ours, all parties could benefit from it.

The legal analyses performed in this paper are not thorough and we make no claims about the fitness of our method in court. We refer to other papers for further discussion [4,5]. While we believe that designing and analysing software systems with our method provides value to the stakeholders, we do not and cannot make any further statements.

9   Conclusion and Future Work

This paper presents a method for automatically verifying purpose limitation in the context of software architectures to comply with the principle of Privacy by Design. The paper also describes a real-world example of a privacy-aware medical register. We use this example as a case study to demonstrate our proposed method. We model the architecture of said register and verify formal purpose limitation properties in two example instances. The paper presents a brief legal discourse on purpose limitation in the context of GDPR and its value in practice. We have shown how Privacy by Design can be implemented with regards to purpose limitation and hope to inspire the modelling of future applications.

Art. 5 of the GDPR names seven "principles relating to processing of personal data." Art. 25 (1), laying down the ideal of privacy by design within the GDPR, accordingly obliges controllers to "implement appropriate technical and organizational measures" to further these principles. The extent of this obligation is subject to, inter alia, the "state of the art." This reflects the GDPR's general ideal of openness to development; an ideal that leaves controllers quite a bit of leeway regarding the "how" of the fulfilment of their obligations and provides them with the possibility of developing solutions and tools that prevail themselves as industry-wide standards. Until enough of these standards have been established, though, the obligation's vagueness and openness lacks the clarification and guidance needed for most controllers. We hope that our tool can contribute to the concretisation of this notion. In this paper we have focussed on purpose limitation. The different principles are closely related. Data minimisation and storage limitation are two principles that can, to a certain extent, already be expressed with the proposed tool. The accountability mentioned in Art. 5 (2) is a new aspect worth looking into. A tool-assisted approach to demonstrate compliance is an interesting field for future work. Similarly, an automated process for the "data protection impact assessment" described in Art. 35 is a promising direction for further research. Finally, it is worthwhile exploring to what extent the management of purposes can be automated.

Acknowledgement. The work is part of the Information Governance Technologies project which is funded by the Behörde für Wissenschaft, Forschung und Gleichstellung.

The IDOMENEO study is funded by the German Joint Federal Committee (Gemeinsamer Bundesausschuss, G-BA) (01VSF16008) and by the German Stifterverband as well as by the CORONA foundation (S199/10061/2015).

References

1. Akinyele, J.A., Lehmann, C.U., Green, M.D., Pagano, M.W., Peterson, Z.N.J., Rubin, A.D.: Self-protecting electronic medical records using attribute-based encryption. Technical report 565, November 2010
2. Ashley, P., Hada, S., Karjoth, G., Powers, C., Schunter, M.: Enterprise privacy authorization language (EPAL). Research report. IBM Research (2003)

3. Bavendiek, K., Adams, R., Schupp, S.: Privacy-preserving architectures with probabilistic guaranties. In: Proceedings of the 16th International Conference on Privacy, Security and Trust, pp. 1–10. IEEE, August 2018

4. Behrendt, C.A., Ir, A.J., Debus, E.S., Kolh, P.: The challenge of data privacy compliant registry based research. Eur. J. Vascul. Endovasc. Surg. **55**(5), 601–602 (2018)

5. Behrendt, C.A., Pridöhl, H., Schaar, K., Federrath, H., Debus, E.S.: Klinische Register im 21. Jahrhundert. Der Chirurg **88**(11), 944–949 (2017)

6. Berman, J.J.: Confidentiality issues for medical data miners. Artif. Intell. Med. **26**(1), 25–36 (2002)

7. Breault, J.L., Goodall, C.R., Fos, P.J.: Data mining a diabetic data warehouse. Artif. Intell. Med. **26**(1), 37–54 (2002)

8. Cios, K.J., William Moore, G.: Uniqueness of medical data mining. Artif. Intell. Med. **26**(1), 1–24 (2002)

9. Drosatos, G., Efraimidis, P.S., Williams, G., Kaldoudi, E.: Towards privacy by design in personal e-Health systems. In: Proceedings of the 9th International Joint Conference on Biomedical Engineering Systems and Technologies, Rome, vol. 5, pp. 472–477, February 2016

10. Fischer-Hübner, S., Ott, A.: From a formal privacy model to its implementation. In: Proceedings of the 21st National Information Systems Security Conference (1998)

11. Forgó, N., Hänold, S., Schütze, B.: The principle of purpose limitation and big data. In: Corrales, M., Fenwick, M., Forgó, N. (eds.) New Technology, Big Data and the Law. PLBI, pp. 17–42. Springer, Singapore (2017). https://doi.org/10.1007/978-981-10-5038-1_2

12. Graf, C., Wolkerstorfer, P., Geven, A., Tscheligi, M.: A pattern collection for privacy enhancing technology, January 2010

13. Haas, S., Wohlgemuth, S., Echizen, I., Sonehara, N., Müller, G.: Aspects of privacy for electronic health records. Int. J. Med. Inform. **80**(2), e26–e31 (2011)

14. Hafiz, M.: A pattern language for developing privacy enhancing technologies. Softw.: Pract. Exp. **43**(7), 769–787 (2013)

15. Karjoth, G., Schunter, M., Waidner, M.: Platform for enterprise privacy practices: privacy-enabled management of customer data. In: Dingledine, R., Syverson, P. (eds.) PET 2002. LNCS, vol. 2482, pp. 69–84. Springer, Heidelberg (2003). https://doi.org/10.1007/3-540-36467-6_6

16. Kaye, J., Boddington, P., de Vries, J., Hawkins, N., Melham, K.: Ethical implications of the use of whole genome methods in medical research. Eur. J. Hum. Genet. **18**(4), 398–403 (2009)

17. Kung, A.: PEARs: privacy enhancing architectures. In: Preneel, B., Ikonomou, D. (eds.) APF 2014. LNCS, vol. 8450, pp. 18–29. Springer, Cham (2014). https://doi.org/10.1007/978-3-319-06749-0_2

18. Mayer-Schönberger, V., Padova, Y.: Regime change? Enabling big data through Europe's new data protection regulation. Columbia Sci. Technol. Law Rev. **17**(315), 21 (2016)

19. Pilyankevich, E., Korchagin, I., Mnatsakanov, A.: Hermes. A framework for cryptographically assured access control and data security. Technical report 200, February 2018

20. Safran, C., et al.: Toward a national framework for the secondary use of health data: an American medical informatics association white paper. J. Am. Med. Inf. Assoc. **14**(1), 1–9 (2007)

21. Schulz: DS-GVO Art. 6 Rechtmäßigkeit der Verarbeitung. Gola, p. 210 (2018)

Commit Signatures for Centralized Version Control Systems

Sangat Vaidya[1], Santiago Torres-Arias[2], Reza Curtmola[1(✉)],
and Justin Cappos[2]

[1] New Jersey Institute of Technology, Newark, NJ, USA
reza.curtmola@njit.edu
[2] Tandon School of Engineering, New York University, New York, NY, USA

Abstract. Version Control Systems (VCS-es) play a major role in the software development life cycle, yet historically their security has been relatively underdeveloped compared to their importance. Recent history has shown that source code repositories represent appealing attack targets. Attacks that violate the integrity of repository data can impact negatively millions of users. Some VCS-es, such as Git, employ *commit signatures* as a mechanism to provide developers with cryptographic protections for the code they contribute to a repository. However, an entire class of other VCS-es, including the well-known Apache Subversion (SVN), lacks such protections.

We design the first commit signing mechanism for centralized version control systems, which supports features such as working with a subset of the repository and allowing clients to work on disjoint sets of files without having to retrieve each other's changes. We implement a prototype for the proposed commit signing mechanism on top of the SVN codebase and show experimentally that it only incurs a modest overhead. With our solution in place, the VCS security model is substantially improved.

Keywords: SVN · Commit signature · Version control system

1 Introduction

A Version Control System (VCS) plays an important part in any software development project. The VCS facilitates the development and maintenance process by allowing multiple contributors to collaborate in writing and modifying the source code. The VCS also maintains a history of the software development in a source code repository, thus providing the ability to rollback to earlier versions when needed. Some well-known VCS-es include Git [10], Subversion [2], Mercurial [18], and CVS [7].

The full version of this paper is available as a technical report [34].

© IFIP International Federation for Information Processing 2019
Published by Springer Nature Switzerland AG 2019
G. Dhillon et al. (Eds.): SEC 2019, IFIP AICT 562, pp. 359–373, 2019.
https://doi.org/10.1007/978-3-030-22312-0_25

Source code repositories represent appealing attack targets. Attackers that break into repositories can violate their integrity, both when the repository is hosted independently, such as internal to an enterprise, or when the repository is hosted at a specialized provider, such as GitHub [12], GitLab [13], or Sourceforge [20]. The attack surface is even larger when the hosting provider relies on the services of a third party for storing the repository, such as a cloud storage provider like Amazon or Google. Integrity violation attacks can introduce vulnerabilities by adding or removing some part of the codebase. In turn, such malicious activity can have a devastating impact, as it affects millions of users that retrieve data from the compromised repositories. In recent years, these types of attacks have been on the rise [16], and have affected most types of repositories, including Git [1,4,17,32], Subversion [5,6], Perforce [15], and CVS [26].

To ensure the integrity and authenticity of externally-hosted repositories, some VCS-es such as Git and Mercurial employ a mechanism called *commit signatures*, by which developers can use digital signatures to protect the code they contribute to a repository. Perhaps surprisingly, several other VCS-es, such as Apache Subversion [2] (known as SVN), lack this ability and are vulnerable to attacks that manipulate files on a remote repository in an undetectable fashion.

Contributions. In this work, we design and implement a commit signing mechanism for centralized version control systems that rely on a client-server architecture. Our solution is the first that supports VCS features such as working with a portion of the repository on the client side and allowing clients to work on disjoint sets of files without having to retrieve each other's changes. During a commit, clients compute the commit signature over the root of a Merkle Hash Tree (MHT) built on top of the repository. A client obtains from the server an efficient proof that covers the portions of the repository that are not stored locally, and uses it in conjunction with data stored locally to compute the commit signature. During an update, a client retrieves a revision's data from the central repository, together with the commit signature over that revision and a proof that attests to the integrity and authenticity of the retrieved data. To minimize the performance footprint of the commit signing mechanism, the proofs about non-local data contain siblings of nodes in the Steiner tree determined by items in the commit/update changeset.

When our commit signing protocol is in place, repository integrity and authenticity can be guaranteed even when the server hosting the repository is not trustworthy. We make the following contributions:

- We examine Apache SVN, a representative centralized version control system, and identify a range of attacks that stem from the lack of integrity mechanisms for the repository.
- We identify fundamental architectural and functional differences between centralized and decentralized VCS-es. Decentralized VCS-es like Git replicate the entire repository at the client side and eliminate the need to interact with the server when performing commits. Moreover, they do not support partial checkouts and require clients to retrieve other clients' changes before

committing their own changes. These differences introduce security and performance challenges that prevent us from applying to centralized VCS-es a commit signing solution such as the one used in Git.

- We design the first commit signing mechanism for centralized VCS-es that rely on a client-server architecture and support features such as working with a subset of the repository and allowing clients to work on disjoint sets of files without having to retrieve each other's changes. Our solution substantially improves the security model of such version control systems. We describe a solution for SVN, but our techniques are applicable to other VCS-es that fit this model, such as GNU Bazaar [3], Perforce Helix Core [19], Surround SCM [22], StarTeam [21], and Vault [27].
- We implement SSVN, a prototype for the proposed commit signature mechanism on top of the SVN codebase. We perform an extensive experimental evaluation based on three representative SVN repositories (FileZilla, SVN, GCC) and show that SSVN is efficient and incurs only a modest overhead compared to a regular (insecure) SVN system.

2   Background

This section provides background on version control systems (VCS-es) that have a (centralized) client-server architecture [3, 19, 21, 22, 27] and on (non-standard) Merkle Hash Trees, which will be used in subsequent sections. We overview the main protocols of such VCS-es, commit and update, which have been designed for a benign setting (*i.e.*, the VCS server is assumed to be fully trusted). Our description is focused on Apache SVN [2], an open source VCS that is representative for this class of VCS-es.

2.1   Centralized Version Control Systems

In a centralized VCS, the VCS server stores the main repository for a project and multiple clients collaborate on the project. The main (central) repository contains all the revisions since the project was created, whereas each client stores in its local repository only one revision, referred to as a *base revision*. The clients make changes to their local repositories and then publish these changes in the central repository on the server for others to see these changes.

Project management involves two components: the main repository on the server side and a *local working copy (LWC)* on the client side. The LWC contains a *base revision* for files retrieved by the client from the main repository, plus any changes the client makes on top of the base revision. A client can publish the changes from her LWC to the main repository by using the "commit" command. As a result, the server creates a new revision which incorporates these changes into the main repository. If a client wants to update her LWC with the changes made by other clients, she uses the "update" command. The codebase revisions are referred to by a unique identifier called a *revision number*. In SVN, this is

PROTOCOL: Commit	PROTOCOL: Update
1: **for** (each file F in the commit changeset) **do**	1: $C \rightarrow S : i$ // C informs S that it wants to retrieve revision i
2: $C \rightarrow S : \delta$ // Client computes and sends δ, such that $F_i = F_{i-1} + \delta$	2: **for** (each file F in the update set) **do**
3: S computes F_{i-1} based on the data in the repository (*i.e.*, start from F_0 and apply skip deltas)	3: $C \rightarrow S : j$ // C sends to S it local revision number for F
4: S computes $F_i = F_{i-1} + \delta$	4: S computes F_j and F_i based on the data in the repository (*i.e.*, start from F_0 and apply skip deltas)
5: S computes $F_{skip(i)}$ based on the data in the repository (*i.e.*, start from F_0 and apply skip deltas)	5: S computes δ such that $F_i = F_j + \delta$
	6: $S \rightarrow C : \delta$
6: S computes Δ_i such that $F_i = F_{skip(i)} + \Delta_i$ and stores Δ_i	7: C computes F_i as $F_i = F_j + \delta$ and stores F_i in its local repository

an integer number that has value 1 initially and is incremented by 1 every time a client commits changes to the repository.

The server stores revisions using *skip delta encoding*, in which only the first revision is stored in its entirety and each subsequent revision is stored as the difference (*i.e.*, delta) relative to an earlier revision [24].

Notation: The VCS (main) repository contains i revisions, and we assume without loss of generality that for every file F there are i revisions which are stored as $F_0, \Delta_1, \Delta_2, \ldots, \Delta_{i-1}$. F_0 is the initial version of the file, and the $i - 1$ delta files are based on skip delta encoding.

We use F_i to denote revision i of the file. We use $F_{skip(i)}$ to denote the skip version for F_i (*i.e.*, the base revision relative to which Δ_i is computed). We write $F_i = F_j + \delta$ to denote that F_i is obtained by applying δ to F_j. Also, we use $C \rightarrow S : M$ to denote that client C sends a message M to the server S.

Commit Protocol: The client C's local working copy contains changes made over a base revision that was previously retrieved from the server S. We refer to the changes that the client wants to commit as the *commit changeset*. Note that changes can only be committed for files for which the client has the latest revision from the server (*i.e.*, i − 1). Otherwise, the client is prompted to first retrieve the latest revision for all the files in the changeset. After C commits the changes, the latest revision at S will become i. After executing the steps described in the **Commit** protocol, the server sends the revision number i to the client, and the client sets i as the revision number for all the files in the commit changeset.

Update Protocol: The client wants to retrieve revision i for a set of files in the repository, referred to as the *update set*. After finalizing the update, the client sets i as the revision number for all the files in the update set.

2.2   Merkle Hash Trees

A Merkle Hash Tree (MHT) [31] is an authenticated data structure used to prove set membership efficiently. An MHT follows a tree data structure, in which every leaf node is a hash of data associated with that leaf. The nodes are concatenated and hashed using a collision-resistant hash function to create a parent node, until the root of the tree is reached. Typically, a standard MHT is a full binary tree. Given the MHT for a set of elements, one can prove efficiently that an element belongs to this set, based on a proof that contains the root node (authenticated using a digital signature) and the siblings of all the nodes on the path between the node to be verified and the root node.

In this work, we will work with sets of files and directories. As a result, we will use non-standard MHTs, which are different than standard MHTs in two aspects: (1) the tree is not necessarily binary (*i.e.*, internal nodes have branching factors larger than two), and (2) the tree may not be full, with leaf nodes having different depths. An internal node is obtained by hashing a concatenation of its children nodes, ordered lexicographically. This ensures that for a given repository, a unique MHT is obtained. We will use MHTs to provide proof that a file or a set of files and directories belongs to the repository in a particular revision.

3   Can Git Commit Signing Be Used?

In this section, we review the commit signing mechanism used in Git [10] and then identify several fundamental differences between centralized and distributed VCS-es that prevent us from using the same solution used to sign commits in Git. Git is a popular decentralized VCS, which stores the contents of the repository in form of objects. When the client commits to the repository, Git creates a *commit object* that is a snapshot of the entire repository at that moment, obtained as the root of an MHT computed over the repository. This commit object is digitally signed by the client, thus ensuring its integrity and authenticity.

We have identified several fundamental differences between Git and SVN in their workflow, functionality, and architecture. These differences make it challenging to apply the same commit signing solution used in Git to centralized VCS-es such as SVN.

Non-interactive vs. Interactive Commits: One important difference is that Git allows clients to perform commits without interacting with the server that hosts the main repository, whereas in SVN clients must interact with the server. A few architectural and functional differences dictate this behavior:

- *Working with a subset of the repository:* Git relies on a distributed model, in which the entire repository (*i.e.*, all files and directories) for a given revision is mirrored on the client side. As opposed to that, SVN uses a centralized model, in which clients store locally a single revision, but have the ability to retrieve only a portion of a remote repository for that revision (i.e., they can retrieve only one directory, or a subset of all the directories). This feature can be useful for very large repositories, when the client only wants to work on a small subset of the repository.

  In such cases, SVN clients do not have a global view of the entire repository and cannot use a Git-like strategy for commit signatures, which requires information about the entire repository. Instead, SVN clients must rely on the server to get a global view of the repository which raises security concerns if the server is not trustworthy and may also incur a significant amount of data transfer over the network.

- *Commit identifier:* SVN and Git use fundamentally different methods to identify a commit. Git uses a unique identifier that is computed by the client solely based on the data in that revision. This identifier is the hash of the commit object, and can be computed by the client based on the data in its local working copy, and without the involvement of the server that hosts the main remote repository. However, in SVN, the revision identifier is an integer which is chosen by the server, and which does not depend on the data in that revision. To perform a commit, the client sends the changes to the server, who then decides the revision number and sends it back to the client. Thus, a Git-like commit signature mechanism cannot be used in SVN, because clients do not have the ability to decide independently the revision identifier. This raises security concerns when the server is not trustworthy.

Working with Mutually Exclusive Sets of Files: SVN allows clients to perform commits on mutually exclusive sets of files without having to update their local working copies. For example, client A modifies a file $F1$ in directory $D1$ and client B modifies a file in another directory $D2$ of the same repository. When A wants to commit additional changes to $F1$, A does not have to update its local copy with the changes made by B. Git clients do not have this ability, as they need the most up-to-date version for the entire repository before pushing commits to the main repository (*i.e.*, they need to retrieve all changes made anywhere in the repository before pushing changes). This ensures that a Git client has updated metadata about the entire repository before pushing changes. As opposed to that, SVN clients may not have the most up-to-date information for some of the files. Thus, SVN clients cannot generate or verify commit signatures in the same way as Git does, and may be tricked into signing incorrect data.

Repository Structure: SVN stores revisions of a file based on the skip delta encoding mechanism, in which a revision is stored as the difference from a previous revision. Thus, to obtain a revision for a file, the server has to start from the first revision and apply a series of deltas. On the other hand, Git stores the entire content for all versions of all files. This difference in repository structure complicates the SVN client's ability to compute and verify commit signatures. For example, a naive solution in which the client signs only the delta difference between revisions may be inefficient and insecure.

4   Adversarial Model and Security Guarantees

We assume that the server hosting the central repository is not trusted to preserve the integrity of the repository. For example, it may tamper with the repository in order to remove code (*e.g.*, a security patch) or to introduce malicious code (*e.g.*, a backdoor). This captures a setting in which the server is either compromised or is malicious. It also captures a setting in which the VCS server relies on the services of a third party for storing the repository, such as a cloud storage provider which may itself be malicious or may be victim of a compromise. Existing centralized VCS-es offer no protection against such attacks.

In addition to tampering with data at rest (*i.e.*, the repository), a compromised or malicious server may choose to not follow correctly the VCS protocols, as long as such actions will not incriminate the server. For example, since commit is an interactive protocol, the server may present incorrect information to clients during a commit, which may trick clients into committing incorrect data.

When a mechanism such as commit signing is available, we assume that clients are trusted to sign their commits. In this case, we also assume that attackers cannot get hold of client cryptographic keys. The integrity of commits that are not signed cannot be guaranteed.

4.1 Attacks

When the VCS employs no mechanisms to ensure repository integrity, the data in the repository is subject to a wide range of attacks as attackers can arbitrarily tamper with data. In this section, we describe a few concrete attacks that violate the integrity and authenticity of the data in the repository. This list is not meant to be comprehensive, but to suggest desirable defense goals.

Tampering Attack. The attacker can arbitrarily tamper with the repository data, such as modifying file contents, adding a file to a revision, or deleting a file from a revision. Such actions may lead to serious security integrity violations, such as the removal of a security patch or the insertion of a backdoor, which can have disastrous consequences. A defense should protect against direct modification of the files. An attacker may also try to delete a historical revision entirely, for example to hide past activity. A defense should link together consecutive revisions, such that any tampering with the sequence of revision is detected.

Impersonation Attack. The attacker can tamper with the author field of a committed revision. This will make it look like developers committed code they never actually did, which can potentially damage their reputation. Thus, a defense should protect the author field from tampering.

Mix and Match Attack. A revision reflects the state of the repository at the moment when the revision is committed. That is, the revision refers to the version of the files and directories at the moment when the commit is performed. However, the various versions of files in the repository are not securely bound to the revision they belong to. When the server is asked to deliver a particular revision, it can send versions of the files that belong to different revisions. A defense should securely bind together the files versions that belong to a revision, and should also bind them to the revision identifier.

4.2 Security Guarantees

SG1: Ensure accurate commits. Commits performed by clients should be accurately reflected in the repository (*i.e.*, as if the server followed the commit protocol faithfully). After each commit, the repository should be in a state that reflects the client's actions. This protects against attacks in which the server does not follow the protocol and provides incorrect information to clients during a commit.

SG2: Integrity and authenticity of committed data. An attacker should not be able to modify data that has been committed to the repository without being detected. This ensures the integrity and authenticity of both individual commits and the sequence of commits. This also ensures accurate updates, *i.e.*, an attacker is not able to present incorrect information to clients that are retrieving data from the repository without being detected.

SG3: Non-repudiation of committed data. Clients that performed a commit operation should not be able to deny having performed that commit.

5   Commit Signatures for Centralized VCS-es

We now present our design for enabling commit signatures by enhancing the standard **Commit** and **Update** protocols. We use the following notation, in addition to what we defined in Sect. 2.1. $CSIG_i$ denotes the client's commit signature over revision i, and $MHTROOT_i$ denotes the root of the Merkle hash tree built on top of revision i. We use $Sign$ and $Verify$ to denote the signing and verification algorithms of a standard digital signature scheme. To simplify the notation, we will omit the keys, but $Sign$ and $Verify$ use the private and public keys of the client who committed the revision. Due to space limitations, the security analysis of these protocols is included in the full version of the paper.

Secure Commit Protocol. We now present the **Secure_Commit** protocol. The client has a commit changeset with changes on top of revision $i - 1$, and wants to commit revision i. The client needs to compute the commit signature over revision i of the entire repository. However, the client's local working copy may only contain a subset of the entire repository (e.g., only the files that are part of the commit changeset). Thus, in order to compute the commit signature, the client needs additional information from the server about the files in the repository that are not in its local working copy. The server will provide this additional information in the form of a proof relative the client's changeset (line

---

PROTOCOL: Secure_Commit

1: // Steps 1-6 are the same as in the standard **Commit** protocol
7: S computes proof P_{i-1} // S uses revision $i - 1$ if the repository to compute a proof relative to the client's commit changeset
8: $S \to C : i, P_{i-1}, CSIG_{i-1}, RevInfo_{i-1}$ // S sends the new revision number i, the proof for the changeset, and the commit signature and revision information for revision $i - 1$
9: **if** $(Verify(CSIG_{i-1}) == invalid)$ **then** C aborts the protocol // C verifies the commit signature using $P_{i-1}, RevInfo_{i-1}$ and revision $i - 1$ of the files in the commit changeset
10: C computes the $MHTROOT_i$ using P_{i-1} and revision i of the files in the commit changeset
11: C sets $RevInfo_i = i, i - 1, ID_{client}$
12: C computes $CSIG_i = Sign(MHTROOT_i, RevInfo_i)$
13: $C \to S : CSIG_i, RevInfo_i$
14: S computes the MHT for revision i using the MHT for revision $i - 1$ and the client's changeset
15: S stores $CSIG_{i-1}, RevInfo_i$ and the MHT for revision i

---

PROTOCOL: Secure_Update

1: // Steps 1-7 are the same as in the standard **Update** protocol
8: S computes proof P_i // S uses revision i of the repository to compute proof P_i relative to the client's update set
9: $S \rightarrow C : P_i, CSIG_i, RevInfo_i$ // S sends the proof for the update set, and the commit signature and revision information for revision i
10: **if** $(Verify(CSIG_i) == invalid)$ **then** C aborts the protocol // C verifies the commit signature using $P_i, RevInfo_i$ and revision i of the files in the update set
11: **for** (each file F in the update set) **do**
12:     C stores F_i in its local repository

---

7). We describe how this proof is computed and verified in Sect. 5.1. After receiving the new revision number, the proof, and the commit signature and revision information for revision $i - 1$ (line 8), the client verifies the validity of the proof (line 9). The client then uses this proof and the files in the changeset to compute the root of the MHT over revision i of the repository (line 10). Finally, the client computes the commit signature over revision i as a digital signature over the root of the MHT and the revision information (which includes the current revision number i, the previous revision number $i - 1$, and the client's ID as the author of the commit) (line 12). Upon receiving the commit signature (line 13), the server recomputes the MHT for revision i and stores it together with the client's commit signature and revision information (lines 14–15).

Secure Update Protocol. The client wants to retrieve revision i for a set of files in the repository, referred to as the *update set*. To allow the client to check the authenticity of the deltas, the server computes a proof for the MHT build on top of revision i, relative to the client's update set (line 8). The server sends this proof to the client, together with the commit signature and revision information for revision i (line 9). The client then verifies this proof (line 10). After finalizing the update, the client sets i as revision number for all the files in the update set.

5.1 MHT-Based Proofs

As described in the previous sections, the commit signature $CSIG_i = Sign(MHTROOT_i, RevInfo_i)$ binds together via a digital signature the root of a Merkle Hash Tree (MHT) with the revision information, both computed over revision i. In the **Secure_Commit** and **Secure_Update** protocols, the client relies on an MHT-based proof from the server to verify the validity of information provided by the server that is not present in the client's local repository. This covers scenarios in which the client works locally with only a portion of the repository. We now describe how such a proof can be computed and verified.

MHT for a Repository. To compute the commit signature, an MHT is built over a revision of the repository. The MHT leaves are hashes of files, which are concatenated and hashed to get the hash of the parent directory. This process continues recursively until we obtain the root of the MHT. Figure 1 shows the directory structure and the corresponding MHT for a revision of repository R1.

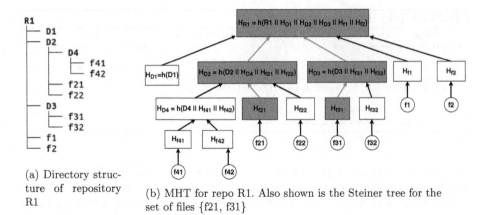

(a) Directory structure of repository R1

(b) MHT for repo R1. Also shown is the Steiner tree for the set of files {f21, f31}

Fig. 1. MHT for a revision of repository R1

MHT-Based Proofs. The client relies on a proof from the server to verify the validity of information received relative to a set of files that it stores locally (*i.e.*, the commit changeset for a commit, or the update set for an update).

The proof of membership for an element contains the siblings of all the nodes on the path between the node to be verified and the root node. For example, consider the MHT for the repository R1 as shown in Fig. 1b. The proof for node H_{f31} is $\{H_{f32}, H_{D1}, H_{D2}, H_{f1}, H_{f2}\}$, whereas the proof for node H_{f21} is $\{H_{D4}, H_{f22}, H_{D1}, H_{D3}, H_{f1}, H_{f2}\}$ We can see that nodes H_{D1}, H_{f1}, and H_{f2} are repeated in the proofs of these two nodes. Thus, when computing a proof of verification for multiple nodes in the MHT, many of the nodes at higher levels of the tree will be common to all the nodes and will be sent multiple times.

To avoid unnecessary duplication and to reduce the data sent from server to client, we follow an approach based on a Steiner tree to compute the proof on the server side. For a given tree and a subset of leaves of that tree, the Steiner tree induced by the set of leaves is defined as the minimal subtree of the tree that connects all the leaves in the subset. This Steiner tree is unique for a given tree and a subset of leaves. The proof for a set of nodes consists of the nodes that "hang off" the Steiner tree induced by the set of nodes (*i.e.*, siblings of nodes in the Steiner tree). Using the same example as earlier, the Steiner tree for the set of nodes $\{H_{f21}, H_{f31}\}$ if shown in Fig. 1b using solid-filled nodes. Thus, the proof is $\{H_{D4}, H_{f22}, H_{f32}, H_{D1}, H_{f1}, H_{f2}\}$.

6   Implementation and Experimental Evaluation

6.1   Implementation and Experimental Setup

We implemented SSVN by adding approximately 2,500 lines of C code on top version 1.9.2 of the SVN codebase. For cryptographic functionality, we used the following primitives from the OpenSSL version 1.0.2g: RSA with 2048-bit keys for digital signatures, and SHA1 for hashing.

We ran experiments with both SVN server and SVN clients running on the same machine, an Intel Core i7 system with 4 cores (each running at 2.90 GHz), 16 GB RAM, and a 500 GB hard disk with ext4 file system. The system runs Ubuntu 16.04 LTS, kernel v. 4.10.14-041014-generic, and OpenSSL 1.0.2g.

Repository Selection. For the experimental evaluation, we wanted to cover a diverse set of repositories with regard to the number of revisions, number of files, and average file size. Thus, we have chosen three representative public SVN repositories: FileZilla [8], SVN [2], and GCC [9], as shown in Table 1.

Overview of Experiments. We have evaluated the end-to-end delay, and the communication and storage overhead associated with the commit and update operations for both SSVN and SVN. We average the overhead over the first 100 revisions of the three selected repositories (labeled FileZilla, SVN, and GCC1). GCC is a large size repository, with over 250K revisions and close to 80K files. Since for GCC the difference between the first 100 revisions and the last 100 revisions is considerable in the size of the repository, we included in our experiments the overhead average over the last 100 revisions of GCC (labeled GCC2). All the data points in the experimental evaluation section are averaged over three independent runs.

6.2   Experimental Evaluation for Commit Operations

End-to-End Delay. The results for end-to-end delay per commit operation are shown in Table 3. Compared to SVN, SSVN increases the end-to-end delay between 12% (for SVN) and 35% (for FileZilla). The overhead is smaller for the SVN repository because the changeset in each commit is small, and thus the corresponding change in the MHT metadata is also small. Even though 35% is a large relative increase for the FileZilla repository, we note that the increase is only 0.06 s per commit. For the GCC repository, the overhead decreases from 20% to 12% as we look at the first 100 revisions compared to the last 100 revisions. This is because the changeset in a commit represents a smaller percentage as the size of the files in the GCC codebase increases. In absolute terms, the increase for GCC remains less than 1 s.

Communication Overhead. Table 2 shows that SSVN adds about 256 bytes to the communication from client to server, which matches the size of the commit signature that is sent by the client with committing a revision. SSVN adds between 0.27 KB to 0.8 KB of communication overhead from server to client. This overhead is caused by the verification metadata sent by server which the client uses to verify the signature over previous commit and to generate the signature for this commit.

Storage Overhead. There is no storage overhead on the client side as the client does not store any additional data in SSVN. On the server side, Table 4 shows that SSVN adds between 0.1 MB–0.16 MB per commit over SVN for FileZilla,

Table 1. Statistics for the selected repositories (as of March 2018). The number of files and the average file size are based on the latest revision in the repository.

	FileZilla	SVN	GCC
Number of revisions	8,738	1,826,802	258,555
Number of files	1,454	2,207	79,552
Average file size	21 KB	18 KB	6 KB
Repository size (all revisions)	29.2 MB	43.9 MB	492.7 MB

Table 2. Network communication for committing one revision (in KBs): from client to server (top two rows), from server to client (bottom two rows).

	FileZilla	SVN	GCC1	GCC2
SVN	35.565	46.672	4.676	20.347
SSVN	35.825	46.934	4.933	20.605
SVN	0.865	1.095	0.539	2.476
SSVN	1.137	1.432	0.962	3.275

Table 3. Commit time per revision (in seconds).

	FileZilla	SVN	GCC1	GCC2
SVN	0.183	0.300	0.385	7.342
SSVN	0.248	0.336	0.459	8.217

Table 4. Server storage per revision (in MBs).

	FileZilla	SVN	GCC1	GCC2
SVN	4.504	0.514	4.263	20.346
SSVN	4.610	0.682	4.415	23.563

SVN, and GCC1. This reflects the fact that the server stores one MHT per revision and the size of the MHT is proportional to the number of files in the repository. We also see the storage overhead increases significantly between GCC1 and GCC2, because the number of files in the GCC repository increases significantly from revision 1 (about 3,000 files) to the latest revision (close to 80,000 files). Since the MHT is proportional to the number of files, the storage overhead for recent revisions in the GCC repository increases to about 3 MB.

6.3   Experimental Evaluation for Update Operations

End-to-End Delay. The results for end-to-end delay per update operation are shown in Table 5. The time needed retrieve a revision in SSVN increases between 11% and 41% compared to regular SVN. Even though 41% looks high, note that the increase is quite modest as an absolute value, at 0.03 s. Even for GCC2, the maximum increase remains modest, at 0.638 s. This increase is caused by the time needed to generate the proof on the server side, to send the proof to the client, and to verify the proof on the client side.

Communication Overhead. Table 6 shows that SSVN adds between 0.24 KB–0.66 KB to the communication from the server to the client. This overhead is

Table 5. Update time per revision (in seconds).

	FileZilla	SVN	GCC1	GCC2
SVN	0.072	0.098	0.150	3.215
SSVN	0.098	0.109	0.182	3.853

Table 6. Network communication for updating one revision (in KBs): from client to server (top two rows), from server to client (bottom two rows).

	FileZilla	SVN	GCC1	GCC2
SVN	1.243	1.328	0.953	10.235
SSVN	1.235	1.548	1.045	11.369
SVN	36.342	49.978	5.782	54.678
SSVN	36.745	50.225	6.245	55.346

caused by the proof that the server sends to the client, which is required on the client side to verify the commit signature for the requested revision.

7 Related Work

Even though an early proposal draft for SVN changeset signing has been considered [23], it only contains a high-level description and lacks concrete details. It has not been followed by any further discussion regarding efficiency or security aspects, and it did not lead to an implementation. Furthermore, the proposal suggests to sign the actual changeset, which may lead to inefficient and insecure solutions, and does not cover features such as allowing partial repository checkout, or allowing clients to work with disjoint sets of files without having to retrieve other clients' changes.

GNU Bazaar [3] is a centralized VCS that allows to sign and verify commits [14] using GPG keys. However, although Bazaar supports features such as partial repository checkout and working with disjoint sets of files, commit signing is not available when these features are used.

Wheeler [35] provides a comprehensive overview of security issues related to source code management (SCM) tools. This includes security requirements, threat models and suggested solutions to address the threats. In this work, we are concerned with similar security guarantees for commit operations, *i.e.*, integrity, authenticity and non-repudiation.

Git provides GPG-based commit signature functionality to ensure the integrity and authenticity of the repository data [11]. Metadata manipulation attacks against Git were identified by Torres-Arias *et al.* [33]. Gerwitz [30] gives a detailed description of Git signed commits and covers how to create and verify signed commits for a few scenarios associated with common development workflows. As we argued earlier in the paper (Sect. 3), several fundamental architectural and functional differences prevent us from applying the same commit signing solution used in Git to centralized VCS-es such as SVN.

Chen and Curtmola [29] proposed mechanisms to ensure that all of the versions of a file are retrievable from an untrusted VCS server over time. The focus of their work is different than ours, as they are concerned with providing probabilistic long-term reliability guarantees for the data in a repository. Relevant to our work, they provide useful insights into the inner workings of VCS-es that rely on delta-based encoding.

8 Conclusion

In this work, we introduce a commit signing mechanism that substantially improves the security model for an entire class of centralized version control systems (VCS-es), which includes among others the well-known Apache SVN. As a result, we enable integrity, authenticity and non-repudiation of data committed by developers. These security guarantees would not be otherwise available for the considered VCS-es.

We are the first to consider commit signing in conjunction with supporting VCS features such as working with a subset of the repository and allowing clients to work on disjoint sets of files without having to retrieve each other's changes. This is achieved efficiently by signing a Merkle Hash Tree (MHT) computed over the entire repository, whereas the proofs about non-local data contain siblings of nodes in the Steiner tree determined by items in the commit/update changeset. This technique is of independent interest and can also be applied to distributed VCS-es like Git in case Git moved to support partial checkouts (a feature that has been considered before) or in ongoing efforts to optimize working with very large Git repositories [25, 28].

We implemented a prototype on top of the existing SVN codebase and evaluated its performance with a diverse set of repositories. The evaluation shows that our solution incurs a modest overhead: for medium-sized repositories we add less than 0.5 KB network communication and less than 0.2 s end-to-end delay per commit/update; even for very large repositories, the communication overhead is under 1 KB and end-to-end delay overhead remains under 1 s per commit/update.

Acknowledgments. This research was supported by the NSF under Grants No. CNS 1801430 and DGE 1565478. We would like to thank Ruchir Arya for contributions to an earlier version of this work.

References

1. Adobe source code breach; it's bad, real bad. https://gigaom.com/2013/10/04/adobe-source-code-breech-its-bad-real-bad/
2. Apache subversion. https://subversion.apache.org/
3. Bazaar. http://bazaar.canonical.com/en/
4. Bitcoin gold critical warning. https://bitcoingold.org/critical-warning-nov-26/
5. Breaching Fort Apache.org - What went wrong?. http://www.theregister.co.uk/2009/09/03/apache_website_breach_postmortem/

6. Cloud source host Code Spaces hacked, developers lose code. https://www.gamasutra.com/view/news/219462/Cloud_source_host_Code_Spaces_hacked_developers_lose_code.php
7. Concurrent versions system. https://www.nongnu.org/cvs/
8. Filezilla. https://filezilla-project.org/
9. GCC. https://gcc.gnu.org/
10. Git. https://git-scm.com/
11. Git commit signature. https://git-scm.com/book/en/v2/Git-Tools-Signing-Your-Work
12. GitHub. https://github.com/
13. GitLab. https://about.gitlab.com/
14. Gnu bazaar GnuPG signatures. http://doc.bazaar.canonical.com/beta/en/user-guide/gpg_signatures.html
15. 'Google' Hackers Had Ability to Alter Source Code. https://www.wired.com/2010/03/source-code-hacks/
16. Internet security threat report, symantec. https://www.symantec.com/content/dam/symantec/docs/reports/istr-23-2018-en.pdf
17. Kernel.org linux repository rooted in hack attack. http://www.theregister.co.uk/2011/08/31/linux_kernel_security_breach/
18. Mercurial. https://www.mercurial-scm.org/
19. Perforce Helix Core. https://www.perforce.com/products/helix-core
20. Sourceforge. https://sourceforge.net/
21. StarTeam. https://www.microfocus.com/products/change-management/starteam/
22. Surround SCM. https://www.perforce.com/products/surround-scm
23. SVN changeset signing. http://svn.apache.org/repos/asf/subversion/trunk/notes/changeset-signing.txt
24. SVN skip deltas. http://svn.apache.org/repos/asf/subversion/trunk/notes/skip-deltas
25. Teach git to support a virtual (partially populated) work directory. https://public-inbox.org/git/20181213194107.31572-1-peartben@gmail.com/
26. The Linux Backdoor Attempt of 2003. https://freedom-to-tinker.com/2013/10/09/the-linux-backdoor-attempt-of-2003/
27. Vault. http://www.sourcegear.com/vault/
28. VFS for Git. https://vfsforgit.org/
29. Chen, B., Curtmola, R.: Auditable version control systems. In: Proceedings of the 21st ISOC Annual Network & Distributed System Security Symposium, February 2014
30. Gerwitz, M.: A git horror story: repository integrity with signed commits. https://mikegerwitz.com/papers/git-horror-story
31. Merkle, R.: Protocols for public key cryptosystems. In: Proceedings of IEEE Symposium on Security and Privacy (1980)
32. Talos: CCleanup: a vast number of machines at risk. https://blog.talosintelligence.com/2017/09/avast-distributes-malware.html
33. Torres-Arias, S., Ammula, A.K., Curtmola, R., Cappos, J.: On omitting commits and committing omissions: preventing git metadata tampering that (re)introduces software vulnerabilities. In: Proceedings of the 25th USENIX Security Symposium (2016)
34. Vaidya, S., Torres-Arias, S., Curtmola, R., Cappos, J.: Commit signatures for centralized version control systems. Technical report, NJIT, March 2019
35. Wheeler, D.A.: Software configuration management (SCM) security. https://www.dwheeler.com/essays/scm-security.html

Towards Contractual Agreements
for Revocation of Online Data

Theodor Schnitzler[1]([✉]), Markus Dürmuth[1], and Christina Pöpper[2]

[1] Ruhr-Universität Bochum, Bochum, Germany
{theodor.schnitzler,markus.duermuth}@rub.de
[2] NYU Abu Dhabi, Abu Dhabi, UAE
christina.poepper@nyu.edu

Abstract. Once personal data is published online, it is out of the control of the user and can be a threat to users' privacy. Retroactively deleting data after it has been published is notoriously unreliable due to the distributed and open nature of the Internet. Cryptographic approaches implementing data revocation address this problem, but have serious limitations when considering practical deployment, and they have not been broadly adopted.

In this paper, we tackle the problem of data revocation from a different perspective by examining how contractual agreements can be applied to create incentives for providers to conform to expiration regulations. Specifically, we propose a protocol to automate the handling of data revocation. We have implemented a prototype smart contract on a local Ethereum blockchain to demonstrate the feasibility of our approach. Our approach has distinct advantages over existing proposals: It can deal with a wide spectrum of revocation conditions, it can be applied retroactively after data has been published, and it does not require additional effort for users accessing the published data. It thus constitutes an interesting, novel approach to data revocation.

1 Introduction

Cloud infrastructures and cheap storage are changing how we handle and share data and how long-lasting data becomes. While promising opportunities go along with this storage and access to data, concerns about the negative impact on ever available data and resources arise. As more and more personal information is shared online, the concepts of *digital forgetting* [19] and *revocation of online data* [28] become increasingly important for protecting our online privacy.

From the jurisdictional and also regulatory perspective, the European Parliament has approved the General Data Protection Regulation (GDPR) [9] on the protection of natural persons with regard to the processing of their personal data and on the free movement of such data. This *Right to be Forgotten* [10] already received considerable attention in 2014 due to a ruling by the European Court of Justice (ECJ) [11]. The ECJ determined that online search engines need to

G. Dhillon et al. (Eds.): SEC 2019, IFIP AICT 562, pp. 374–387, 2019.
https://doi.org/10.1007/978-3-030-22312-0_26

provide an interface and procedures for EU citizens to request the removal of their personal information from search results.

Technical approaches to implement the Right to be Forgotten apply cryptographic erasure, i.e., publishing data only in encrypted form and making it irretrievable after expiration by a suitable key management [5,12,23,24,26]. In this work, we explore a mechanism for the revocation of online data that does not purely limit the availability of data in technical terms but provides monetary incentives such that providers take appropriate measures to support data revocation and to comply with expiration conditions. In particular, we propose to conduct an agreement between a user, who owns a specific data item, and a platform provider, who offers access to the data. The agreement defines the expiration conditions for the data items. In contrast to existing approaches, our technique can be applied not only for data at the point of publishing but also for data that has already been made available in the past. Technically, we explore how smart contracts, as available in certain cryptocurrencies such as Ethereum [4], can be used to realize agreements between users and providers. Compared to other forms of reaching formal agreements, e.g., digitally signed PDFs, smart contracts allow a high degree of automation. Using a distributed ledger system is attractive, as it comprises a trusted third party in a decentralized manner. Our approach is designed to handle the majority of agreements on data expiration. Violations and disagreements in particular cases can still be handled in the jurisdictional system. Data owners might even refer to the contract as a piece of evidence in a potential legal action.

We assume that services featuring a revocation mechanism can also be attractive from the provider's perspective. The broader adoption of services such as Snapchat [29] and Signal [21] over several years shows that there is a demand for online interaction granting a certain level of privacy.

In short, this paper makes the following contributions:

- We design a data revocation scheme based on contractual agreements that can be applied both to new data and to data that has been published in the past. Users can take action both during the lifetime of data and after its intended expiration. To the best of our knowledge, this is the first approach to cover all these aspects at the same time.
- Our proposal is particularly efficient for fast data access without unnecessarily burdening users legitimately reading the published data.
- We provide an extensive overview of the design space of solutions for the revocation of online data based on cryptocurrencies and contractual agreements.
- We provide insights into an instantiation of our protocol: We have implemented a prototype as an Ethereum smart contract to demonstrate the general feasibility of our approach.

2   Solution Overview

We will now provide an overview of the basic ideas of our approach, before giving detailed insights into our protocol in Sect. 3.

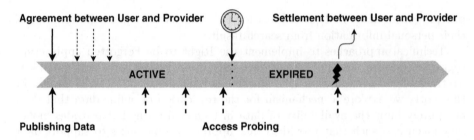

Fig. 1. Timeline of our approach. User and provider conclude the agreement at the time of publishing or later. During its lifetime, the data is freely accessible on the provider's platform. A successful access after expiration leads to a settlement process.

There are three main entities in the system: the *user U*, who initially holds and owns the data and wants to publish it online. The data should be published with a *provider P*, who makes it publicly available on a large *online platform* in the *WWW*, such as a social network, or image sharing website.

Whereas previous work has tried to enforce automatic deletion of data, we pursue an idea where both the user and the provider establish a contractual agreement. This agreement states that P will take measures to limit the distribution of the data (e.g., by enforcing a limited lifetime on the data) and that U is entitled to compensation if the provider violates this agreement.

For this process, we distinguish the following four actions: (1) registering the agreement, (2) publishing the data, and (3) probing if the data is still available. When a violation is recognized we (4) enter a settlement process. We illustrate the chronology of these actions in Fig. 1.

(1) First, the user and the provider need to reach a formal agreement. This agreement will include (i) the data d it concerns, by means of a unique identifier, (ii) the expiration condition on which the data should become unavailable, e.g., after a specific time t or an event e, and (iii) the penalty p for the provider in case the agreement is violated by the data being available after time t or event e. Similar agreements can be reached in the form of a traditional contract, but enforcing such agreements incorporating very small penalties, e.g., in the range of a few cents, is prohibitively expensive.

(2) The actual publishing of data typically involves sending the data to the provider, who will make it publicly accessible. One interesting property of this approach is that Steps (1) and (2) can be initiated in arbitrary order. In other words, it is also possible to reach such an agreement even after the data was published. While most other approaches require deciding on using protective measures and even fixing specific parameters such as the expiration time upon publishing, our system can also be invoked retroactively.

(3) After publishing and before expiration by meeting the expiration time t or the event e, the data is freely available. Accessing the data requires no additional tools or measures. The user can easily probe whether the data is still accessible before and after expiration.

(4) If the user detects a violation, i.e., the provider fails to handle the data correctly in making it available other than specified in the initial agreement, a penalty mechanism is triggered. A typical settlement could be a small financial compensation.

Adversary Model

The security notion for data revocation considered in previous works is security against a *retrospective attacker*: Basically, the retrospective adversary becomes active only after the data has been revoked. The attacker should not be able to reconstruct the original data after the expiration. This strong notion can typically only be achieved when the users are entirely honest and refrain from re-uploading the accessed data without protection to a different location. One can argue that re-uploading the data while still available constitutes an explicit act of archiving, in which case the data should indeed be available over time [19].

Additional aspects consider how the protocol can be exploited by the participants to obtain an advantage over the opposite party. The use of financial compensations might tempt the user to claim violations in cases in which the provider complied with the agreement. The provider is therefore interested in a guarantee that the data have actually been accessible at the time of the claim. At the same time upon violation claim, the provider might be interested in false statements, i.e., contending that the data do not exist. Thus, the user is interested in a decision finding that can objectively determine whether the data exists and that is resilient to a provider tampering with it.

3   Revocation Contract Scheme

We use Ethereum smart contracts [4] as a convenient method to specify and validate agreements between the data owner and the service provider on the expiration of data. However, Ethereum smart contracts comprise certain limitations, such as accessing data outside the blockchain and strictly enforcing payments between parties. For the specification of our protocol, we therefore require (i) the presence of a data feed for accessing external data and (ii) that a fraction of a potential penalty is placed as a deposit in the contract. The interactions in the stages of our protocol are illustrated in Fig. 2.

(1+2) Registration. The registration process incorporates the first two steps as proposed in Sect. 2, since publishing data is a key subject of the agreement registration. A user U can publish a data item d along with an associated expiration condition exp_d on the platform of the provider P and will receive an identifier id_d for the data from the provider.

The user, who is identified by the address $addr_U$, can then initiate a transaction invoking the registration function of the smart contract C, passing id_d and exp_d to the contract.

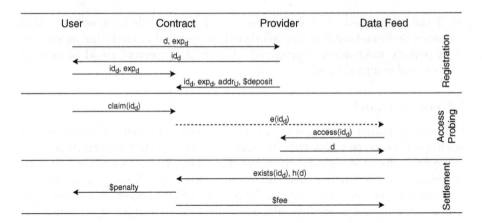

Fig. 2. Protocols of the registration, access probing, and violation settlement processes.

The registration process is finalized as soon as the provider also invokes the contract, providing id_d, exp_d and $addr_U$. Along with the confirmation transaction, the provider places a deposit in the contract, from which a potential penalty can be paid out. We assume that the deposit can be significantly smaller than the prospective penalty, obviously separated for each provider since we expect that contract violations and the resulting settlement processes will only occur in exceptional cases. Deposits are not bound to specific data items, but instead, the contract balance should cover penalties for a certain proportion of all data registered in the contract. In practice there is no fixed order for the last two steps, i.e., it is also possible that the provider is the first party to pass the registration information to the smart contract.

During the registration process, the contract may check the conditions for general plausibility. In case the expiration condition is, e.g., a time t, it should be checked whether t lies in the future. The registration process does not necessarily have to be completed immediately after the publication of the data. It can generally be initiated at any time after the data has been published on the provider's platform.

(3) Access Probing. After its publication, the data is freely accessible on the provider's platform. There is no need to encrypt the data prior to legitimately reading it during its lifetime. In addition, the owner can probe access in direct interaction with the provider without the contract being involved. We can also delegate this task to a third party, e.g., the owner can assign the task of access probing to a trusted service. This requires that the data is publicly available during its lifetime – our scheme does not cover non-public data.

(4) Violation Settlement. Ideally, the provider would have revoked the data according to the expiration condition specified in the contract. Nonetheless, if the owner detects that the data is yet available, a transaction can be initiated to

notify the smart contract about the violation, passing the corresponding identifier along with the function call. Violations can be reported not only by the user
but by everyone who has access to the data. Thus, the user can even conclude
out-of-band agreements at will with dedicated services to check the availability.
These services regularly observe the provider's compliance with the contract and
initiate settlement on behalf of the user in the case of a contract violation. When
the contract receives such a violation claim, it must be checked whether the data
is available online. Therefore, the smart contract will request an external data
feed service S to retrieve the data from the platform of the provider. Finally, S
will initiate a transaction notifying the smart contract that the access attempt
has been successful and will provide a hash $h(d)$ of the retrieved data. As the
last step of the violation settlement, the smart contract will pay out the penalty
to the user and a small fee to reward the data feed service for taking part in the
protocol.

4 Protocol Design Space

Next, we sketch how extensions of our protocol and alternatives in its design
influence specific properties of the protocol.

4.1 Data Identification

In a basic approach, identifiers assigned by the providers are stored in the smart
contract for data identification. Usually, this identifier does not change during
the lifetime of data, regardless where it is referenced. To improve user privacy,
the identifier should not be stored in plain text, but in a cryptographically secure
hashed form. However, uploading the data to an external website or even under
a new identifier is sufficient to circumvent the protection.

Alternatively, data can be identified using hashes of the original data. Robust
hashes [32,33] are indented to tolerate minor modifications of data, such as re-
scaling, compression artifacts, and similar in the case of images. The use of such
alternative identifiers is generally feasible within our proposed protocol, it only
requires the location information as additional input in the first step of the
settlement process.

Such extensions appear more desirable for the user, but put a significantly
stronger burden on the provider. Location-independent data identification raises
additional challenges with regard to the accountability of contract violations.
Moreover, by storing images while they are available and re-uploading them at
a later point, a malicious user may be able to force compensation payments by
the provider, even though the provider has behaved correctly. However, users
are not entirely protected from malicious providers or third parties. If the script
is transparently available, an adversary can gradually introduce slight changes
into the data until they are classified as different to circumvent protection.

4.2 Data Feeds

Our system requires the presence of a data feed to incorporate real-world events from outside the blockchain. However, the actual request processing, contacting external resources such as a website, raises concerns regarding the trust required in these services and challenges in validating the delivered results.

First, the response delivered by the data feed should be correct, i.e., the data feed should process the data as they are present at the time of the request. Such a response must be time-bound, because it is not sufficient to show that the data under consideration have existed on the website at any point in time before. Second, transactions performed on the blockchain are publicly visible and irrevocable. The data feed cannot simply access a website and write its contents (i.e., the data) to the blockchain, as this is effectively opposed to the goals of data revocation. We require the data feed to access the data from the provider website, perform an additional step such as determining whether the retrieved data comprise an image, and write the result to the blockchain.

Oraclize [22] and Town Crier [35] both use cryptographic means such as *TLS Notary Proofs* or *Software Guard Extensions (SGX)* to deliver a proof that certain data exist on a source website. From the perspective of the provider, such cryptographic attestations are attractive to prevent spurious claims from adversarial users. However, the use of a single service appears to be susceptible to malicious providers, as they could identify a particular data feed and deliver false responses to them to avoid the settlement process.

The concept of ChainLink [7] comprises multiple data feeds that access data independently from each other and are capable of performing computations on the retrieved data before responding. Its consensus mechanism for aggregating a result from the distributed responses (which can be built on trusted hardware such as SGX on an individual basis) reduces the trust required in the individual data feeds. Moreover, the use of a reputation mechanism that records false responses and a deposit-based penalty mechanism incentivize data feeds to deliver true responses. We imagine this as a crowd-sourcing approach in which even end-users can serve as verifiers in individual cases. Thus, massive misbehaving from the provider appears impossible, which ensures a high level of availability for such a system.

4.3 Complex Revocation Conditions

While the majority of prior work on data revocation simply use time-based expiration mechanisms, we suppose that smart contracts can be used to create almost arbitrary expiration conditions. However, these conditions need to be objectively observable through reliable data sources providing appropriate interfaces. Such scenarios would require an additional step in the verification step of our protocol, whereas the general procedure would be similar and security requirements would not be stricter than for the data access verification.

In a naïve approach, the expiration condition is represented as an expression in natural language. There is a need for reliable information sources that are capable of processing the information in their represented form, e.g. providing an interface for natural-language processing. The trust in these services (e.g., Google, Wikipedia, Wolfram Alpha) can be reduced through the use of decentralized verification as proposed by ChainLink, assuming that the service cannot distinguish verification attempts from regular requests to their services. In addition, the stability of information must be taken into account, i.e., the verification must be resilient to short-time changes, e.g., malicious users manipulating the information to their advantage and then triggering the settlement process. This is feasible, if Wikipedia is used as a knowledge base, but can be prevented by also considering the information history.

However, with more complex expiration conditions, we also see the possibility that data items can become valid again after they have expired. Examples are scenarios in which data are supposed to be available only on specific days of the week, or have a daily access limit (under the assumption that the number of accesses can be verified reliably). This property makes the use of smart contracts a much more powerful instrument than previous approaches.

4.4   Financial Reserve Model

In order to guarantee the payout of penalties, the provider needs to place a deposit in the contract that is paid out when the contract is violated. However, in the case of large-scale application of our system, this would lead to large amounts of provider capital locked in the smart contract. Thus, we propose that only a fractional reserve (e.g., 1%) has to be deposited in the contract, penalties are paid from the pool of all deposits, and that the provider has the possibility to withdraw money from the contract as long as the total amount is above the reserve threshold. This threshold is determined by the number of data items covered by the contract. If an item has expired and no violation has been reported for an adequate period of time, this item can be taken into account with less weight for calculating the threshold.

Large-scale violations that exceed the contract capacity determined by the total amount of deposits can still be handled resorting to the jurisdictional system. In this case, the contract can even be used as a piece of evidence to support that user and provider have agreed on the expiration of data beforehand.

5   Prototype Implementation

In this section, we describe our prototype implementation and evaluate the transaction cost incurring in its use and the scalability in terms of the numbers of data items that can be protected. Our contract employs time-based expiration conditions for data items that can be identified with a unique ID. We have implemented our prototype system using Ethereum with a local blockchain using the *Go Ethereum (Geth)* client. We have initialized Ethereum accounts representing a user, a provider, and the external service, as well as a smart contract in which items can be managed. For experimental interactions, i.e., registering or checking items with the smart contract, we utilized the *Ethereum Mist Wallet* application.

5.1 Smart Contract

Our Smart Contract consists of roughly 200 lines of Solidity code, is deployed on a local private blockchain, and available on GitHub[1]. The registration process consists of two steps that can be executed in arbitrary order. Both owner and provider have to commit information to create a valid entry. A user can register data in `addItem()`, providing as inputs its identifier and the remaining time it is intended to be available. The provider approves registration by using the function `confirmItem()`, also passing the identifier, the time left, and the owner's Ethereum address to the contract. We aim to ensure that the contract balance covers a minimum proportion of all registered data. After both parties have committed to the registration, the item agreement has become valid.

Data owners can be remove their data from the contract by calling the function `removeItem()`, which represents a cancellation of the agreement. Likewise, the provider can also withdraw confirmed items from the contract, as long as the owner has not added it.

The function `verifyAccess()` initiates the verification, whether data with a given identifier is available. In general, this function can be invoked by anyone and at any time, but the verification will only be triggered if the expiration date as stated in the record has been reached. However, access verification must be conducted by a data feed service external to the blockchain. For our prototype, we used an external script providing basic functionality. If the check is successful, the service can invoke the contract function `itemFound()`, which will initiate the compensation process. Thus, the contract will send the specified amount of Ether from its balance to the picture owner. In exceptional circumstances, i.e., if there is a widespread distribution of violations and many settlement processes are initiated at the same time, the Ether transfer may fail due to a low contract balance. The function `claimPending(id)` allows the user to initiate the compensation retroactively when the accessibility after expiration has already been verified before.

5.2 Evaluation

The feasibility of our approach mainly depends on the transaction cost arising from the use of the contract and the number of agreements that can be achieved in total.

Transaction fees in Ethereum, referred to as *gas*, generally depend on the complexity of the transaction, but can also be specified by the transaction sender. If a lower fee is selected, it may take a longer time-span for the transaction to be processed. As of December 2018, waiting times for transaction processing have been 45 s [8] on average. This time-span seems acceptable, as our application is not time-sensitive in terms of a few minutes.

---

[1] https://github.com/theoschnitzler/data-revocation-contract/.

Table 1. Cost of contract execution. We assume gas fees of 1 GWei and an Ethereum exchange rate of $100.00.

Transaction	Gas	USD	Actor
Create	82,446	$0.0082	User
Confirm	35,107	$0.0035	Provider
CheckExpiry	27,199	$0.0027	User
VerifyAccess	28,621	$0.0029	User
ItemFound	22,071	$0.0022	Data Feed
ClaimPending	21,922	$0.0022	User

In Table 1, we illustrate the cost for the six transactions offered by our contract and the actor who needs to provide the fee on triggering the transaction. We assume an exchange rate of $100.00 per Ether and a transaction fee of 1 GWei. At the time of writing, miners of 30% of total blocks have accepted this or even a lower fee [8]. However, these numbers are constantly changing, due to the current network load. Registering a new data item to the contract requires a user to provide a transaction fee of 0.8 cents. For a user who uploads one item per day, this results in a total amount of roughly $3.00 per year. For each item confirmation, the provider has to bear costs of 0.35 cents. The cost of the other transactions will only incur in case of a contract violation. When we assume that the penalty for the violation is significantly higher (even in the range of a few dollars), the additional transaction cost will be negligible.

The amount of gas consumed by all transactions to be included in a new Ethereum block is limited to 8 million. When the contract registration requires 117,553 gas (see Table 1), and a new block is created every 15 s on average, this allows roughly 390,000 new registrations each day (2.75 million per week) under the assumption that the overall Ethereum network is not used for any other means. In 2017, Google has received roughly 2000 removal requests under European privacy law on a weekly basis [14], which is well below 0.1% of the limit that could be achieved within Ethereum.

6 Discussion

In this section, we discuss the effect of our proposal on trust assumptions and privacy aspects of practical implementations.

Trust Requirements. Whereas we have introduced smart contracts for data revocation to reduce the trust required in providers when users upload personal data on their platforms, the use of data feeds still requires a certain level of trust. We consider the data feed a neutral adjudicator and, therefore, trust requirements in these entities are less critical than in providers who have—due to their business models—interest in making user content accessible as long as possible. However, trust in particular services is reduced by the presence of several alternatives users can ideally pick from, and also an approach based on crowd-sourcing. With regular other users serving as verifiers, it becomes harder for providers to make false attestations to data feeds to circumvent a violation detection, as regular accesses cannot be distinguished from accesses for verification purposes.

Metadata Privacy. The data accessible to smart contracts is stored on the blockchain, a distributed and publicly readable and irrevocable data structure. This may lead to two challenges:

First, the arising database might be a valuable target for an adversary who aims to collect all the sensitive data prior to expiration. This mainly depends on the number of data items covered by the contract and can particularly become a problem when our system is only used for sensitive data. However, if there is also sufficient non-sensitive data covered by the contract, which we advocate for, an attacker cannot directly reason that the data is sensitive just from its presence in the contract.

Second, the contract data can reveal information about the privacy preferences of individual users. If many data items registered in the contract belong to the same owner, it is evident that this user attaches importance to privacy in common and data revocation in particular. On the other hand, we can also conclude that this user also shares a lot of data publicly. We assume that the blockchain interactions involving a specific user provide pseudonymity, without direct linkability to a concrete person. Thus, information derived from the contract does not comprise an additional privacy leak.

Provider Participation. We see the use of smart contracts to register user data and specify expiration conditions similar to establishing a regular contract. Thus, a fundamental requirement is the willingness of service providers to support such a mechanism and active participation in the protocol. For a successful registration of a particular agreement, the provider must commit information to the contract. We cannot ultimately prevent that the provider might refrain from entering the agreement in specific individual cases. However, such a misbehavior can be observed, as there will be open registration requests by users, which are publicly visible on the blockchain. This makes it possible for a user to check upfront whether a provider actually complies with the system, before finally uploading new data.

Recent studies [3,20] have found out that users of online social networks are actively employing privacy preferences as offered by the providers for their data. Therefore, we assume that applying a service for data revocation as we propose can make social networks more attractive to users, especially to those who are generally more concerned about privacy issues. Moreover, our scheme can be utilized as a reputation mechanism, in that providers use the system to demonstrate that they take user privacy seriously as they comply with the preferences defined in the contract. If more providers apply the system, they can even compete with each other in reaching the lowest violation rate. From a regulatory point of view, our approach can be considered as a technical instantiation for establishing the users' right to be forgotten. In the future, technical revocation mechanisms, but also our proposal for contractual agreements can provide directions towards the automated handling of such removal requests. This seems desirable not only for the users but also for providers, in that it renders the manual check of revocation requests and dealing with individual cases unnecessary.

7 Related Work

Cryptographic erasure mechanisms to enforce digital forgetting have been well studied. One of the first approaches was Ephemerizer by Perlman [23]. Vanish [12] uses distributed hash tables for key distribution EphPub [5] and Neuralyzer [34] use the Domain Name System as infrastructure for key distribution, and Reimann and Dürmuth [26] leverage the continuous change of website contents for key expiration. Amjad et al. [1] propose an approach raising the effort required to access data to prevent large-scale adversarial data collection during the lifetime of data. Chen et al. [6] propose a method to delete not only data but also remaining structural artifacts on a hardware level.

Opposed to data revocation scenarios, approaches exist to keep data secret after its publication until a certain release condition is fulfilled. Li and Palanisamy [17] present a scheme leveraging distributed hash tables that is built on the cryptographic foundations of time-lock encryption [18,27]. In this context, Jager [15] shows that blockchain technology can be leveraged to emulate real-world-time in a computational model.

Studies have explored how much users are willing to pay to recover data that have become inaccessible [30] and for improvements in data privacy in an online shopping scenario [31]. González et al. [13] present a tool for computing the value of personal data uploaded to Facebook based on the financial revenue generated from the data, which can be useful to determine appropriate penalty amounts in our contract.

Integrating blockchain technology with privacy scenarios raises challenges in handling sensitive personal information. In this context, approaches to change the blockchain history without omitting advantages such as public verifiability have been proposed [2,25]. Kosba et al. [16] introduce a framework for privacy-preserving smart contracts by developing a blockchain model of cryptography.

8 Conclusion

In this paper, we have developed an approach for the support of data revocation. Different from previous work, we did not use cryptographic measures, but combined both technical and regulatory aspects in order to incentivize the provider to delete data as determined by its owner. Based on this idea, we have developed a protocol for the specification of revocation conditions in smart contracts and implemented a prototype that supports time-based revocation conditions and is processed on a local Ethereum blockchain. The contract incorporates a penalty mechanism for data that remains available deviating from the conditions in the contract. With our approach, users can take action both proactively in defining expiration conditions for data they have published, and also retroactively in that they can get compensation in case the data provider has failed to delete data as specified.

Acknowledgments. This research was supported by the BMBF InStruct project under grant 16KIS0581.

References

1. Amjad, G., Mirza, M.S., Pöpper, C.: Forgetting with puzzles: using cryptographic puzzles to support digital forgetting. In: CODASPY 2018, pp. 342–353. ACM (2018)
2. Ateniese, G., Magri, B., Venturi, D., Andrade, E.: Redactable blockchain–or– rewriting history in bitcoin and friends. In: EuroSP 2017. IEEE (2017)
3. Ayalon, O., Toch, E.: Retrospective privacy: managing longitudinal privacy in online social networks. In: SOUPS 2013, pp. 4:1–4:13. ACM (2013)
4. Buterin, V.: A next-generation smart contract and decentralized application platform (2014). https://github.com/ethereum/wiki/wiki/White-Paper. Accessed 28 Feb 2019
5. Castelluccia, C., De Cristofaro, E., Francillon, A., Kaafar, M.A.: EphPub: toward robust ephemeral publishing. In: ICNP 2011, pp. 165–175. IEEE (2011)
6. Chen, B., Jia, S., Xia, L., Liu, P.: Sanitizing data is not enough!: towards sanitizing structural artifacts in flash media. In: ACSAC 2016, pp. 496–507. ACM (2016)
7. Ellis, S., Juels, A., Nazarov, S.: ChainLink - A Decentralized Oracle Network (2017). https://crushcrypto.com/wp-content/uploads/2017/09/LINK-Whitepaper.pdf. Accessed 28 Feb 2019
8. ETH Gas Station: ETH Gas Station (2017). https://ethgasstation.info/. Accessed 28 Feb 2019
9. European Parliament: Regulation (EU) 2016/679 (General Data Protection Regulation). Official Journal of the European Union 59 (2016)
10. European Union: Factsheet on the "Right to be Forgotten" Ruling (C-131/12) (2014)
11. European Union Court of Justice: Judgment in Case C-131/12. Press release No. 70/14 (2014). https://curia.europa.eu/jcms/upload/docs/application/pdf/2014-05/cp140070en.pdf. Accessed 28 Feb 2019
12. Geambasu, R., Kohno, T., Levy, A.A., Levy, H.M.: Vanish: increasing data privacy with self-destructing data. In: USENIX Security 2009, pp. 299–316. USENIX (2009)
13. González Cabañas, J., Cuevas, A., Cuevas, R.: FDVT: data valuation tool for Facebook users. In: CHI 2017, pp. 3799–3809. ACM (2017)
14. Google Inc.: European Privacy Requests for Search Removals. Transparency report (2017). https://www.google.com/transparencyreport/removals/europeprivacy/. Accessed 28 Feb 2019
15. Jager, T.: How to build time-lock encryption. Cryptology ePrint Archive, Report 2015/478 (2015). http://eprint.iacr.org/2015/478
16. Kosba, A., Miller, A., Shi, E., Wen, Z., Papamanthou, C.: Hawk: the blockchain model of cryptography and privacy-preserving smart contracts. In: SP 2016, pp. 839–858. IEEE (2016)

17. Li, C., Palanisamy, B.: Timed-release of self-emerging data using distributed hash tables. In: ICDCS 2017, pp. 2344–2351, June 2017
18. Mahmoody, M., Moran, T., Vadhan, S.: Time-lock puzzles in the random oracle model. In: Rogaway, P. (ed.) CRYPTO 2011. LNCS, vol. 6841, pp. 39–50. Springer, Heidelberg (2011). https://doi.org/10.1007/978-3-642-22792-9_3
19. Mayer-Schönberger, V.: Delete: The Virtue of Forgetting in the Digital Age. Princeton University Press, Princeton (2011)
20. Mondal, M., Messias, J., Ghosh, S., Gummadi, K.P., Kate, A.: Forgetting in social media: understanding and controlling longitudinal exposure of socially shared data. In: SOUPS 2016, pp. 287–299. USENIX (2016)
21. Open Whisper Systems: Signal (2010). https://signal.org/. Accessed 28 Feb 2019
22. Oraclize Ltd.: Oraclize - Blockchain Oracle Service, Enabling Data-Rich Smart Contracts (2017). http://www.oraclize.it. Accessed 28 Feb 2019
23. Perlman, R.: The Ephemerizer: making data disappear. Technical report SMLI TR-2005-140, Sun Microsystems Laboratories, Inc. (2005)
24. Pöpper, C., Basin, D., Capkun, S., Cremers, C.: Keeping data secret under full compromise using porter devices. In: ACSAC 2010, pp. 241–250. ACM (2010)
25. Puddu, I., Dmitrienko, A., Capkun, S.: μchain: how to forget without hard forks. Cryptology ePrint Archive, Report 2017/106 (2017). http://eprint.iacr.org/2017/106
26. Reimann, S., Dürmuth, M.: Timed revocation of user data: long expiration times from existing infrastructure. In: WPES 2012, pp. 65–74. ACM (2012)
27. Rivest, R.L., Shamir, A., Wagner, D.A.: Time-lock puzzles and timed-release crypto. Technical report (1996)
28. Shein, E.: Ephemeral data. Commun. ACM **56**(9), 20–22 (2013)
29. Snap Inc.: Snapchat (2011). https://www.snapchat.com/. Accessed 28 Feb 2019
30. Spiekermann, S., Korunovska, J.: Towards a value theory for personal data. J. Inf. Technol. **32**(1), 62–84 (2017)
31. Tsai, J.Y., Egelman, S., Cranor, L., Acquisti, A.: The effect of online privacy information on purchasing behavior: an experimental study. Inf. Syst. Res. **22**(2), 254–268 (2011)
32. Venkatesan, R., Koon, S.M., Jakubowski, M.H., Moulin, P.: Robust image hashing. In: ICIP 2000, pp. 664–666. IEEE (2000)
33. Yang, B., Gu, F., Niu, X.: Block mean value based image perceptual hashing. In: IIH-MSP 2006, pp. 167–172. IEEE (2006)
34. Zarras, A., Kohls, K., Dürmuth, M., Pöpper, C.: Neuralyzer: flexible expiration times for the revocation of online data. In: CODASPY 2016, pp. 14–25. ACM (2016)
35. Zhang, F., Cecchetti, E., Croman, K., Juels, A., Shi, E.: Town crier: an authenticated data feed for smart contracts. In: CCS 2016, pp. 270–282. ACM (2016)

Correction to: ESARA: A Framework for Enterprise Smartphone Apps Risk Assessment

Majid Hatamian, Sebastian Pape⬤, and Kai Rannenberg

Correction to:
Chapter "ESARA: A Framework for Enterprise Smartphone Apps Risk Assessment" in: G. Dhillon et al. (Eds.): *ICT Systems Security and Privacy Protection*, IFIP AICT 562, https://doi.org/10.1007/978-3-030-22312-0_12

By mistake the originally published version of this chapter did not include the acknowledgement text. This has been corrected so that the updated version of the chapter now contains the following acknowledgement: This research was supported by the European Union's Horizon 2020 Research and Innovation program under the Marie Skłodowska-Curie "Privacy&Us" project (GA No. 675730).

The updated version of this chapter can be found at
https://doi.org/10.1007/978-3-030-22312-0_12

Author Index

Printed in the United States
By Bookmasters